W9-BIE-659

FOURTH EDITION

PUTTING FOOD BY

Decent mention of all the distinguished food scientists who helped with this Fourth Edition would be a **Who's Who** *in home food preservation in the United States and Canada. They are in Washington, D.C., and Ottawa; in each state's Cooperative Extension Service, and in each province's Ministry of Agriculture and Food; and continue out to all county agents, Extension Home Economists and Master Food Preservers, and informal groups that reach into our homes.*

We, and all like-minded people everywhere, are beholden to them and salute them.

—JG

FOURTH EDITION, NEWLY REVISED

PUTTING FOOD BY

JANET GREENE
RUTH HERTZBERG
BEATRICE VAUGHAN

A JANET GREENE BOOK

THE STEPHEN GREENE PRESS
LEXINGTON, MASSACHUSETTS

THE STEPHEN GREENE PRESS, INC.
Published by the Penguin Group
Viking Penguin Inc., 40 West 23rd Street, New York, New York 10010, U.S.A.
Penguin Books Ltd, 27 Wrights Lane, London W8 5TZ, England
Penguin Books Australia Ltd, Ringwood, Victoria, Australia
Penguin Books Canada Ltd, 2801 John Street, Markham, Ontario, Canada L3R 1B4
Penguin Books (N.Z.) Ltd, 182-190 Wairau Road, Auckland 10, New Zealand
Penguin Books Ltd, Registered Offices: Harmondsworth, Middlesex, England

This fourth edition of *Putting Food By* first published in 1988 in simultaneous
hardcover and paperback editions by The Stephen Greene Press, Inc.
Published simultaneously in Canada
Distributed by Viking Penguin Inc.

Copyright © The Stephen Greene Press, Inc., 1973, 1975, 1982, 1988
All rights reserved

Publishing history of *Putting Food By*:
First edition published January 1973
Second edition published July 1975
Third edition published July 1982

Some parts of Chapter 21, ''Drying,'' appeared in different form in the August
1981 issue of *Blair & Ketchum's Country Journal*.

Photographs credited to Jeffery V. Baird are copyright © 1987 by Jeffery V.
Baird and reproduced with permission.
Designed by Deborah Schneider
Printed in the United States of America
by Haddon Craftsmen
Set in Times Roman and Futura by Maryland Composition
Produced by Unicorn Production Services, Inc.

Library of Congress Cataloging-in-Publication Data

Greene, Janet
  Putting food by.

  Hertzberg's name appears first on the previous ed.
  ''A Janet Greene book.''
  Includes bibliographical references and index.
  1. Food—Preservation. I. Hertzberg, Ruth.
II. Vaughan, Beatrice. III. Title.
TX601.G74 1988   641.4   87-34262
ISBN 0-8289-0644-0
ISBN 0-8289-0645-9 (pbk.)

CONTENTS

CONTENTS

1
WHAT IS IT?

To "put by" is an old, deep-country way of saying to "save something you don't use now, against the time when you'll need it." Putting food by is simply food preservation.

Who does it? Millions of households in the United States and Canada, for openers. Preserving food at home is prehistorical, though: drying and fermenting are the earliest known means, followed in very short order by salting and brining (pickling). Preserving with sugar came next, but much later. Home-canning is less than two centuries old, and deliberate deep-freezing is the youngest method used at home.

Nowadays every nation on earth practices its own forms of preservation family-by-family, but chances are that North Americans do more of it on a wider scale than householders in any other major region. One statistic will give an idea: in the last year that figures were available (at this writing, the year was 1985) experts in the field of preserving food at home estimated that in the United States alone *$310 million* was spent on canning jars, lids, sugar, salt, pectin, vinegar, and spices.

There is no hocus-pocus about food preservation, no touchstone, no luck, no mystery. Food preservation is the protection of food from spoilage—period. Spoilage can mean unattractively over-the-hill, on to downright nasty, to finally—and most dangerously because sometimes it is not self-evident—deadly.

This book has been written to tell the lay person how to control spoilage or prevent it entirely, so that a full program/menu of foods can be harvested in a time of plenty and treated to be wholesome and available in a time of need.

Whether we are cooking dinner or canning tomatoes, we first clean the food, ridding it of external spoilers like dirt, blemishes, or infestations. We do the same thing in starting to preserve it. Next, we treat the unseen

1

causes of deterioration, chief among them being the enzymes, those re-
markable substances programmed to make the food fulfill its ordained life
cycle.

And finally we deal with the greater trouble-makers: the micro-organisms that can poison our food. These are stopped dead or destroyed outright by reducing the oxygen that most microbes need; by applying heat
or radically decreasing the temperature (heat kills them, freezing holds
them immobilized); by increasing acid (because virtually no microbial
action occurs in strong-acid mediums); by decreasing the available water
that they require.

George York of the University of California, Davis, gives the following
examples of how these microbial controls are applied, thereby demon-
strating the beauty of home-preservation of food. The preserving methods
are: (1) making jams and jellies—lowering the available water, removing
oxygen, applying heat, adding acid; (2) canning fruits and tomatoes—
reducing oxygen (sealing jars), applying heat, relying on natural or added
acid; (3) drying—taking away most of the available water; (4) canning
vegetables and meats (both low-acid)—removing oxygen again, and applying *intense* heat; (5) freezing—inhibiting enzymatic action and radically lowering temperature; and (6) pickling—greatly increasing acidity
beyond the tolerances of deadly micro-organisms.

PFB has said earlier, and says again, that anyone with the drive to
preserve food has the gumption to want to do it right. No big deal: anyone
who can ride herd on a backyard gas barbecue can follow the ways to
preserve food safely.

2
WHY FOODS SPOIL

Everything in this chapter applies to every safe method of preserving food at home, as you will see as *PFB* describes each process for each individual food. Meanwhile here follow the basic principles that are amplified throughout the rest of the book. They are corralled here for ready reference because we feel so strongly that any newcomer to canning, freezing, making preserves, drying, root-cellaring, or curing can always keep the *How* of food safety in mind if the *Why* is clear and handy.

Four kinds of things cause spoilage in preserved food: (1) *enzymes,* which are naturally occuring substances in living tissues, and are necessary to complete growth and reproduction; and three types of microorganisms—(2) *molds,* (3) *yeasts,* and (4) *bacteria.* All these microorganisms are present in the soil, water, and air around us. They can be controlled adequately by care in choosing ingredients, scrupulous cleanliness in handling them, and faithful good sense in following the established safe procedures for putting food by at home.

CONVERSIONS FOR PUTTING FOOD BY

Do look at the conversions for metrics (with workable roundings-off) and for altitude—both in Chapter 3—and apply them.

How Enzymes Act

Nature has designed each plant or animal with the ability to program the production of its own enzymes, which are the biochemicals that help the organism to ripen and mature—in short, enzymes promote the organic

3

changes necessary to the life cycle of all growing things. However, their action is reversible: they can turn around and cause decomposition, thereby causing changes in color, flavor, and texture, and making food unappetizing.

Their action slows down in cold conditions, increases most quickly between around 85 to 120 degrees Fahrenheit/29 to 49 degrees Celsius, and begins to be destroyed at about 140 F/60 C. However, the heat resistance of enzymes in a vinegar-loaded—and hence strong-acid—food like pickles can be stiffer than the opposition inherent in bacteria or molds or yeasts in the same type of food. The natural enzymes in cucumbers are especially tough.

How Molds Act

Molds are microscopic fungi whose dry spores (seeds) alight on food and start growing silken threads that can become slight streaks of discoloration in food or cover it with a mat of fuzz.

It used to be felt that "a little mold won't hurt you," but modern research has disclosed that only the mold introduced deliberately into the "blue" cheeses like Roquefort, Gorgonzola, or Stilton is trustworthy. The others, as they grow in food, are capable of producing substances called mycotoxins, and some of them can be hurtful indeed.

In addition, molds eat natural acid present in food, thereby lowering the acidity that is protection against more actively dangerous poisons— but we'll have more to say about this in a minute.

Molds are alive but don't grow below 32 F/Zero C; they start to grow above freezing, have their maximum acceleration at 50–100 F/10–38 C, and then taper off to inactivity beginning around 120 F/49 C; they die with increasing speed at temperatures from 140–190 F/60–88 C.

How Yeasts Act

The micro-organisms we call yeasts are also fungi grown from spores, and they cause fermentation—which is delightful in beer, necessary in sauerkraut, and horrid in applesauce. As with molds, severe cold holds them inactive, 50–100 F/10–38 C hurries their growth, and 140–190 F/ 60–88 C destroys them.

How Bacteria Act

Bacteria also are present in soil and water, and their spores, too, can be carried by the air. But bacteria are often far tougher than molds and yeasts are; certain ones actually thrive in heat that kills these fungi, and in some foods there can exist bacterial spores which can make hidden toxins.

TEMPERATURE *vs.* THE SPOILERS

F = Fahrenheit/C = Celsius

(At sea level to 1000 feet/305 meters altitude)

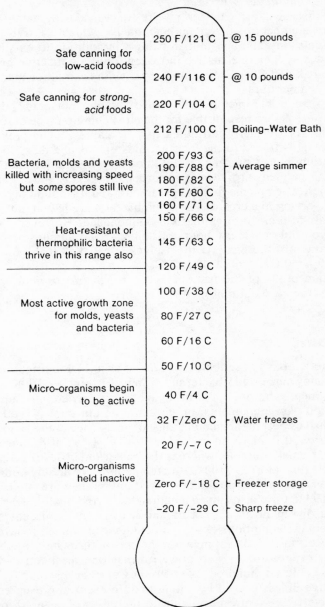

Safe canning for low-acid foods	250 F/121 C	@ 15 pounds
Safe canning for *strong-acid* foods	240 F/116 C	@ 10 pounds
	220 F/104 C	
	212 F/100 C	Boiling–Water Bath
Bacteria, molds and yeasts killed with increasing speed but *some* spores still live	200 F/93 C	Average simmer
	190 F/88 C	
	180 F/82 C	
	175 F/80 C	
	160 F/71 C	
	150 F/66 C	
Heat-resistant or thermophilic bacteria thrive in this range also	145 F/63 C	
	120 F/49 C	
Most active growth zone for molds, yeasts and bacteria	100 F/38 C	
	80 F/27 C	
	60 F/16 C	
	50 F/10 C	
Micro-organisms begin to be active	40 F/4 C	
	32 F/Zero C	Water freezes
	20 F/−7 C	
Micro-organisms held inactive	Zero F/−18 C	Freezer storage
	−20 F/−29 C	Sharp freeze

(Drawing by Irving Perkins Associates)

These spores will be destroyed in a reasonable time only if the food is heated at 240–250 F/116–121 C—*at least 28 degrees higher than the boiling temperature for water AT SEA-LEVEL CLASSIFICATION,* and *obtainable only under pressure.*

Of the disease-causing bacteria we're concerned with mainly, the most fragile are members of the genus Salmonella, which are transmitted by pets, rodents, insects, and human beings, in addition to existing in our soil and water. Salmonellae live in frozen food, are inactive up to 45 F/ 7 C, and are killed when held at 140 F/60 C, with destruction much quicker at higher temperatures.

The transmittable bacteria that cause "staph" poisoning, the *Staphlococcus aureus,* are responsible for the most common type of food-borne illness, the sort formerly attributed to "ptomaines." The prevalence of the poisoning is traced to the fact that most meats and dairy foods carry the bacterium but its presence is not known unless the foods are allowed to sit unrefrigerated at room temperature; it is specially widespread during the summer, when picnic food is spread out, and it is found in meats kept warm for serving to a crowd. The staph bacteria are halted in their tracks fairly easily at temperatures that kill Salmonellae. However, their toxin is destroyed only by many hours of boiling or 30 minutes at 240 F/116 C; the growth of the bacteria themselves is checked if the food is kept above 140 F/60 C or below 45 F/7 C.

Most dangerous of the bacteria is the tough *Clostridium botulinum,* which deserves a section of its own.

Botulism

The scientific books describe *C. botulinum* as a "soil-borne, mesophilic, spore-forming, anaerobic bacterium." Which, translated into everyday language, means that it is present in soil that is carried into our kitchens on raw foods, on implements, on clothing, on our hands—you name it. Next, it thrives best in the middle range of heat—beginning at about room temperature, 70 F/21 C, on up to 110 F/43 C. Next, it produces spores that are extremely durable: whereas the bacterium is destroyed in a relatively short time at 212 F/100 C, the temperature of briskly boiling water at sea level, the *spores* are not destroyed unless they are subjected to at least 240 F/116 C for a sustained length of time. And finally, the bacterium lives and grows in the *absence* of air (and also in a very moist environment; these combined conditions exist in a container of canned food).

This description does not mention the poison thrown off by the spores as they grow: the toxin is so powerful that one teaspoon of the pure substance could kill hundreds of thousands of people.

The grave illness caused by eating toxin present in preserved food is comparatively rare—rare in relation to the cases of "staph" or salmonellosis—but, unlike them, it is often fatal unless life-support care is given right away. The symptoms are blurred vision, slurred speech, inability to hold up the head, and eventual respiratory arrest unless the victim is given help to breathe until the body can reverse the progress of the illness with the aid of medication. Between 1899 and 1949 the case-fatality rate of food-borne botulism was 60 percent. Since then it has declined markedly and steadily, thanks to quicker diagnosis and great improvements in intensive care at the outset. The *case* rate took a brief upward spurt in the mid-seventies, however, because a whole generation of people went in for food preservation, especially home-canning, without having the information or equipment needed for a safe product.

The botulinum toxin can be destroyed by brisk boiling in an open vessel. This is why, throughout this book, we warn people to boil hard—for the time demanded by its density and acidity—any home-canned food *about which there is the slightest safety problem.*

C. botulinum is held inactive at freezing and comes into its own at room temperatures, as mentioned earlier. Type E, the comparatively rare food-borne strain that is found on sea- and freshwater seafoods, begins to grow at refrigeration temperatures.

As with all micro-organisms, a moisture content of below 35 percent directly inhibits its growth.

DEALING WITH THE SPOILERS

Enter *pH*—the Acidity Factor

The strength of the acid in any food determines to a great extent which of the spoilage micro-organisms can grow in each food. Therefore acidity is a built-in directive that tells us what temperatures are necessary to destroy these spoilers within it, and make it safe to eat. (Heat alone, not natural acidity, controls the action of enzymes.)

Acid strength is measured on the *pH scale,* which starts with strongest acid at 1 and declines to strongest alkali at 14, with the Neutral point at 7, where the food is considered neither acid nor alkaline. The *pH* ratings appear to run backward, since the larger the number, the less the acid, but it may help to think of the ratings as like sewing thread: size 100 cotton thread is smaller (finer) than size 60.

The term "*pH*" is an abbreviation in chemistry for "potential of hydrogen," traced by a sophisticated laboratory device. Crude indications can be made by litmus-like test papers gettable from scientific supply houses, in the Yellow Pages.

---

pH RATINGS

This listing indicates in a general way the natural acid strength of common foods on the *pH* scale. Different varieties of the same food will have different ratings, of course, as will identical varieties grown under different conditions. The Food and Drug Administration now uses a *pH* rating of 4.6 as the dividing point that dictates the preserving techniques to use: most notably, which canning method is right. Thus foods rated 2.2 (some lemons) on up to 4.6 (virtually every tomato, but see Chapter 8) *PFB* calls *strong-acid*—inserting "strong" as a sharper workable distinction than merely "acid," the customary designation. Foods over 4.6 up to Neutral 7 are always specified as *low-acid*. Foods over Neutral are *alkaline*.

Lemons	2.2–2.8	Beets	4.9–5.8
Gooseberries	2.8–3.0	Squash	5.0–5.4
Plums	2.8–4.0	Beans, string, green, wax	5.0–6.0
Apples	2.9–3.7	Spinach	5.1–5.9
Grapefruit	3.0–3.7	Cabbage	5.2–5.4
Strawberries	3.0–3.9	Turnips	5.2–5.6
Oranges	3.0–4.0	Peppers, green, bell	5.3
Rhubarb	3.1–3.2	Sweet potatoes	5.3–5.6
Blackberries	3.2–4.0 +	Asparagus	5.4–5.8
Cherries	3.2–4.2	Potatoes, white	5.4–6.0
Raspberries	2.8–3.6	Mushrooms	5.8–5.9
Blueberries	3.3–3.5	Peas	5.8–6.5
Sauerkraut	3.4–3.7	Tuna fish	5.9–6.1
Peaches	3.4–4.0 +	Beans, Lima	6.0–6.3
Apricots	3.4–4.4 +	Corn	6.0–6.8
Pears	3.6–4.4 +	Meats	6.0–6.9
Pineapple	3.7	Salmon	6.1–6.3
Tomatoes	4.0 + –4.6	Oysters	6.1–6.6
Pimientos	4.6–5.2	Milk, cow's	6.3–6.6
Pumpkin	4.8–5.2	Shrimp	6.8–7.0
Carrots	4.9–5.4	Hominy	6.8–8.0

The Importance of Sanitation

Even though the spoilage micro-organisms in a food are rendered inactive by cold, they kick in with a vengeance as soon as they warm up.

Or, adequately preheated food can be contaminated by air-borne spores while it, or its unfilled container, stands around unprotected; and, unless it is cooled quickly or refrigerated, the new batch of spoilers can start growing—and growing fast. In this connection, it's interesting to trace back the notion that one must not put hot food in a refrigerator in order to cool it quickly. This idea is a holdover from the days of the wooden ice chest, which was kept cool by a big block of ice: of course the hot food warmed the inside of the ice chest, and there was a long lag before the ice (or what was left of it) could reduce the internal temperature of the cabinet to a safe-holding coolness. With modern refrigerators, though, a container of warm food merely causes the thermostat to kick on, and the cooling machinery goes to work immediately.

Another good way to deal with the spoilers is simply to wash them off the food as carefully as possible, and to keep work surfaces and equipment sanitary at every stage in the procedure. Food scientists refer to the "bacterial load" in describing the almost staggering rate of increase in bacteria if the food is handled in an unsanitary manner or allowed to remain at the optimum growing temperature for the spoilers. The procedures described in this book have been established after years of research by food scientists, and naturally there is an allowance made for extra micro-organisms that must be destroyed. But in many cases the food would have to be processed almost beyond palatability if the bacterial load had been allowed to increase geometrically to an enormous extent; the alternative of course would be that the treatment was not intensified, so the food spoiled after all.

Household Disinfectants

Boiling water can destroy many organisms *if* it is indeed boiling and *if* it has long enough contact with a contaminated surface after it has dropped a hair below the boiling point. Cleaning work surfaces with boiling water is a cumbersome exercise, though; better to use a good household disinfectant, which, for our money, is the liquid chlorine bleach on hand in most kitchens. An extremely strong solution of such bleach is 1:4, or 1 part bleach to each 4 parts water; this should do a virtually instantaneous job of destroying bacteria—the drawback is that if the water is hot the fumes are strong enough to be unpleasant, even dangerous in large amounts. For a less strong yet efficient mixture, see Bacterial Load in Chapter 4.

Chlorine dissipates after a while: when you can smell it in the air, it is dissipating. (The best way to rid big-city tap water of too much chlorine is merely to draw a pitcher/bottle of water and let it sit uncovered in the refrigerator for several hours.)

The carbolic-acid-based disinfectants, generally pine-scented, and always milky in solution, are likely to leave their flavor on surfaces.

Heat + Acidity = Canning Safety

Combining (1) the temperatures that control life and growth of spoilage micro-organisms with (2) the *pH* acidity factor of the foods that are particularly hospitable to certain spoilers, gives the conscientious home-canner this rule to go by: *it is safe to can strong-acid foods at 212 F/100 C (at sea level to 1000 feet) in a Boiling–Water Bath, but low-acid/nonacid foods must be canned at 240 F/116 C (again, at sea-level zone)—a temperature possible only in a Pressure Canner at 10 pounds pressure—if they are to be safe.*

Processing times vary according to acidity and density of the food concerned. Adequate temperature and length of processing time are given in the specific instructions for individual foods.

3
ALTITUDE AND METRICS

This chapter is designed to provide the *Whys* and *Hows* for all altitude adjustments or metric conversions stipulated in the variety of preserving methods found in the chapters that follow. It is meant to quell any notions that home preservation is at all daunting as far as the numbers and "math" involved are concerned. For your altitude adjustments, especially, refer to the information here as often as is necessary—and make a note of any adjustments/conversions in the specific how-to's as you go along especially in the Canning chapters (6–12).

CORRECTING FOR ALTITUDE

More North Americans who preserve food in some manner live below 1000 feet/305 meters—in the sea-level zone—than live higher. But there remain a great many householders who, to ensure that their home-preserved foods are safe and appetizing, must compensate for the perceptible *decrease* in atmospheric pressure as the altitude *increases*. It's one thing to have a roast pork undercooked and yet dry; it's another thing to court food poisoning in a canner-ful of green beans.

Before we go any further, we must consider altitude.

Boiling: What Is It?

Boiling is the process by which a liquid (for our purposes here the liquid is water) becomes a vapor, and this conversion is produced by heat. We put water into a pot, and put the pot over a source of heat, and soon the water boils. In boiling, bubbles of vapor are created at the bottom of the vessel near the heat; and they rise to the surface and break and waft into

11

the air as steam, which is the gaseous form of water. As they rise, the bubbles agitate the water: gently as in a simmer, or violently as in a full boil.

When water is boiling—and by this we mean a thrashing, rolling boil—it looks the same at different altitudes *but it is not boiling at the same temperature*. The temperature is different because the blanket of air around us, called atmospheric pressure, is lighter when we live higher; and *the lighter the air the lower the boiling point of water*. At sea level our blanket of air weighs 14.7 pounds per square inch (psi), but at 10,000 feet/3048 meters the atmospheric pressure is only 10.1 psi; this difference of 4.6 psi will be the basis for the special thumbnail tables in the section Canning at High Altitude, coming soon.

However, back to boiling. Water at sea level, and for simplicity's sake up to 1000 ft/305 m, boils at 212 F/100 C; at 5000 ft/1524 m, boiling point is rounded to 203 F/95 C; and at 10,000 ft/3048 m, water boils at a mere 194 F/90 C. Right away you see that this makes it harder for us to apply a heat treatment that will deal with bad micro-organisms in the food we intend to can. Say that *Bacterium xyz* is destroyed if the container of strong-acid food (*pH* of 4.6 or lower) is held at 212 F/100 C for 35 minutes—but we live above the sea-level zone. What do we do?

We add processing *time*. Up to a feasible, sensible point, that is.

What if the food is low-acid (*pH* higher than 4.6) and of course we use a Pressure Canner to achieve the heat needed to destroy *Clostridium botulinum* spores, but we live well up in the mountains?

We add *Pressure*.

ALTITUDE AND WATER'S BOILING POINT
(WITH METRICS)

ALTITUDE FEET/ METERS	WATER BOILS FAHRENHEIT/ CELSIUS	ALTITUDE FEET/ METERS	WATER BOILS FAHRENHEIT/ CELSIUS
Sea Level: 0+/0+	212.0/100		
1000/305	210.0/98.9	5000/1524	202.6/94.8
1500/457	209.1/98.4	5500/1676	201.7/94.3
2000/610	208.2/97.9	6000/1829	200.7/93.7
2500/762	207.1/97.3	6500/1982	199.8/93.2
3000/914	206.2/96.8	7000/2134	198.7/92.6
3500/1067	205.3/96.3	7500/2286	198.0/92.2
4000/1219	204.4/95.8	8000/2438	196.9/91.6
4500/1372	203.4/95.2	10,000/3048	194.0/90.0

Time Out for "Simmer" and "Pasteurize"

There's no easy way to put these below-boiling terms later, so this interruption does get them recorded for future reference, and indexing.

The descriptions of *simmer* range from the French cook's "making the pot smile" on up to the one we mean when we say to simmer: which is 185–200 F/85–93 C at sea level. In this range small bubbles rise gently from the bottom of the pot, and the surface merely quivers, instead of dancing as in the full boil at 212 F/100 C.

We mention this now because we use a *Hot*–Water Bath in processing a number of canned fruit juices. The individual instructions specify the simmer temperature needed—180 F/82 C to 190 F/88 C—but again this is at sea-level, so at higher altitudes increase Hot–Water Bath processing time for such foods.

Pasteurization is the *partial* sterilization of food, and it was devised in the nineteenth century by Louis Pasteur as a means of controlling the fermentation of wine. As applied to milk, one method raises the temperature to 142–145 F/61–63 C, holds it there for 30 minutes, then quickly reduces the temperature to well below 50 F/10 C, where is is held for storage. The other way—called the "flash" method and used commercially by dairies—raises the temperature to 160–165 F/71–74 C for a mere 15 seconds, followed by rapid cooling to well below 50 F/10 C.

In addition we use the term *pasteurizing* in connection with the *Hot*–Water Bath for processing canned fruit juices (Chapter 7) and to ensure a safe seal for pickles, etc. in Chapter 19. Note that PFB recommends a "finishing" treatment by dry heat at 175 F/80 C for foods dried in open-air/sun.

Canning at High Altitude

The thumbnail tables on page 14 were given to *PFB* by Dr. Gerald Kuhn of Pennsylvania State University, who has organized, at the Center for Excellence, research for and publication of *The Complete Guide to Home Canning*. They simplify greatly the way that researchers have dealt with the idiosyncrasy of the dial *versus* the deadweight gauges of 12-quart Pressure Canners used for research for the Guide. Amplifying suggestions for applying its research will be published by the Penn State scientists. It is these tables that householders are referred to when advised to insert their own altitude adjustments in the instructions for specific foods in Chapters 7 through 12.

PFB has no intent to gainsay any part of the presentation of research done for *The Complete Guide to Home Canning* (1988/89) when we point out recommendations made over the years by food technologists who work closely with the public at very high altitudes. In the Canadian and American Rockies it has been the accepted custom to add *time* to

15-psig instructions, because it is not feasible to suggest canning at pressures between 15 and 20 psig: the usual householder is leery of pressures taken deliberately higher than 15 psig. Betty Wenstadt of Presto points out that much research was done by University of Minnesota food scientists for her company, and a copy of related material may be had by writing to her at National Presto Industries, Inc., 3925 North Hastings Way, Eau Claire, Wisconsin 54701.

DR. KUHN'S RULES-OF-THUMB FOR CANNING AT ALTITUDE

(If your county's Cooperative Extension Service has no recommendations for your particular area, ask your State University CES for help.)

If the *base* Boiling–Water Bath time, always given as for the sea level zone, is 20 minutes or less, increase the time by increments of 5, thus (using Hot-packed pints of stewed rhubarb as our example):

0–1000 ft/0–305 m = 10 minutes
1001–6000 ft/305–1829 m = 15 minutes
6001–9000 ft/1829–2743 m = 20 minutes

If the *base* Boiling–Water Bath time is more than 20 minutes (as, say, for a pint of Hot-packed quartered tomatoes with an added 1 tablespoon of bottled lemon juice):

0–1000 ft/0–305 m = 35 minutes
1001–3000 ft/305–914 m = 40 minutes
3001–6000 ft/914–1829 m = 45 minutes
6001–(9000) ft/1829–(2743) m = 50 minutes

For Pressure-processing low-acid foods in a *Dial*-Gauge Pressure Canner ("psig" = pounds per square inch by gauge)—using as our example the ever-present green/snap/wax bean, Hot-packed in pints and processed for 20 minutes:

0–1000 ft/0–305 m = 11 psig for 20 minutes
1001–3000 ft/305–914 m = 12 psig for 20 minutes
3001–5000 ft/914–1524 m = 13 psig for 20 minutes
5001–7000 ft/1524–2134 m = 14 psig for 20 minutes
7001+ ft/2134+ m = 15 psig for 20 minutes

For Pressure-processing low-acid food in a Deadweight (i.e., weighted) Gauge Canner—using the same example of green beans processed for 20 minutes:

0–1000 ft/0–305 m = 10 psig for 20 minutes
1001–9000+ ft/305–2743+ m = 15 psig for 20 minutes

Summarizing: Ways to Deal with Altitude

Because water boils at lower temperatures the higher we go above sea level—and therefore is less effective as a destroyer of micro-organisms with every increase in altitude—we should:

1) Add processing time in a Boiling–Water Bath as the USDA advises (given recently in this section), *but*—and this recommendation is our own, based on interview material—*only up to 2500 ft/762 m*. There and higher, go to Pressure-processing. Updated instructions for using your Pressure Canner will help here. For some rules-of-thumb from makers, see notes for this chapter in the Appendix. *PFB* arrived at this opinion after talking with experts whose jobs are to handle everyday problems posed by everyday people. Research scientists have our wholehearted respect; our feeling for workers out in the field comes from the difference between "making do" as opposed to what's called in the deep country "having everything to do *with*."

2) Remember that gas (steam) expands more at higher altitudes, so we allow extra room for it inside our canning jars. This is called headroom, and it's shown in General Steps in Canning, in Chapter 6. If we did not increase headroom, the contents of the jars would erupt out between the lid and the sealing rim. Bad business.

3) The virtues and drawbacks of dial gauges compared with deadweight/weighted gauges are matters for later discussion. But unless you'd bet your life on the accuracy of your dial—and maybe you are, at that—it's safer and easier to go along with the 5–10–15 psig of the deadweight gauge at high altitudes.

Other Bothers with Altitude

Jelly-making and cooking syrups for candy and cake frosting are affected by lowered boiling points as you go higher. What happens is that water in the juice/syrup evaporates faster as the altitude increases. So if you are counting on the jelly stage being reached at 8 F/4.4 C above the boiling point of water at sea level—or a reading of 220 F/104.4 C—you'd have some mighty stiff jelly. To deal with this, Pat Kendall says to lower the final cooking temperature by around 2 F/1.1 C for each 1000 ft/305 m in elevation. Quickly from Fahrenheit, then, the temperature would be 218 at 2000 feet, 216 at 3000, 214 at 4000, and so on.

Maybe the Sheet Test (Chapter 18, "Jellies, Jams, and Other Sweet Things") is your best bet.

Just plain boiling food to cook it will take longer—and the pot might be in danger of going dry: again, the problems are lower boiling points plus greater evaporation. These factors also must be considered in cooking by steaming.

Blanching before freezing or drying. Hard to make an across-the-board recommendation here, because so much depends on the food (certainly all vegetables) being treated—age, tenderness, cut size. Leaf vegetables are usually better boiled than steamed, because they tend to mat in the steam basket, but can roll around if gathered loosely in cheesecloth and popped into water at a thrashing boil. Dr. Pat Kendall of Colorado State University Extension (Fort Collins) offers a good rule-of-thumb for the "Centennial State," where average altitude is 4000 ft/1219 m above sea level. She says that dwellers at 5000 ft/1524 m should preheat vegetables for *one minute longer* than the sea level zone requirement—and should *not* increase blanching time further at greater altitude.

Baking: yeast-bread doughs rise very quickly, thus allowing less time for flavor to develop; Professor Kendall suggests letting yeast-leavened mixtures rise twice. The gases from baking powder also expand more, as does the trapped air in beaten egg whites.

Roasting at high altitudes can result in an expensive chunk of meat's being dry outside and woefully underdone inside.

Deep-frying: temperature of the fat must be reduced the higher you go, she says. Above 3000 ft/914 m, temperature of the fat could be lowered 3 F/1.7 C for each added 1000 ft/305 m as a way to avoid outsides too dark and insides not fully cooked.

Using a double-boiler can be tricky because water boils at so much lower a temperature that many starchy thickeners are not activated to do their job. Better, Pat Kendall says, to use a fine, heavy, enameled saucepan over well-managed direct heat.

METRIC CONVERSIONS

In her crisp, commonsense fashion, Canada has brought off "going metric," thereby letting her people join householders from most other major powers in figuring everyday weights and measures by the universal system of metrics. The notable holdout is still the United States. True, we have moved along a bit in the past decade: Americans get soda pop by the liter now (with auxiliary labeling to tell quarts and fluid ounces); and, though milk may be sold by the half-gallon, volume also is given in liters, with milliliters noted as percentages (which is fair enough). More important is the appearance of metric measurements alongside the old-style ones in the new cookbooks—especially those with an international flair.

Metric Conversions for Temperature

In 1714 Gabriel Daniel Fahrenheit created a thermometer using a scale based on human body temperature and set Zero as the freezing point of a mixture of *ice and salt,* with 32 F (*Fahrenheit*) the freezing point for *pure water,* and 212 F as the boiling point of pure water. Soon—only 28 years later—Anders Celsius created his scale whereby pure water freezes at Zero and boils at sea level at 100 C (*Celsius,* or also called Centigrade). The two scales intersect only at *minus 40 degrees* both *F* and *C*. And the boiling points of pure water on both scales depend on a reading at sea level. This insistence on *sea level* has great bearing on variations in boiling temperatures when altitude increases (as we saw earlier in this chapter—and will apply in all canning specifics later on).

Because the usual American householder's measuring equipment is not precise enough to reflect the mathematical computations in the temperatures that follow, we indicate *in italics* the rounded-off figure that offers a reasonable equivalent for everyday use.

FAHRENHEIT AND CELSIUS (CENTIGRADE)
(At Sea Level to 1000 Feet/305 Meters Altitude)

F	C (ROUNDED)	F	C (ROUNDED)	F	C (ROUNDED)
− 35	− 37	80	27	220	104
− 30	− 34	85	29	225	107
− 25	− 32	90	32	228	109
− 20	− 29	100	38	230	110
− 15	− 26	110	43	235	113
− 10	− 23	120	49	240	116
− 5	− 21	130	54	245	118
0	− 18	140	60	250	121
5	− 15	142	61	255	124
10	− 12	145	63	260	127
15	− 9	150	66	265	130
20	− 7	160	71	270	132
25	− 4	165	74	275	135
30	− 1	170	77	280	138
32	0	175	80	285	141
35	2	180	82	290	143
36	2	185	85	295	146
37	3	190	88	300	149
38	3	195	91	325	163
39	4	198	92	350	177
40	4	200	93	375	191
45	7	203	95	400	205
50	10	205	96	425	218
60	16	208	98	450	232
65	18	210	99	475	246
70	21	212	100	500	260

The Arithmetic

The list of equivalents includes the temperatures cited frequently in this book. To convert other temperatures, you can use the following formulas.

If you know Fahrenheit, you can find *Celsius* by subtracting 32, then multiplying by 5/9 (i.e., multiply by 5 and divide by 9). For example, 200 F = (200 − 32) × 5 ÷ 9 = 93 C.

If you know Celsius, you can find *Fahrenheit* by multiplying by 9/5 then adding 32. For example, 93 C = 93 × 9 ÷ 5 + 32 = 200 F.

Metric Conversions for Volume

For ready reference, *PFB* uses ovenproof glass volume measures, one side marked for fluid ounces and cups, the other in milliliters (also in *deciliters,* or 1/10-liter steps—which so far haven't cropped up much; but not in *gills,* a rarely used amount that equals ¼ pint, or ½ U.S. cup). The comparison set of spoons offers double ends, one side going from ¼ teaspoon, to ½, to 1 teaspoon, to 1 tablespoon, and the other side giving the workable metric equivalent.

The terms *teaspoon, tablespoon, gill, cup, pint, quart,* and *gallon* in the list that follows—and indeed throughout the book—are U.S. measurements. Although their volumes differ in some cases from those of similar British or European designations, their metric equivalents provide the means for translating recipes from around the world. Further (but marginal): Americans are used to seeing the term *cubic centimeter* for volume in pharmaceutical prescriptions and solutions, etc., so note that *1 cubic centimeter (cc) = 1 milliliter (mL).*

The first list below—The (Fairly Pure) Arithmetic—shows how to figure back and forth between units of current standard volume and metrics, because it's important to know the method, and you can always fall back on it. Because the householders' means for measuring volume metrically are so much less precise than are the available thermometers, diet scales, and foot/centimeter rules, we must round off volume measurements quite roughly ("Now Rounded Off"). They're workable for now, though; and they often come on measuring cups and canning jars.

The (Fairly Pure) Arithmetic

Start with the U.S. fluid ounce, and work back for spoonfuls:

One fluid ounce (fl oz) = 29.574 milliliters (mL).
There are 2 tablespoons in 1 fl oz, so 1 tablespoon = 14.787 mL.
There are 3 teaspoons in 1 tablespoon, so 1 teaspoon = 4.929 mL.
To get ½ teaspoon, divide 4.929 by 2 = 2.465 mL.
To get ¼ teaspoon, divide 4.929 by 4 = 1.232 mL.
To get ⅛ teaspoon (usually "a pinch"), divide 4.929 by 8 = 0.6161 mL.

Now work forward for cups, etc.:

One fluid ounce (fl oz) = 29.574 mL.
To get ¼ cup (2 fl oz), multiply 29.574 by 2 = 59.148 mL.
To get ⅓ cup (2⅔ fl oz), multiply 29.574 by 2.67 = 78.96 mL.
To get ½ cup (1 gill; but for use here, 4 fl oz), multiply by 4 = 118.30 mL.
To get ⅔ cup (5⅓ fl oz), multiply 29.574 by 5.33 = 157.74 mL.
To get ¾ cup (6 fl oz), multiply 29.574 by 6 = 177.44 mL.
To get 1 cup (8 fl oz) multiply 29.574 by 8 = 236.59 mL.
To get 1 pint (2 cups, 16 fl oz), multiply 29.574 by 16 = 473.18 mL.
To get 1 quart (4 cups, 32 fl oz), multiply 29.574 by 32 = 936.36 mL.
To get 1 gallon (4 quarts, 128 fl oz), multiply 29.574 by 128 = 3785.47 mL.

Now Rounded Off

When America goes convincingly metric, our measuring cups and spoons will be much closer to the real figures we have just made. With the correlations possible with our current utensils, though, we round them as best we can: 10 or so milliliters when we're dealing with cupfuls in the kitchen is not a killing matter.

⅛ teaspoon = *0.62 mL*	⅛ cup = 1 fl oz = *30 mL*
¼ teaspoon = *1.25 mL*	¼ cup = 2 fl oz = *60 mL*
½ teaspoon = *2.5 mL*	⅓ cup = 2⅔ fl oz = *80 mL*
¾ teaspoon = *3.75 mL*	½ cup = 4 fl oz = *125 mL*
1 teaspoon = *5 mL*	¾ cup = 6 fl oz = *200 mL*
1¼ teaspoons = *6.25 mL*	1 cup = 8 fl oz = *250 mL*
1½ teaspoons = *7.5 mL*	2 cups = 16 fl oz = *500 mL*
1¾ teaspoons = *8.75 mL*	1 (U.S.) pint = 16 fl oz = *500 mL*
2 teaspoons = *10 mL*	1 quart = 32 fl oz = *1000 mL/1 liter (L)*
2½ teaspoons = *12.5 mL*	4 quarts = 128 fl oz = *4 L*
3 teaspoons = ½ fl oz = *15 mL*	1 gallon = 128 fl oz = *4 L*
1 tablespoon = ½ fl oz = *15 mL*	(1 deciliter = *0.10 L*)
2 tablespoons = 1 fl oz = *30 mL*	(1 decaliter = *10 L*)

Metric Conversions for Weight

If you know ounces avoirdupois (oz av), you can find grams (g) by multiplying by 28.35.

If you know pounds (lb: 16 oz av), you can find kilograms (kg: 1000 g) by multiplying by 0.4536.

If you know grams, you can find ounces avoirdupois by multiplying by 0.0353 (1 divided by 0.4536 = 0.0353: some find it easier to set up a multiplication than a division).

If you know kilograms (kg), you can find pounds by multiplying by 2.205 (1 divided by 0.4536 = 2.20458 = 2.205).

Because even our good spring-type scales are not refined enough to reflect mathematical figuring in the amounts below, we show *in italics* a rounded-off figure that offers a reasonable equivalent for everyday use.

1 oz ($\frac{1}{16}$ lb)	= *28 g*	2 lbs	= *0.91 kg*
2 oz ($\frac{1}{8}$ lb)	= *57 g*	5 lbs	= *2.27 kg*
4 oz ($\frac{1}{4}$ lb)	= *113 g*	10 lbs	= *4.54 kg*
8 oz ($\frac{1}{2}$ lb)	= *227 g*	15 lbs	= *6.80 kg*
12 oz ($\frac{3}{4}$ lb)	= *340 g*	20 lbs	= *9.07 kg*
16 oz (1 lb)	= *454 g/0.454 kg*	25 lbs	= *11.34 kg*

Metric Conversions for Length

Automobile mechanics and machinists are luckier than homebodies when it comes to metrics, because so much of their equipment has meticulous conversions that allow workers to deal with the measurements of foreign supplies and specifications.

If you know inches (in.), you can find centimeters (cm: 1/100 meter) by multiplying by 2.54. (*PFB* doesn't deal with millimeters—mm: 1/1000 meter—but if you know inches, you can find millimeters by multiplying by 25.40).

If you know centimeters, you can find inches by multiplying by 0.3937 (1 divided by 2.54 = 0.3937).

If you know millimeters, you can find inches by multiplying by 0.0394 (1 divided by 25.40 = 0.03937).

Because the usual U.S. householder's measuring equipment is not precise enough to reflect the mathematical computations for the amounts below, we show *in italics* a rounded-off figure that offers, as best it can for the time being, a reasonable equivalent for everyday use.

Now Rounded Off

$\frac{1}{8}$ in.	= *0.31 cm*	2 in.	= *5.1 cm*
$\frac{1}{4}$ in.	= *0.63 cm*	3 in.	= *7.5 cm*
$\frac{1}{2}$ in.	= *1.25 cm*	4 in.	= *10 cm*
$\frac{3}{4}$ in.	= *1.90 cm*	6 in.	= *15 cm*
1 in.	= *2.50 cm*	8 in.	= *20 cm*
$1\frac{1}{2}$ in.	= *3.75 cm*	12 in.	= *30 cm*

4
FAIR WARNING

With the older editions of *PFB,* and their successive printings, has come a permanent and ongoing collection of oddments that go from polite information to hair-raising tales. Here are updates on some concerns that have stayed alive or multiplied over the past six years or so.

IRRADIATION

Technically, we should ignore the matter, because there is no chance that the North American householder will start treating his put-by food with little shots of gamma rays, or whatever. This is just a casual summing-up to date.

First, something is treated *with* radiation, or it is treated *by ir*radiation.

No doses of radiation currently allowed on items in our food supply are intended to sterilize, because radiation is "dose dependent." This means that one or more specific targets are chosen to be dealt with, usually with the idea of prolonging shelf-life. A good example is subjecting pork in the United States to radiation to kill the trichina worms and eggs that infest pork meat. Or it may be used on certain fruits and vegetables, where a protein substance that's part of an enzyme is targeted in order to slow down a perfectly natural progression in the cycle of food spoilage. Grains and their by-products have been treated with radiation for some years, to kill off infestations of insects that attack grains in storage.

Irradiated food is not radioactive. The example given most often is that human beings are not rendered radioactive by a dentist's X-Ray machine.

Some forty nations use radiation in treating their food supplies. The United States is slower to apply radiation than most of them.

SO WHAT'S A LITTLE BIT OF MOLD . . .

A mold is a fungus, and the fungi include mushrooms like truffles or the elegant morel, and yeasts, plus the specially cultured growths that trace blue veins in a noble Stilton cheese and or give us the means of producing streptomycin and penicillin. Surely these molds can't be said to hurt you.

But then there are the fungi that embrace toadstools and mildew, and the mold on fruits and vegetables and grains and nuts and seeds, and on food too long in the refrigerator, and in foods improperly canned or sealed. Inhale spores from some of these molds and you can develop allergies or suffer injury to your bronchia and lungs. As they grow, some molds throw off poisons called mycotoxins (much like the way spores of *Clostridium botulinum* make the deadly toxin that causes botulism); of the mycotoxins, the one with the most and worst publicity is aflatoxin. Aflatoxin occurs naturally on nuts and peanuts and corn and wheat and millet and rice and cottonseed among other things, and none of the usual means of destroying a poison like, say, botulinum toxin is able to faze it.

Mold grows in warmish temperatures on foodstuffs exposed to moisture and oxygen. These are the reasons why grains must be stored dry and kept cool, and fresh vegetables and fruit must be kept cool, and canned food must be kept sealed after it has been processed (and processed correctly in the first place, to destroy mold spores in the food and in the containers). In addition to attacking plant materials, molds in a refrigerator will invade preserves and ham and cold cuts and dairy products and cake and bread; they just take longer to do so in cool storage.

So far molds have been a matter for routine good handling. *PFB* is especially harsh about molds because they can get into an apparently well-sealed jar of canned food in the 4.5 or 4.6 *pH* range—that is, in foods rated as strong-acid—and they can consume just enough acid to make the contents tip over into the low-acid category, thus giving any undestroyed spores of *C. botulinum* a medium in which they can flourish. And as these spores multiply, they make the dreaded poison; and as they make their poison, they can produce a different sort of acid in minuscule amounts. But apparently these amounts can add up to enough acidity to restore the *pH* rating that should guarantee safety IF the food had been canned with reliable procedures in the first place.

This is not a scenario made up to be scary. It actually occurred in the early 1970's when tomatoes were canned by the old open-kettle method, long in disrepute. Canned juice, even canned pears, also caused an outbreak of botulism. In the last case it was considered that specks of food lodged on the sealing-rim of the jar, preventing a true seal—and thereby giving mold a chance to eat its way into the contents. Or maybe a bad sealing-ring was used by a cook who wasn't much bothered by mold.

USING SPACE-AGE PLASTICS

In Chapter 21, "Drying," *PFB* warns against using any plastic for storing food of any kind unless the wrap/bag/box is rated as *food-grade*. So how to find out if one of these plastic things is O.K. to use with food, when the labels don't tell you NOT to do it?

First, really examine the label. Just as certain wraps and bags are quick to tell you that they're "ideal for freezing" or that they're intended for use in freezers in the first place, so will wrappings that are food-grade tell you to use them to protect food.

If the blurb or the label does not mention food, consider that the plastics are *not O.K.'d* for use with food. Such plastics are those used in making waste bins or trash cans, in liners for waste-collectors, and lawn/leaf/trash bags. The best-known makers of plastics for such use are the chemical division of Mobil, and Union Carbide. Both these giants take the trouble to warn *against* using their trash plastics for food. Neither company has asked the Food and Drug Administration to approve these particular plastic products for use with food; we understand that they're not about to, either. Both, along with Dow, make highly publicized lines of food-grade plastic containers and wraps.

Trash plastics contain a chemical toughener and usually a substance called a "plasticizer" added to the ingredients to allow the goop to extrude readily in the form desired (as a sheet, or as toothbrush handles, or as cat boxes, or whatever). These plasticizer chemicals often contain cadmium—definitely not a thing for human beings to mess with, much less ingest, since all its soluble compounds are poisonous. As are the garbage bags with built-in germicides.

BACTERIAL LOAD

Everyone careful in handling food knows that the best way to have safe food is to start with clean food; and that the utensils and boards and holding vessels must be immaculate or they will supply contamination.

Over the years, *PFB* has been trying to come up with a truly simple, easy-to-make-from-household-stuff germicide—something that will do a competent job of dealing with gross bacterial load, and deadly microorganisms. Now we have it. And it's from one of our favorite publications, *The Morbidity and Mortality Weekly Report* from the Centers for Disease Control, in Atlanta.

In a round-up for health professionals on dealing with surfaces possibly contaminated by the AIDS virus, the recommendation was 1:10 household bleach and water. Earlier this same publication had cited 1 part of

5.25 percent sodium hypochlorite (good old household bleach) to 9 parts of water. This 1:9 solution, then, certainly would disinfect cutting boards, sinks, counters, you name it, after we have been dealing with Salmonellae in our cut-up fryers, etc.

PFB is pleased to pass this along.

After food—and food utensil—cleanliness, the food-handler knows that it's not merely O.K., its *absolutely necessary* to refrigerate food that's not going to be held at 140 F/60 C or above until it is eaten. *PFB* is leery of any steam-table unless it is supervised by a professional, and generally we deplore casual "keeping something warm" at home. For several generations now, the refrigerator has been capable of kicking its motor on automatically to cool any warm food put into it; certainly it's no trick to reheat quickly any dish to be served. (A case of botulism was traced to a frozen pot pie that was not completely cooked when the oven was turned off; it stayed in the oven until the next day, as we recollect the report; and then it was reheated and eaten by a hungry teenager—who recovered after intensive care in the hospital.)

PFB also is stiff-necked about refreezing thawed food, especially poultry and seafood—actually, any low-acid thing that has not been thoroughly cooked. When you look at the thermometer-chart in Chapter 2, you see the range where bacteria begin to grow and where they thrive. When the package goes back into the freezer, the micro-organisms don't die—they just become inactive. Then if that piece of food is thawed on a countertop before being cooked, and the way it finally is cooked doesn't take care of the manifold increase of bacteria . . .

Salmonella food poisoning is nothing like the killer that *C. botulinum* can be, but it's all too commonplace in poultry and eggs, etc., etc. *PFB* removes eggs from their carton and wipes each one with a clean dishcloth wrung out in the 1:9 solution, before stashing them in their holders in the refrigerator.

HOME-CURING PORK

Professionals should do the slaughtering of meat animals, and the carcasses should be correctly hung and bled and chilled. *PFB* got a letter, restrained under the circumstances, from a reader who'd followed our directions for preparing the cure, and applying it, and holding the hams and bacon properly and at 38 F/3 C. She was clear and definite about the temperature; they'd used an old refrigerator, kept a thermometer inside, and checked it often. Despite such care the meat spoiled: it discolored and a nasty liquid collected in the bottom of the container.

We took the problem to various Agriculture Department experts, all of whom agreed that our directions were correct, and they offered these reasons for the spoilage:

The carcass had not been bled immediately or fully.

The carcass had been improperly chilled. This can mean either that the flesh had been allowed to freeze to some extent, thereby making it less capable of absorbing salt; or it had not been cooled adequately and quickly enough, and so the bacterial load had increased to such an extent that the cure could not offset its progress.

Perhaps the cure had contained ingredients that are not food-grade. Standard commercial mixtures are food-grade (even though some consumers jib at the nitrites), but an ice-melting type of salt might have been introduced into the cure.

Perhaps sea salt—which could mean solar salt, dried in beds (see Chapters 5 and 20)—*had been part of the cure.* Even if it is food-grade and O.K. to ingest, apparently sea salt can contain substances that react poorly with other natural chemicals in the raw meat (of special interest, because we'd not been told this before). Salt sold commercially as *solar* is loaded with impurities.

5
COMMON INGREDIENTS AND HOW TO USE THEM

Water

There's hardly any method of putting food by that does not involve water somewhere along the line, beginning with the first washing of raw materials, continuing to the various water-based solutions used in canning and freezing and curing—which includes brining of all sorts, to ferment or to preserve—and even extending to some steps in drying.

The most important single thing about water used in any process for preserving food is this: *The water must be fresh and at least of drinking quality*. A staggering number of spoilage micro-organisms are added to food by impure water. Therefore:

Don't assume that "it's-going-to-be-cooked-anyway" will counteract *all* the extra contamination. Remember that an excessive bacterial load can tax your preserving method beyond the point where it is effective. (Nor is there any sure-fire, non-toxic sterilizing substance that you can tuck into a container of food before you process it—never mind what folkloric compounds keep surfacing in descriptions of ye olde-tyme methods.)

In any step of preserving food, don't use any water that you cannot vouch for as safe to drink. In the late 1970's an outbreak of botulism from home-canned food was traced to the soil-borne bacteria in a family's water supply; the canning procedures were not able to deal with them.

And change wash-water often, or wash under running tap water you would drink.

Dealing with Minerals in Water

"Hard" water has above-average mineral content (calcium is often an offender here). Hard water can shrivel pickles or toughen vegetables.

You can check for hardness by shaking a small amount of soap—*not* detergent—in a jar of water: if it makes a good head of suds in your water, hardness is not a problem. Or ask your municipal water department to tell you the composition of the water that comes from your tap. In rural areas, your health officer can tell you how to have your private water supply tested for mineral content as well as for bacterial count.

Where there's *no dangerous air pollution* ("acid rain," etc): if your water is hard, and you can't get distilled water, collect rainwater *in the open*—not as run-off from dusty roofs—and strain it through layers of cheesecloth.

Or if you know that hardness is caused by calcium or magnesium carbonates, boil the water for 20 to 30 minutes to settle out the mineral salts; then pour off and save the relatively soft water, taking care not to disturb the sediment.

If you plan to make pickles with whole small cucumbers, another means of dealing with very hard water is to boil it for 15 minutes, then let it stand carefully covered for 24 hours; remove the expected light film of scum from the top, and do not disturb the sediment. Pour off the water carefully into a clean, large container, again not disturbing the sediment; add 1 tablespoon white vinegar for each gallon of boiled water before you use it.

Or add ½ cup vinegar to each 1 gallon of water, so the acid will cause precipitation of calcium and other minerals; then pour off the treated water, as above.

Iron compounds in your water will darken the foods you put by.

In preparing fish or shellfish for processing or freezing, do not wash in sea water or use sea water for precooking, etc.

Don't use water that contains sulfur if you can avoid doing so. The sulfates in water do settle out with boiling, but become more concentrated as the water evaporates. Sulfur will darken foods.

Salt

Only about 2 percent of all the salt—the common salt all of us know, which is sodium chloride (NaCl)—used in the United States is used for food, a statistic indicating that great amounts of available salt are not food-grade. In this book we refer often to salt, and always mean either table salt or a pickling salt that is fit to put in our food. We offer a quick description of the various kinds of salt that are offered at retail to householders in order to clear up some misunderstandings that probably stem from simple misnaming.

The differing "cuts" of popular canning salt, and how their bulk (volume) is affected. *PFB* prefers the Sterling. (*Photograph by Jeffery V. Baird*)

Use of any salt is optional: added salt to canned foods is only a flavoring, because it is included in amounts so small that it has no effect as a preservative. It is up to individual choice to avoid foods that need a good deal of salt as a preservative (certain pickles, cured meats, fish, vegetables—Chapters 19, 20, and 21).

Table salt is finely ground; is either plain or with an added iodine compound (if it is iodized the legend on the box will say so); and it has a non-caking agent as a filler. This "free-running" additive that prevents caking is not soluble, and therefore gives a cloudiness to a canning or pickling liquid. The very small optional amounts in canning are not enough to produce noticeable cloudiness, but this effect would be quite apparent in pickling liquids that contain much greater amounts of salt.

Any iodine in the table salt would discolor or darken pickled foods; this is why we specify plain pickling salt for such food.

Characteristically, table salt is in the form of tiny cubes. Its density is such that 1 pound of this salt will equal as near to 1⅓ cups as makes no difference in household use.

Cooking and canning salt usually comes in 5-pound containers. It is pure sodium chloride with nothing added to it—no agent to prevent caking, no iodine compound. Its density equals that of the table salt above, so 1 pound of it will also be about 1⅓ cups.

Kosher salt is so named because it is used in the ritual cleansing of food in the Jewish religion, and it can come in granules coarser than table salt, or as "gourmet" *flaked* salt (i.e., made by rolling out granulated salt). The density of this particular flaked salt from the Morton company and sold in the blue, rectangular, 3-pound box is such that 1 pound makes about 1½ cups. Another kosher salt, this one from the Diamond Crystal company (orange carton) and made by the Alberger process, is even fluffier: it takes about 1⅔ cups to equal 1 pound.

Dairy salt, though it is mentioned in many cookbooks as an acceptable food-grade salt sold at retail, is not sold under this name by Farm Bureau outlets or by the large regional and national seed-and-feed suppliers in *PFB*'s area. Instead, the salt used in making butter, etc., by farm households is table/pickling salt sold in 25-pound bags, and is clearly designated as *food-grade* and containing no anti-caking or iodine compounds. (NOT TO BE USED with food for human beings are the trace-mineral salts designed to be, and sold as, supplements for animal feed.)

Sea salt comes in many grades. One is sold in gourmet and natural-food stores and is considered to be purified enough for use with food; it is quite expensive compared with regular table/canning/pickling salt, but is used presumably because it contains minerals from sea water (the minerals can affect the color of preserved foods). It often is coarse, and is ground in a hand-mill at the table. Its cost and its mineral content are likely to discourage its use as a preservative for food.

Solar salt sometimes is referred to as "sea salt," but it is a far cry from the gourmet variety just mentioned, and is *not food-grade,* is *not labeled food-grade,* and has never been promoted by its manufacturer as food-grade. Instead, this salt is produced mainly for use in water-purifying systems, and its coarse, rather soft pellets are discolored by the organic residues of dead aquatic life. It is called "solar" because it is evaporated in open ponds; among the sources of it in North America are the Great Salt Lake, San Francisco Bay, Baja California around the Sea of Cortez, Great Anagua in the Bahamas (legendary Turks Island salt comes from this chain of islands in the Caribbean).

The 50-pound bag *PFB* bought on our comparison-shopping spree cost less than 10 cents per pound, an attractive price for a bad ingredient.

(Apropos Turks Island salt, which was a favorite several generations ago, we bought some at 25 cents per pound, and 1 pound made $1\frac{3}{4}$ cups. The salt was bought in bulk by the store, which did not guarantee its being food-grade, but just said that "the older folks used to use it a lot, but we seldom have a call for it.")

Triple Warning: never use (1) solar sea-salt in curing, *especially* in curing meat and similar low-acid foods—not food-grade, it contains substances that interact with protein to cause spoilage. Never use (2) halite salt, the sort used to clear ice from walkways—it is not food-grade, and therefore could contaminate the food it was supposed to protect. Finally (3), never use a salt substitute as a seasoning in preparing food that is going to be heat-processed—it can have an unpleasant aftertaste from canning or even just cooking.

Brining

Brine is salt dissolved in liquid, and for the purposes of this book there are two kinds of brine. One comes from adding pure pickling salt proportionate to the volume of water. The other is the solution that results when dry salt is added to plant or animal material to draw out the juices from the tissues, and the salt combines with the juices.

When we describe brine as being of a certain *percent,* we are referring to the proportion of weight of salt to the volume of liquid—*NOT* to the sophisticated salinometer/salometer reading used in laboratories or in industry (and occasionally confused in some bulletins with the simple weight/volume rule-of-thumb that is adequate for use in putting food by at home). Which is: 1 part of salt to 19 parts of water (totaling 20 parts) = 5 percent, allows some benign fermentation; 1 part of salt to 9 parts of water (totaling 10) = 10 percent, the strongest solution generally used in pickling at home, prevents the growth of most bacteria.

Brine of $2\frac{1}{2}$ percent actually encourages benign fermentation; brine of 20 percent controls the growth of even salt-tolerant bacteria (see Chapter 20).

These percentages are offered for your ready reference, but are not really needed here: at every stage in every procedure involving brine, we tell you how much salt and how much liquid you need to achieve the result desired. Above, we gave volume-per-pound equivalents for the main types of salt used in pickling and preserving just as a help in figuring these proportions.

Sweeteners

The relatively small amount of sugar (sucrose) or alternative natural sweetener used in canning or freezing fruit helps keep the color, texture, and flavor of the food, *but it is optional.*

The sugar in jams and jellies (Chapter 18) *helps the gel to form,* points up the flavor, and, in the large amount called for, acts as a preservative. As a preservative and for the gel sugar is *not* optional. (The so-called sugarless or "diet" confections rely on gelatin, not pectin; are made with artificial non-nutritive sweeteners; and must be refrigerated or frozen.)

A sweetener is also an ingredient in some vinegar solutions used for vegetable pickles, where it enhances the flavor (Chapter 19).

In curing meats like ham, bacon, etc., a relatively small amount of sugar (perhaps brown) is combined with the salt; but the sweetener is added more to feed flavor-producing bacteria than to provide flavor on its own (Chapter 20).

In canning, freezing, and drying fruits, sweetening is in the form of a syrup of varying concentrations of sugar; in addition, dry sugar is used for some types of packs of fruit for freezing.

The "Natural" Choices

The usual granulated table *sugar* (sucrose) is white and refined from cane or beets. It is the sweetening implied in almost all the instructions that do not specify another type of sweetener. Other sweeteners may be substituted for it; where the volume ratio of the substitution will make a difference in the product this fact is noted below. One teaspoon (5 mL) granulated white sugar contains 18 calories.

Brown sugar, semi-refined and more moist than white—is called for in some directions for curing meats or cooking. It also may be used in place of white sugar where its color and flavor will not affect the looks and taste of the put-by food in a way you might dislike (it would impart its characteristic taste to canned or frozen fruit, for example). Substituting it for white sugar: measure for measure, but pack it down well; 1 teaspoon (5 mL) has 16 calories.

Corn syrup comes in light and dark forms; use light only in substituting for sugar—and then replace only up to 25 percent of the sugar with the corn syrup. Corn syrup increases the gloss and jewel-like color of jellies and jams, and helps canning syrups to cling to the fruits. One teaspoon (5 mL) has about 20 calories.

Fructose was the glamour sweetener of the end of the 1970's, but by 1982 it began suffering backlash from being over-promoted as natural (the process for extracting it is no more natural than the means of refining table sugar), and as a boon to diabetics and to those wanting to lose weight. Fructose occurs naturally in fruit, but most of what is used in supermarket products comes from corn. Industry uses high-fructose syrup (HFS), a substance far from easily available to the householder looking for a substitute for sucrose (sugar); fructose sold in tablet and liquid form as a sweetening agent is not the concentrated sweetener HFS.

Granulated 100 percent fructose is sold in specialty food stores; *PFB*'s sample 8 ounces cost $2.38. Fructose has the same calories per teaspoon as white table sugar but, because on the average its ½ *again* as sweet (150 percent) as sugar, 10 calories' worth will do the job of 15 calories of sucrose.

Fructose is notably sweeter on cold foods, and especially on fresh fruit; its sweetening power seems noticeably less on hot foods. In cooking with fructose it is a good idea to lower the temperatures slightly, because it tends to caramelize more quickly than sucrose.

Honey has nearly twice the sweetening power of white sugar. Because of a distinctive taste, use mild-flavored honey. You will get best results by replacing only part (no more than ½) of called-for sugar with honey.

Maple syrup is usually too hard to come by for routine use in the kitchen but should replace only about ¼ of the required sugar, because of its pronounced flavor and color.

Sorghum and *molasses* are not recommended for most food-preservation because of their strong flavors.

Artificial Sweeteners

Also called "non-nutritive" sweeteners, these are available in liquid or tablets in the United States.

Although amounts of artificial sweetener to use are given in the recipes, it is sensible also to read carefully the equivalent-to-sugar measurements on the label of the sweetener. Sweetening power may vary from one brand name to another, and between liquid and tablet and granulated forms. What you want is the effect, not volume, in the non-nutritive sweetener you use.

Aspartame won final approval from the U.S. Food and Drug Administration in July 1981 (approval had been withdrawn pending studies about a side-effect that resembled a problem with some glutamates), and thereby gave the American public an alternative to saccharin. Aspartame is *180 times* sweeter than sugar and does not have the bitter aftertaste of saccharin. See the Appendix for where to find more information on aspartame.

Cyclamates were banned in 1970 and are under continued investigation by the FDA and therefore are not on the market.

Saccharin is 10 times sweeter than aspartame—which makes it 1800 times sweeter than sugar. Saccharin is prohibited as a sweetener in commercial food products in Canada, and home economists in the Dominion recommend strongly against its use in food prepared and served at home. The thorough and highly regarded Canadian studies helped to prompt FDA action against saccharin in the food supply of America. Following a general outcry against the U.S. action, saccharin is used in foods sold commercially, but the products are obliged to carry on their labels this legend: "Use of this product may be hazardous to your health. This product contains saccharin which has been determined to cause cancer in laboratory animals."

Anti-discoloration Treatments

Special treatments are given the cut surface of certain fruits to prevent oxidizing in the air and turning brown when they are canned, frozen, or dried. Vegetables are treated to prevent discoloration from enzymatic action when they are frozen or dried. Some cured meats may be treated to retard the inevitable loss of their appetizing pink color during storage. Canned shellfish are given a pre-canning treatment to prevent discoloration from the natural sulfur in their flesh. Fatty fish are given a special treatment to forestall some of the oxidation that causes them to turn rancid in the freezer after awhile.

C-enamel and R-enamel tin cans (see Chapter 6) have special linings that prevent naturally sulfur-y or bright-colored foods from changing color in contact with the metal. If glass jars of red or bright fruits are not stored in the dark or wrapped in paper, the light will bleach the contents. Fruit canned with too much headroom or too little liquid is likely to darken at the top of the jar; for the same reason, keep fruit submerged in its juice or syrup during freezer storage.

Specifically for Fruits

Ascorbic acid is Vitamin C, and volume for volume it is the most effective of the anti-oxidants. You should be able to get pure crystalline ascorbic acid at any drugstore and perhaps at any natural-food store; certainly either is likely to have Vitamin C in tablet form. There are about 3000 milligrams of ascorbic acid in 1 teaspoon of the fine, pure crystals, so buy 400-mg or 500-mg tablets to get the maximum amount of Vitamin C with the minimum of filler, then crush them between the nested bowls of two spoons. Ascorbic acid dissolves readily in water or juice (both of which should be boiled and cooled before the solution is made).

It is used most often with apples, apricots, nectarines, peaches, and pears; in a strong solution, it is a coating for cut fruit waiting to be processed or packed.

Generally speaking, the ascorbic-acid solution is strongest for drying fruits (Chapter 21), less for freezing them (Chapter 14), and least strong when they're canned (Chapter 7).

The crystals also may be added to the canning syrup or to the Wet packs of frozen fruit.

Citric acid is known to Kosher cooks as "sour salt" and to Greek cooks as "lemon salt": so designated, it is sold in specialty or ethnic food stores—or occasionally in gourmet sections of a metropolitan supermarket—as coarse crystals. *PFB* is still working with some lemon salt bought at Ratto's in Oakland, California, and with some sour salt bought at Zabar's on Manhattan's West Side. It is much more expensive from your drugstore, where it is finely granulated; special-order it in advance if you have to. The large crystals are easily pulverized between the nested bowls of two spoons or in a mortar and pestle. (See also Acids to Add for Safety later in this chapter.)

Volume for volume, it's about $\frac{1}{3}$ as effective as ascorbic acid for controlling oxidation (darkening), and therefore enough to achieve the same result could mask delicate flavors of some fruits.

Lemon juice contains both ascorbic and citric acids. Average acid-strength of fresh lemons is about 5 percent (also the labeled strength of reconstituted bottled lemon juice; some strains of California lemons are less strongly acid, however).

Being in solution naturally, it's about $\frac{1}{6}$ as effective volume for volume as ascorbic acid for preventing darkening. Even more of a flavor-masker than citric acid, it also adds a distinctive lemony taste to the food. It is used primarily to augment a food's natural acidity.

Commercial color-preservers are gettable at supermarkets alongside paraffin wax and commercial pectins; most often comes in 5-ounce tins. The best-known brand has a sugar base, with ascorbic acid and an anti-caking agent. Expensive to use because of the relatively small proportion of ascorbic acid in the mixture. The label tells how much to use for canning or freezing.

Another brand contains sugar and citric acid, plus several other ingredients, but no ascorbic acid. Moral: read labels to learn what you're paying for.

Both preparations, and similar ones, also have directions for using with fresh fruit cocktail, etc., that is made well in advance and chilled.

Mild acid-brine holding bath is one of the choices for treating apples particularly, but also apricots, nectarines, peaches, and pears while they wait to be packed in containers and processed. It's a solution in the proportions of 2 tablespoons salt and 2 tablespoons white (distilled) vinegar to each 1 gallon of water. Cut fruit is held for no longer than 20 minutes in the acid-brine (which does leach nutrients), then is well rinsed (to remove salt taste), drained, and packed.

A similar treatment before freezing these fruits, particularly apples, omits the vinegar. We recommend against using any such holding bath before drying because it adds some liquid to fruit you're trying to take the moisture out of.

Never use the regular cider vinegar or a wine vinegar. Either could add its own color and, perhaps, sediment to the fruit.

Steam-blanching is sometimes used before freezing fruits (especially apples) likely to darken when cut, and always before drying (even though sulfuring is recommended standard treatment in addition to blanching for fruit that is to be open-air/sun dried). Blanch in a single layer held in strong steam over briskly boiling water, 3 to 5 minutes depending on size of pieces. (See Chapters 14 and 21.)

Microwave-oven blanching. Each make or model comes with particular instructions for its use, so do read the directions that came with your oven. A rule-of-thumb would be, however: (1) of course use only an appropriate vessel to hold the food in the oven; (2) blanch only in 1-pound batches, with very little water; (3) rearrange the food once during the blanching process; and (4) depending on the density of the food's tissues, blanch for $2\frac{1}{2}$ to 6 minutes.

If the food is to be frozen, dunk it immediately in ice water to stop the cooking action and to chill it; pat it dry; package and freeze.

Syrup-blanching is again a pre-drying treatment to hold color, but the syrup is so heavy—1 cup sugar or equivalent natural substitute to each 1 cup water—and the fruit is in it so long (about 30 minutes all told) that you end up with a crystallized (or "candied") confection.

Syrup holding bath is simply dropping each piece of cut fruit into the syrup it will be canned or frozen in, and to which you may have added ascorbic acid. Fruit held this way before canning is usually packed Hot (precooked); if Raw pack, the fruit must be fished out, put in the containers, and covered with the syrup after it has been brought to boiling.

Sulfuring before drying can be used with most fruits in order to slow enzymatic action. All fruits and berries (except grapes) that are to be dried in open-air/sun are sometimes exposed to sulfur dioxide (SO_2), which is the fumes from burning pure sulfur.

Get "sublimed" sulfur (also called "sulfur blossoms") from the drugstore; it's pure, is a soft-yellow powder, and a 2-ounce box will sulfur 16 to 18 pounds of prepared fruit. Chapter 21 tells how to burn it. (*Don't use fumigating compounds* even though they contain some sulfur.)

Note: sulfites in solution have been banned for use on displays of produce for sale, and on "salad bar" offerings.

Specifically for Vegetables

Blanching is used before freezing or drying all vegetables (except for sliced onions or sweet green bell peppers) in order to slow down enzymatic action and produce the side-effect of helping to protect natural color and some texture. Blanching may be done in boiling water, in steam, or in a microwave oven (as in treating fruits, earlier). Specific times are given in instructions for individual foods later.

Before canning, white potatoes are held in a mild salt solution; salsify (oyster plant) is held in the mild acid-brine bath we mentioned earlier for certain cut fruits.

Specifically for Meats: The Nitrates/Nitrites

Potassium nitrate and *sodium nitrate* have been called *saltpeter* for generations, and for even longer have been used in the salt-curing of meats. Most simply, nit*rates* are changed into nit*rites* by metabolism when we eat them, or by the action with the protein of raw meat being cured. The nitrites, in turn, help to make nitrosamines—and these latter substances have been found to cause cancer in laboratory animals. It is this fear of carcinogens in human beings that has caused a continuing controversy about the use of nitrites in preservation of food, and they are under scrutiny by the FDA and the USDA.

Meanwhile it has been established that nitrites do help prevent the formation of the dreaded *C. botulinum* toxin—aside from stabilizing the appetizing pink color of ham, cured meats, frankfurters, etc., that may or may not be smoked following their cure in salt and nitrates/nitrites. Because they are considered able to help reduce the possibility of botulism poisoning, nitrites currently are tolerated in small amounts by food scientists. Nitrates can mask food spoilage, so read labels for expiration dates.

The answer to their use is continued vigilance and restriction to minimum proportions in commercially prepared foods. It is important not to add more nitrates/nitrites than are given in the recipes in Chapter 20 or to increase the amounts of commercially prepared cures containing these substances and which are sold in hardware and farm supply stores, and some supermarkets.

It is interesting that most Americans ingest at least as much nitrite-forming substances in leafy and green vegetables and other highly approved foods as they get in the cured meats.

Ascorbic acid. Good old Vitamin C again. It will hold the color of cured meat—not so brightly and certainly more briefly than saltpeter does. *It will not prevent botulism.* Use ¼ teaspoon pure crystalline ascorbic acid for every 5 pounds of dressed meat to be cured; add it to the salt mixture or to the brine.

Specifically for Seafood

Citric acid (and lemon juice) can be used for canning: the picked meat of crabs, lobsters, shrimp, and clams is given a brief dunk in a fairly tart solution of citric acid or lemon juice as a way to offset the darkening action of minerals naturally present in such foods (otherwise the meats would be likely to discolor during processing and storage). The dip lasts about 1 minute, and the meat is pressed gently to remove excess solution. (See Chapter 11.)

White (distilled) vinegar may be used too, but it might contribute a slight flavor of its own.

Ascorbic acid can be used for freezing fatty fish: a 20-second dip in a cold solution of 2 teaspoons crystalline ascorbic acid dissolved in 1 quart of water will lessen the chance of rancidity during freezer storage. The fish to be treated include mackerel, pink and chum salmon, lake trout, tuna, and eel, plus all fish roe.

Acids to Add for Safety

Two acids—citric acid in dry form or in lemon juice, and acetic acid in vinegar—are added to foods in this book for reasons that have little to do with the cosmetic purpose of helping to control oxidation or color changes in put-by food (see Anti-discoloration Treatments, above).

Adding Acids to Preserves and Pickles

We add acid to enhance flavor, making it brighter or more tangy, in condiments like ketchup or chili sauce or chutney. We add it to help create the balance that makes a gel in combination with pectin and sugar, in cooked jellies and jams. And we add it much more lavishly to aid preservation of a number of pickles served as garnishes. How much of which particular acid is added to all these foods is given in the specific instructions in Chapters 18 and 19.

Adding Acids to Natural Foods

The relationship between a food's natural acidity (*pH* rating) and its ability to provide hospitable growing conditions for spoilage micro-organisms is discussed at some length in the first part of Chapter 2. In Chapters 7 and 9, small amounts of acid are added for canning specific foods: figs, berry juices, the nectars and purées of apricots, peaches, and pears; sweet green bell peppers and pimientos; and in canning tomatoes (Chapter 8).

Acid is added *only* to the foods cited immediately above before canning them by the method specified and for the specified processing time. The added acid *does not allow any short cut* for any step in safe canning procedure, and it *does not permit any fiddling with canning methods*.

Citric acid. Pure crystalline citric acid, U.S.P. (meaning "United States Pharmacopoeia" and therefore of uniform stability and quality) is the acid that USDA Extension Service nutritionists allowed (in April 1975) added in canning tomatoes. It is sold by weight, is gettable at most drugstores, is not expensive—especially when you consider that 4 ounces (the consistency is like finely granulated sugar) will do about 45 quarts or slightly more than 90 pints of tomatoes. (As an anti-oxidant it is cheaper than ascorbic acid, though less effective.)

Citric acid is preferred for increasing acid-strength of foods because it does not contribute flavor of its own to food (unlike lemon juice and vinegars, which can alter flavor if used in large enough amounts).

Fine citric acid may be substituted for a 5-percent acid solution (the average for store-bought vinegar or for the juice of most lemons) *whenever the called-for measurements of the solutions are by the spoonful,* in this general proportion: $\frac{1}{4}$ teaspoon citric acid powder = a generous 1 tablespoon of 5-percent lemon juice/vinegar; $\frac{1}{2}$ teaspoon citric acid powder = a generous 2 tablespoons of the vinegar or lemon juice. (The equivalents actually are $\frac{1}{4}$ = 4 teaspoons, and $\frac{1}{2}$ = 8 teaspoons, but 1 and 2 tablespoons are easier measurements to make in the usual household's kitchen.)

To reverse the coin and make a 5-percent solution of citric acid, use the rule-of-thumb for making salt brines: dissolve 1 part fine citric acid in 19 parts of boiled (and cooled) water. Translated into measurements used in the average kitchen, this means dissolving 2 tablespoons fine citric acid in 1 pint (2 cups) of boiled water; or, if you want to be metric, dissolving 30 mL of fine citric acid crystals in $\frac{1}{2}$ liter (500 mL) of boiled water. Either translation will produce a solution around 6 percent instead of 5—but the result will serve the purpose we're after.

Lemon juice. This is recommended over the other acids for use in canning fruit juice in Chapter 7, in augmenting the innate acidity of tomatoes (Chapter 8), and for increasing the acidity of fruit juices to ensure a good gel for jellies and jams in Chapter 18. In canning tomatoes and some vegetables, use commercially bottled lemon juice: as the label attests, it is guaranteed to be of 5 percent acidity.

Vinegar. Vinegar is acetic acid, and all vinegars corrode metal, so when you use them in larger-than-spoonful amounts—in making pickles and relishes—make sure your kettle or holding vessel is enameled, stainless steel, ceramic, or glass. Acetic acid reacts badly with iron, copper and brass, and with galvanized metal (which we don't like to use with *any* food because of the possibility of contamination from cadmium in connection with the zinc used in galvanizing; and zinc itself can be toxic). Vinegared foods corrode aluminum.

The cider vinegar bought in supermarkets usually runs about 5 to 6 percent acid. Its pronounced flavor can be an asset with spicy condiments and pickles, but a drawback elsewhere. Its color is unimportant for dark relishes, but hurts the looks of light-colored pickles. Because of its flavor and color, it should not be used with fruits or bland foods. The minerals present in any vinegar that has not been distilled can react with compounds in the water or the foods' tissues to produce undesirable color changes. All of which boils down to: *do not use* cider (or malt) vinegar to reduce the *pH* rating of the specific fruits and vegetables cited above— unless it's a last resort, and unless you're prepared for changes in color and flavor.

Use *white* (*distilled*) *vinegar* for decreasing *pH* rating of foods mentioned above if you don't have citric acid or lemon juice—but double the measure used for lemon juice. Reason: the acetic acid in vinegar is more volatile than the citric acid in lemon juice.

Avoid using "raw" or "country" vinegar for the purposes cited here— it's likely to have sediment, and its flavor is pronounced. And save wine and special herbed vinegars for dressing salads or vegetable dishes.

Firming Agents

Having raw materials in prime condition and perfectly fresh, plus handling them promptly and carefully add up to the best *natural* means of ensuring that your put-by foods are firm and appetizing.

Salt. In canning, freezing, or curing fish for smoking, a short stay in a mild brine that is kept ice-cold will not only draw diffused blood from the tissues but also will firm the flesh and result in a better product.

In drying fish (Chapter 21), a larger concentration of salt is generally used to draw moisture from the tissues and make the flesh firm.

In pickles, the original "short-brine" firms the vegetables. Be sure to shave off the vestigial remains of the blossoms of pickling cucumbers, because this is where enzymes concentrate; and unless enzymatic action is halted, such foods will soften.

Cold. Refrigerating meats, poultry, and produce, and holding seafoods on ice until they are prepared for processing, will do much to ensure food with good, firm texture.

Calcium hydroxide is also called *slaked lime* (slaked, because liquid has been added to it, whereby it became heated, and the residue remains) and *pickling lime*. It is sold in drugstores as a firming agent for pickles; it also is gettable as one of the products of a line of preservatives and jelly-making products whose manufacturing headquarters is in Tupelo, Mississippi (see Appendix). Calcium hydroxide is preferred by some nutritionists over alum and calcium chloride (both below), but *PFB* takes a dim view because it can zap *pH* in foods.

Pectin, a starch found in large amounts in green apples and the white spongy layer under the thin skin of citrus fruits, swells to form a clear, thick substance when heated with sugar (usually sucrose) and acid. There is more about pectin in Chapter 18, "Jellies, Jams, and Other Sweet Things"; plus a discussion of low-methoxyl pectin, which creates a gel with less-than-standard amounts of sugar, but which requires a calcium compound as an aid to firming.

Alum is any of several allied compounds, and was called for in old cookery books to make pickles crisp—usually watermelon or cucumber chips. If a fairly large amount is eaten, it often produces nausea and even severe gastro-intestinal trouble. Alum is not an ingredient in any of the pickle recipes in this book.

Calcium chloride, of course, food-grade. Some people find this more acceptable than alum, but we do not include it in any pickle recipe or canning instruction in this book. It is an ingredient often used by commercial canners, especially in tomatoes. If you feel impelled to use it, get it from a drugstore in a food-pure form—*not* as sold at farm and garden supply centers for settling dust on roads or for dehumidifying closets, etc., or for fireproofing. And, because too much of it could leave a bitter aftertaste, never substitute it measure-for-measure for regular salt (sodium chloride). Instead, figure how much salt you'll need for a batch of, say, tomatoes, and in advance mix *not more than 1 part calcium chloride* with 2 parts regular salt. Then add the mixture in the amount of optional salt seasoning that the canning instructions call for.

Thickeners

Waxy Maize (Corn) Flour

Gravies and sauces thickened with all-purpose wheat flour, with potato flour—or even with that old stand-by, cornstarch—are likely to separate in freezing; wheat-flour sauces require thinning after mere refrigeration.

Since *PFB* discovered mochiko (below) some years ago, and advocated its use wholeheartedly, we have only recently been introduced to a waxy cornstarch called Clearjel A, by its maker, National Starch and Chemical Company of Bridgewater, New Jersey. We give it as an ingredient in our fruit-pie fillings in Chapter 12, and we have used it at home as a household staple in gravies, sauces, and "made" dishes for freezing: we believe it is the best thickener we have ever handled. Unfortunately, it must be bought in 500-pound bags from its manufacturer, but see the Appendix for how to encourage its sale in one-pound packets.

If you do get some, handle like run-of-the-mill cornstarch. Unlike such cornstarch, though, it won't become runny if it is long-cooked or processed in canning.

Mochiko: "Sweet" Rice Flour for Frozen Sauces

Mochiko (mo'-chi-ko) is a special rice flour that is often termed "waxy," and is known as "sweet" rice flour by the Japanese (it's not sweet in itself, but gets its name from being used in cakes and other confections). Its virtue for our purposes is that it "stretches," and, like Clearjel A, makes a smooth sauce. It does not equal the corn product in producing silkiness, however, and it does not perform well in puddings.

Mochiko is relatively easy to get at any oriental food store in fairly large cities, and routinely in supermarkets that serve communities with Indonesian or Japanese heritage.

6
THE CANNING
METHODS

Home-canning emerged as an industry with the patenting in 1858 of the mason jar. The usual method was simple. It was called "open-kettle," whereby hot jars were filled to brimming with hot, fully cooked food, then the lids were clapped on, and a vacuum was expected to form an airtight seal (proof against contamination) as the contents cooled. In its youth, open-kettle was used primarily for the home-canning of fruit and tomatoes (botanically a fruit). Because of this generally observed limitation, the method continued to be sanctioned, but only for strong-acid foods, until World War II. By the mid-forties, though, microbiologists and food scientists had started chipping away at its reputation for safety. Now, open-kettle is held to be inadequate for making a safe product.

Acidity and heat penetration dictate how a food must be processed when it is being canned at home. With these factors in mind we choose the Boiling–Water Bath or Pressure-processing.

Note: the processing times for food in glass mason jars using both Boiling–Water Bath and Pressure Canning given in this edition of *PFB* are based on recommendations by the USDA's *Home Canning Guide* (1988). The processing times for home canning in *metal* containers are from National Food Processors Association bulletins.

Before we start on the safe, right ways, though, hear words from two former professors notable in the field of food science:

From Ruth Klippstein—"Klippstein of Cornell"—concerning material sent out to the county agents in home economics across New York State: "[apropos *very* old-time instructions from a number of sources] they are

41

either very imprecise, or use what is now an expensive ingredient, or result in a product which by modern standards is not acceptable."

And from Isabel D. Wolf, of the University of Minnesota, about home-canning recipes: "avoid following the home-canning advice of celebrities, old cookbooks, 'back to nature' publications, and out-of-date home canning leaflets. Some potentially dangerous instructions can be found in old official publications, even those of this state!"

CONVERSIONS FOR CANNING

Do look at the conversions for metrics (with workable roundings-off) and for altitude—both in Chapter 3—and apply them.

The two methods for canning everything else at home are the *Boiling–Water Bath* and *Pressure-processing:* the first is for strong-acid foods; the latter, for low-acid ones. Together, these procedures make canning the means most often used in North America by householders preserving perishable foodstuffs.

For its effectiveness, canning relies on applied heat and the exclusion of air. Between them, these functions destroy the dangerous, targeted things in the food that cause spoilage or poisoning, and drive out air from the contents; thus is created a condition that will form the vacuum that seals the containers against outside contamination during storage.

How much heat depends on the acidity of each particular food we intend to can, as was discussed at some length in Chapter 2. In our kitchens this means that every carefully prepared food with a *pH* rating of 4.6 or below may be canned safely in a Boiling–Water Bath at 212 F/100 C at the sea-level zone (which includes up to 1000 ft./305 m; we'll deal with higher altitudes in a minute). Every carefully prepared food with a rating of higher than 4.6 *pH* must be processed in a Pressure Canner—which at sea level produces temperatures ranging from 1 degree hotter than the boiling point of water on up to 250 F/121 C, and beyond.

The *type of heat* is vitally important, because the ability to transfer heat varies between wet heat and dry heat. Illustration: hold your hand in a steady, strong flow of steam from the spout of a teakettle for one minute, and it will be burned enough to blister badly; hold it in the dry air of an oven at the same temperature-reading for a minute and it will be pleasantly warmed. To carry the idea further, the 240 F/116 C steam in a Pressure Canner at 10 pounds psig (pounds per square inch by gauge) at sea level has a much greater effect on the heat transfer process than does the atmosphere of an oven operating at 240 F/116 C.

So Never Forget This: only under *GREATER THAN ATMOSPHERIC PRESSURE* can you produce *wet heat that is hotter than the boiling point of water at your altitude.*

And you will need wet heat under pressure to reach deep enough inside a container of low-acid food to destroy the spoilers that can make it nasty, even dangerous, to eat.

If you remember this *Why,* you'll always be able to keep track of the *How* in canning food at home.

The rate of heat penetration and the acidity of the food (discussed in Chapter 2) are the criteria for determining the length of time needed to process food safely when it's canned at home.

How Heat Penetrates a Container of Food

Heat from outside the jar or can comes through the wall of the container and moves deep into the contents either by *conduction,* i.e., by being passed from one particle of food to another particle next to it; or by *convection*—being carried, almost swirled, by currents of liquid within the container. Usually both types of heat are involved, of course in varying proportions, when adequate processing occurs.

Conduction

In dense, closely packed food, and with relatively little fluid, heat is passed from the surface of the container inward from top, bottom, and sides, leaving a spot in the center as the last place that will heat. There is extra worry when the food is low in acid. Strained pumpkin and mashed winter squash are extreme examples of low-acid food that would process

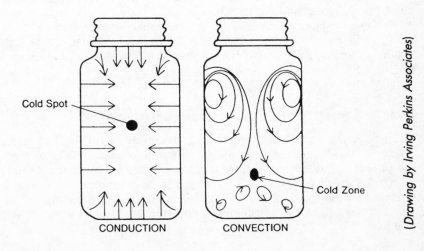

Cold Spot

CONDUCTION

Cold Zone

CONVECTION

(Drawing by Irving Perkins Associates)

mainly by conducted heat—and in Chapter 9, *PFB* recommends NOT to try canning them at home. Reason: the outer areas of the pack would be sadly overcooked in order to make certain that the part in the "cold" area would be heated enough—and long enough—to be safe. Raw meat ground fine and packed without liquid offers the same sort of problem.

Convection

Cut vegetables with added liquid and fruits in syrup of their own lavish juice depend greatly on convection for safe processing. In such packs, the heat seems to spiral inward from the sides, the swirls growing larger as they circle down and back up and closer toward each other, until eventually they touch and expand down to deal with the cold zone; this cold area, in convection, has lain in the center of the bottom one-fourth of the container.

Other Factors in Heat Penetration

There are still more considerations to be dealt with before the *length of the heat* can be specified.

1) How big is the container—and is it slender, or squat? These are questions concerning procedure, not prettiness. Naturally a big jar or can will take longer to heat through than a smaller one; and a short, chunky shape may have a cold zone that takes longer to heat than does a tall, slim one with greater actual volume.

2) How densely or loosely is the food packed inside the container? The less liquid in the container, the more worry you have that the food will be processed safely. (Remember the strained pumpkin that's not for home-canning.)

3) There is a limit on the size of container that can be processed successfully at home, and we're coming to that soon.

4) If you're canning meat or a combination/convenience food: is it as lean and free of fat as possible? Fat insulates against heat just as it does against cold, and the more fat there is in a food, the slower the penetration by any heat.

Basic Equipment for Home-Canning

Some specialized canning equipment is essential for turning out *safe* and attractive products in return for your efforts, and it includes large canners for processing filled containers, and the containers (with their fittings), which are made to withstand the required heat treatments, and to seal well. These few special items—and their *Whys*—are described fully in the following pages; for all the rest, your regular kitchen utensils will be adequate.

A deep-enough, Water–Bath Canner—or a big stockpot—for processing strong-acid foods.

A 12-quart steam Pressure Canner for processing low-acid foods: with gauge(s) of tested accuracy.

6- to 8-quart enameled or stainless steel kettle for precooking or blanching foods to be canned.

Jars or "tin" cans in prime condition, with lids/sealers/gaskets ditto.

Sealing machine (hand-operated), if you're using cans.

Alarm clock, to time processing longer than 1 hour—*plus* a minute-timer with warning bell (more accurate than an alarm clock and better for processing less than 1 hour)—*plus* a clock with a sweep second-hand (if you go in for quite brief 15-psi processing).

Pencil-shaped glass food thermometer (you'll need it for meats, poultry, and seafoods in jars, and for all foods in cans).

Shallow pans (dishpans will do).

Wire basket or cheesecloth to hold foods for blanching.

Ladle or dipper.

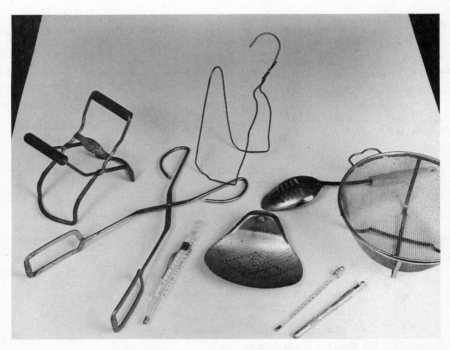

Clockwise from upper left: jar lifter; coat hanger bent to lift lids (works!); slotted spoon; strainer; exhausting thermometer (and cover); antique steel cream skimmer (great for stock and jelly); thermometer for Hot–Water Bath, jelly, etc.; tongs for filled jars. (*Photograph by Jeffery V. Baird*)

Perforated ladle or long-handled slotted spoon, for removing food from its
precooking kettle.
Wide-mouth funnel for filling jars.
Jar-lifter.
Food mill, for puréeing (or a blender; a food processor is ultimate luxury).
Sieve or strainer.
Colander, for draining.
Large measuring cups, and measuring spoons.
Muslin bag, for straining juices.
Plenty of clean, dry potholders, dish cloths, and towels.
Long-handled fork.
Household scales.
Large trays.

If you're looking for a new cookstove and like the idea of a double-
oven range whose upper oven partly overhangs the cooking-top, check
the clearance above the burners before you buy. Some models don't have
room for a really big double-boiler—much less a canner tall enough to
process quart jars on the back *or* front burners.

THE BOILING–WATER BATH

The Boiling–Water Bath has limitations: it is suitable *only for canning
strong-acid foods*—vinegared things, and fruits, which include tomatoes—and for "finishing" pickles, relishes, etc., and cooked sweet fruit
garnishes like jams on through butters to conserves. With such foods, the
B–W Bath does these things *if it is used correctly:*

- In raw or blanched (just partly precooked) strong-acid foods, it destroys
  yeast and molds and the bacteria that cannot live at the temperatures
  reached in the middle of the containers of food. Thus it deals with the
  food, as well as all inner parts of the containers that could have got
  contaminated by airborne spoilers while waiting to be filled and capped.
- Drives out the air naturally present in the tissues of the foods and in the
  canning liquid (air bubbles trapped in the container as it was filled are
  removed before capping). Air can prevent a perfect seal and permit
  spoilage.
- Creates a vacuum that enables the *jars to seal themselves*. (As we'll
  see, *cans* are sealed by hand after the air in the food's tissues and in
  the added liquid is exhausted.)

The B–W Bath Canner

It is round, usually made of heavy enameled ware (also called "granite"
ware).
It must have a cover.

It has a rack to hold containers off the bottom of the kettle, thus letting boiling water circulate under them as they process. *PFB* prefers a simple rack in place, because of cranky experiences with the folding-basket racks that allow filled jars to bump each other.

There are many B–W Bath canners offered for sale, and it should be easy to get one that's right—but often it isn't.

It's Got to Be Deep Enough

The most popular sizes are billed as 21-quart and 33-quart, and *PFB* uses both of them. These capacities actually mean loose contents; which is fair enough, since the labels also say, of the respective sizes, "7-jar rack" and "9-jar rack."

But this 7-jar/9-jar bit is *not* fair, because the reasonable inference is that they hold 7 pints or quarts, or 9 pints or quarts, and *even the 33-quart one is not deep enough to process quart jars correctly*—meaning, for our money, *safely*. When you're in a store looking at canners, and if the salesperson has enough experience with canning to point out that a particular kettle is too shallow for certain jars, pay attention—and ignore what the manufacturer's label implies.

The canner is too shallow for the quart jar (even though it was sold as O.K. for quarts): there's only 1 inch between jar top and kettle rim. (*USDA photograph*)

Here's the arithmetic. Starting at the bottom of the 33-quart canner, you should have ¾ to 1 inch between the holding rack and the bottom of the kettle—call it 1 inch for simplicity. Then you must have between 1 and 2 inches of boiling water covering the tops of your containers: 2 is better, so call it 2. Now you have 3 inches accounted for.

Then you must have a *minimum* of 1 more inch of "boiling room" between the bubbling surface of the water and the rim of the kettle. If you don't, the briskly boiling water will keep slopping out onto your stove-top and drenching the well of your heating element (making a mess even if it doesn't extinguish a gas flame, and offering a temptation to reduce your boil to a polite, and inadequate, simmer). Be cagey and remember that the close-fitting cover on your canner will increase its tendency to slop over: allow $1\frac{1}{2}$ inches for headroom to boil in. Now you have $4\frac{1}{2}$ inches—and your containers haven't even gone in yet.

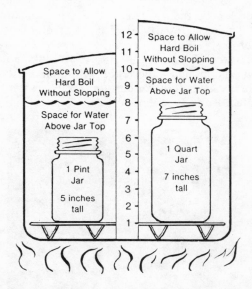

The cutaway drawing shows the room needed above jar tops for correct processing in a Boiling–Water Bath. (*Drawing by Norman Rogers*)

The height of a quart jar (we'll use jars here, as the manufacturer and the great majority of home-canners do) is 7 inches if it's a modern mason with a two-piece screwband lid, or $7\frac{1}{2}$ inches if it's the old bailed type with a domed glass lid.

Add $4\frac{1}{2}$ inches and 7 to $7\frac{1}{2}$ inches and you get $11\frac{1}{2}$ to 12 inches.

But the inside depth of our 33-quart Boiling–Water Bath canner is only a scant 10 inches. Therefore—though it's O.K. for pints, which are 5 to $5\frac{1}{2}$ inches tall, and of course for the even shorter $\frac{1}{2}$-pint jars—*we cannot use it for processing quart jars.*

There are canners around, but you have to insist first on depth when you're looking for a B–W Bath kettle. Know the height of your jars, add $4\frac{1}{2}$ to 5 inches, and measure the height of every canner you see. Stick to your guns, and you'll find the one that will do the job right.

Other B–W Bath Canner Possibilities

You can always sail right on past the inadequate B–W Bath canners and buy a stockpot. There is a heavy-duty aluminum stockpot that would do a good job for safe *processing* of strong-acid foods; it could not double as a preserving kettle for brined or vinegared things because, being aluminum, the metal would react with the acid. Most glorious of all are the stainless steel stockpots. They're relatively expensive, but they do come in many sizes; and they are impervious to acid. Invest in one—or two.

A third idea is to use a Pressure Canner for a B–W Bath (for how, see page 54).

Two stockpots don't hog the cooking top (and they're taller, too). (*Photograph by Jeffery V. Baird*)

How to Use the B–W Bath

Put your canning kettle on the stove and fill it halfway with water; put a rack in the bottom; turn on the heat. If the food you're canning takes relatively little time to prepare for packing in the containers, start heating a large teakettle of extra water now too.

Prepare the food for canning according to the individual instructions, packing it in clean, scalded containers, and putting on lids as directed.

Jars of *raw* (cold) food *must not be put in boiling water,* lest they crack; you may need to add cool water to the canner. However, if the food in the jars is very hot, the filled jars may go into nearly boiling water without fear of breakage. (Cans will always be hot, because they've just been exhausted. See pp. 71)

Process in each load only *one type of food,* in *one size of container.*

If you like your canner's lift-out rack, fill it with capped containers, and lower the whole thing bodily into the hot water. If your batch does not have enough jars to fill the canner, use "dummy" jars of the same size, filled with hot water and capped. These prevent food-jars from knocking together.

Or use your jar-lifter to lower jars/cans individually onto the simple rack at the bottom of the kettle. Place them away from the sides of the canner and about 1 inch apart, so the boiling water will be able to circulate freely around them. Don't jam them in or even let them touch: jars might break as they expand ever so slightly while processing. If your batch is too small to make a full canner-load, submerge mayonnaise jars, or whatever, in the empty spaces to keep your capped jars of food from shifting around as they boil.

If your canner is tall enough to let you process two tiers of small jars/cans at a time, use a wire cake-cooling rack to hold the upper layer of containers. Ensure adequate circulation by staggering them as you do when baking on several racks in your oven.

Pour enough more hot water around the jars/cans to bring the level 2 inches above the tops of all containers; be careful not to dump boiling water smack on top of jars.

What to Do About Racks

The racks that come in store-bought B–W Bath kettles can be infuriating. Made of heavy-gauge wire, they may resemble a flimsy wagon wheel or they look like the skeleton of a basket; at any rate the "spokes" or dividers are so wide apart that even pint jars sometimes fall between them, instead of resting on them. Some lack support strong enough to hold containers high enough off the bottom of the canner. Many actually cause jars to teeter and topple—and you have broken glass to contend with.

So buy several large round cake-cooling racks. Their wire is more fragile, but for support toward the middle, cluster on the bottom of the

Old Boiling–Water Canners are monsters—and we don't like the baskets (jars sit lopsided, and thus can roll and crack). (*Photograph by Jeffery V. Baird*)

kettle five or six screwbands that are no longer fit to use on jar lids. Measure the bottom diameter of your B–W Bath canner (not the top, which usually is about an inch larger), and choose your racks accordingly: $\frac{1}{2}$ inch of leeway all around is about right.

Hot–Water Bath (Pasteurization)

We have bent over backward in referring to the *Boiling*–Water Bath because it's necessary to distinguish clearly between the B–W—which maintains a real 212 F/100 C boil at sea-level—and the *Simmering* Hot–Water version, which is a form of pasteurizing.

The Hot–Water Bath is recommended *only for certain sweet, acid fruit juices,* and as a *finishing*—as against a complete *processing*—procedure to help seal preserves and pickles/relishes and sauerkraut.

Despite the casual swapping of these terms in some older manuals, the processes themselves are not interchangeable.

THE *WHY* OF PRESSURE CANNING

Even in the sea-level zone, every plain vegetable except tomatoes, every meat, every seafood that is canned at home—and almost every mixture containing these—MUST BE CANNED IN A PRESSURE *CANNER.*

Or put it this way: *Pressure Canning is the ONLY process* that is able to destroy the tough spores of bacteria like *Clostridium botulinum* which can grow and produce deadly poison in jars or cans of *any* low-acid food.

Or this way: dangerous spoilers can live through even a day-long Boiling–Water Bath at 212 F/100 C in containers of low-acid food with a natural acidity rating of a *pH* higher than 4.6, but *they are killed by the higher temperatures reached only in a Pressure Canner.*

Please look at the *pH* ratings of selected foods and the section called Dealing with the Spoilers on p. 7.

And before we talk about how to use a Pressure Canner, we'll make three points:

First, the Pressure Canner we're talking about is not to be confused with a *steamer*—either the old-style arrangement that swirled steam at Zero pounds/212 F/100 C around containers of food, or today's compartmented kettles for cooking clams or lobsters, or the "atmospheric canner" described later in this chapter.

Second, the Pressure Canner we're talking about is an honest-to-goodness, regular, conventional *big* Pressure *Canner*—not the 4- , 6- or 8-quart pressure saucepan that you may already use for cooking loose food. *PFB* is leery of processing any containers of low-acid food in such a saucepan. (It's also a worry that manufacturers, and even some food specialists, keep using the term "pressure *cooker*" for both the Canner and the saucepan—adding to the chances for a slip-up.) We're distrustful because:

Third, being so much smaller and thin-skinned, pressure *saucepans* (usually less than 4- to 6-quart loose capacity) are not designed for processing times researched by food scientists for true *canners*. The saucepan's heating-up and cooling-down times are too short to do the job without adding processing time. However, if one of these saucepans is the only thing you have for canning low-acid foods, go ahead and use a pressure saucepan—*with these stipulations:*

> It is 4- to 6-quart loose contents, and is controlled by a deadweight gauge, using 15 psig.
>
> Process nothing larger than a 1-pint jar or a No. 303 can.
>
> Don't let the containers touch each other inside the pan; and of course the jars/cans will be on a rack that holds them up from the bottom of the pan.

Let the pan vent steam for 5 minutes—see The Pressure Canner at Work a few pages later on. Then close the vent and start counting the minutes required for safe processing when the gauge indicates 15 psig.

Add 10 minutes to the processing times given in later chapters for individual foods being Pressure-canned (again, this is to compensate for shorter heating/cooling times in the smaller pot).

Consult Correcting for Altitude in Chapter 3. *If you live higher than 3000 ft/914 m, the two leading makers of the small pressure pots warn against processing low-acid food in one of these saucepans. Reason: water needed for extra processing time at altitude will evaporate before the food is adequately treated.*

Fourth, gauges, vents, and gaskets of *saucepans* may not be checked so carefully or so often, as the sensible householder will do with a Pressure Canner during the season. Less than perfect operation while cooking a Swiss steak is no big deal, but a hitch during Pressure-canning could mean food poisoning when the contents of the containers are eaten several months later.

Many a cook has cooled a pressure *saucepan* quickly under running water in order to inspect the progress of loose food being cooked inside. Doing this with canning jars inside will make a grade-A mess.

Buying a Pressure Canner

Pressure Canners are usually made of cast aluminum. Like B–W Bath kettles, they come in sizes measured according to loose-contents capacity; unlike B–W Bath kettles, though, their diameters are all roughly the same: it's their height that varies. The most popular sizes are 16-quart (actually taking 7 quart jars or 9 pint jars), and 22-quart (actually taking 7 quarts, or 18 pints in two layers, or 34 ½-pints stacked in three tiers).

They're expensive, but they do their job for years and years if you take good care of them and keep their sealing rims and pressure gauges and safety vents in good working order.

The Anatomy of a Pressure Canner

The base, or kettle, is covered with a tight-fitting lid that contains the controls. The lid is fastened down with clamps or a system of interlocking ridges and grooves, bayonet style. It may or may not have a rubber gasket.

Controls are: (1) a pressure gauge—a dial, or a system of measured deadweight that sits on the vent; (2) an open vent to let air and steam exhaust before processing time begins, and is closed with a petcock or separate weight (if gauge is a dial) or with the deadweight gauge, to start raising pressure; and (3) a safety valve/plug that blows if pressure gets unsafe.

Left to right, front to back: 8-quart deadweight gauge Pressure Canner; 12-quart dial gauge Pressure Canner; 12-quart deadweight gauge Pressure Canner; 22-quart combination of dial and weighted gauges. (*Photograph by Jeffery V. Baird*)

The canner also will have a shallow removable rack to keep jars/cans from touching the kettle bottom, or a strong wire basket that serves as a rack and lets you lift all the hot containers out in one fell swoop.

Pressure Canner for a B–W Bath

Put in enough water to come up to the shoulders of the jars you'll process; start the water heating. Put in the filled jars, taking care lest they touch the sides or each other; of course they're on a rack that holds them off the bottom of the pot. Then *lay* the heavy lid on the sealing rim of the canner, but *do not lock it.* Leave the vent open. (These terms come up in the complete description of the operation of the Pressure Canner soon; meanwhile this much of them belongs here because we're talking about B–W Baths, not Pressure-processing.) After a strong flow of escaping steam is established, start timing as for the B–W Bath recommended for the strong-acid food you are treating.

If you lock the lid, you will get 1 to 1½ pounds of pressure, even with the vent open. At *PFB*'s altitude, this is good. We use it only to ensure that our B–W Bath is hot enough; *it's not any alternative to correct Pressure-processing.*

Servicing Your Pressure Canner

First, read the operating manual that comes with it.

Next, clean the canner according to directions, to remove any factory dust or gunk. Do *not* immerse dial in water.

Check the openings of the vent, safety valve, and pressure gauge to make sure they're unclogged and clean. Take a small sharp-pointed tool (like a large darning needle or a bodkin) to the openings if they need it; clean the vent by drawing a narrow strip of cloth through it.

Be sure the sealing surfaces of kettle and lid are smoothly clean, so they'll lock completely tight and not allow any pressure, in the form of steam, to escape. If your canner has a rubber gasket, replace it when it shows signs of losing its gimp or getting hard or tired; it should be as limber as a good jar rubber. The kind of hardware store that has replacement dial gauges is likely to have gaskets as well, so take your canner lid with you to make sure of the right size (and don't forget a fresh safety plug if yours has hardened and cracked).

Checking the Dial Gauge

Each year well before canning season, or more often if you put by large quantities, you should have the dial gauge of your Pressure Canner checked for accuracy against a master gauge. Just because the dial may rest at 2 pounds pressure when the canner is not in use does not necessarily mean that the gauge is simply 2 pounds high, or that it is uniformly 2 pounds off throughout its range: *have it checked*.

Telephone your county Agricultural Extension Service offices. The County Agent in home economics there is likely to have the equipment to do it for you—or at least she can tell you where to go. The director of the EFNEP (Expanded Food and Nutrition Education Program) center for your area might be able to offer similar information.

Presto will check your gauge. See the Appendix for where and how to send it to them.

Dealing With a Faulty Gauge

In earlier editions of this book we relayed faithfully the recommendations from older official publications about how to compensate for a dial gauge that misrepresented the actual psig inside an operating Pressure Canner. Now we edit this advice ruthlessly, because newer research and equipment are asking for more refined measurements.

The Pressure Canner at Work

Here are the mechanics of using a Pressure Canner. And a commonsense rule: don't leave your canner unattended while it's doing its job.

Water In

Put about 1½ inches of warm water in the thoroughly clean canner. If you know that your canner leaks steam slightly when the vent is closed, add an extra inch of water to ensure that the canner will not boil dry if the processing time is more than 30 minutes.

Start It Heating

Place the uncovered kettle (bottom section) over heat high enough to raise the pressure quickly *after* the lid is clamped tight.

Cover on Tight

When the batch is loaded, put the lid in position, matching arrows or other indicators. Turn the cover to the Closed position or tighten knobs, clamps, etc., to fasten the closure so no steam can escape at the rim.

Let It Vent

It is important to consider the *Why* of this next step, because sometimes it isn't stressed in the instruction booklet that comes with your canner—and your processing could be thrown off whack.

Air that is trapped in your Pressure Canner will expand and exert extra pressure—that is, pressure in addition to that of the steam—and your gauge will give you a false indication of the actual temperature inside your canner. Therefore, you must make sure that *all air is displaced by steam before you close your petcock or vent.*

For canners with either type of gauge, leave the vent open on the locked-in-place lid until steam has been issuing from it in a *strong, steady*

Two "musts": time carefully, and open lid away from you. (*USDA photograph*)

stream—*7 minutes* for small canners, *10 minutes for larger ones*. Use your minute-timer here. When your canner has vented, close the vent by the means specified for dial or weighted gauge.

Manage the Controls

For a canner with a *dial gauge,* close the vent and let pressure rise quickly inside the kettle until the dial registers about ½ pound *under* the processing pressure that is called for. At this point, reduce the heat moderately under your canner to slow down the rate at which the pressure is climbing: in a minute the gauge will have reached the exact poundage you want, so you adjust the heat again to keep pressure steady at the correct poundage.

For a canner with a *deadweight-weighted gauge,* set the gauge on the vent and then carry on as for the canner with a dial gauge—but of course keep track of the frequency of the jiggles as the maker's instructions say to do.

Watch the Gauge

Pressure, once the right level is reached, *must be constant.*

You're using a Pressure Canner so your food will be safe, so if the pressure sags you should bring it back up to the mark and start timing again from scratch. Don't guess—turn your timer right back to the full processing period. Your food will be overcooked, of course—but next time you'll keep your eye on the pressure, right?

Fluctuating pressure also can cause liquid to be drawn from jars. And this in turn can prevent a perfect seal.

And Watch the Clock

Count the processing time from the moment pressure reaches the correct level for the food being canned. *Be accurate:* at this high temperature even a few minutes too much can severely overcook the food.

Reduce to Zero

At the end of the required processing time turn off the heat or remove the canner from the stove. It's heavy and hot, so take care.

If you're using *glass jars,* let the canner cool until the pressure drops back to Zero. Then open the vent very slowly.

CAUTION: if the vent is opened suddenly or before the pressure inside the canner has dropped to Zero, liquid will be pulled from jars, or the sudden change in pressure may break them.

If you're using *tin cans,* the pressure does not need to fall naturally to Zero by cooling: open the vent gradually when processing is over and the canner is off heat, to let steam out slowly until pressure is Zero.

New Pressure Canners are designed to keep their covers locked so long as there is any pressure still built up inside.

Lift the Lid

Never remove the lid until after steam has stopped coming from the vent.

Open lid clamps or fastenings.

Raise the farther rim of the cover first, tilting it to direct remaining heat and steam *away* from you.

Take Out Jars or Cans Promptly

If the canner doesn't have a basket rack, use a jar-lifter and *dry* potholders (wet ones get unbearably hot in a wink) to remove jars/cans promptly. Promptness is important. Certain bacteria *like* heat, remember, and some types of spoilage are thermophilic.

Cool *cans* quickly in cold water.

Ignore the "after-boil" still bubbling in jars (this means a vacuum has already formed!) and complete seals if necessary. Stand jars on a wood surface, or one padded with cloth or paper, to cool away from drafts.

Test seals when containers have cooled overnight.

Dangerous Canning *Don'ts*

There's nothing like the *Morbidity and Mortality Weekly Report* for discouraging nostalgic ways of canning food at home, because in almost every account of an outbreak of food-borne botulism, the Editorial Note deduces that "inadequate processing" or "inadequate heating" allowed the toxin to form, with the help of a bad seal.

Spelling it out, this means that low-acid foods that should have been Pressure-processed were merely given a Boiling–Water Bath; and that strong-acid foods—which should have been given a B–W Bath to sterilize container and heat contents adequately—were canned instead by "open-kettle," or worse.

No to Old "Open-Kettle"

The first page of this chapter describes the origin of what was called open-kettle canning, a method whereby hot food—presumably fully cooked—was put into hot jars that once were sterile and were capped with hot lids that also once were sterile, and then usually a vacuum was formed as the contents cooled, helping to form the seal.

You see the problems. Jars, lids, sealing-rims probably were sterilized carefully by being filled with, or lying in, boiling water. Fine. But then they sat for seconds or minutes in the open air of the kitchen while the contents were ladled in, and air-borne spoilers, some potentially deadly, contaminated the inside of jars and covers. The hot food simply was not able to re-sterilize the container; and indeed it may not have cooked quite long enough to destroy the bacteria that *like* heat approaching 212 F/100 C at the sea-level zone.

Mold is one of the leaders in the air-borne danger brigade, and it can settle on the underside of a canning lid, and grow. In the process of growing it can metabolize the safe margin of acid just enough to allow surviving *C. botulinum* spores to develop and throw off their wicked toxin. So your jar of supposedly "safe" open-kettle-canned tomatoes—or dill pickles or jams or condiments or pears or peaches, all of which traditionally have been regarded as strong-acid enough to be protected—may contain a deadly threat. And aside from botulism, there could be mycotoxins from mold itself.

And *No* to Old "Oven-Canning"—or Microwaving

The folklore of homemaking has another bad method that continues to surface: trying to process foods by baking them. Please look back at How Heat Penetrates a Container of Food, early in this chapter. Dry heat just plain cannot produce the same effect as boiling water, either at atmospheric pressure or under extra pressure, can.

In addition to the danger of inadequate processing, there is also the likelihood that a jar of food will burst in your face as you open the oven door.

Quite early in the marketing of microwave ovens to the general public, one maker touted his product as able to can fruits. In very short order the manufacturer discontinued such a claim for his oven. Now all responsible makers of microwave ovens warn against using them for canning. (Blanching produce as a preliminary treatment before freezing or drying may be done in a microwave.)

And *No* to Old "Steamer-Canning"

As far as we can discover, steamer-canning (NOT to be confused with Pressure-processing—*please*) was carried out in a tank arrangement that let plain steam waft around the jars placed on racks inside it. The outfit was not on the market very long: apparently too many canning failures from letting jars of food putter along in the steamer. ("But steam is *steam,* isn't it?" Nope. It isn't. As any youngster taking practical science in grade-school can tell us.)

And *No* to Canning in an Automatic Dishwasher

(Scout's honor, *PFB* keeps getting asked about this cockeyed theory.) It would be interesting to chart the thought processes behind such a procedure, since deluxe models reach an atmospheric heat of 140+ F/60+ C only when they're set for extra hot cycles. In a home dishwasher, the temperature reached inside a container of food could only help produce a bacterial load that nobody needs.

And *No* to Canning in a Crockpot

We don't know who could have advocated this, since we have never seen any such claims in print by the makers of Crockpots. But it has come up several times in letters to *PFB*, usually from writers who say they hope it's safe, because they felt that slow-cooking was more likely to preserve nutrients than subjecting the food to heats like 240 F/116 C. All we need to say about a Crockpot is that if it's on a long Low-heat setting, chances are that the nastier micro-organisms will be encouraged to grow and breed like mad, and what else should you expect?

It is interesting that the people asking about Crockpot canning do not think in terms of pasteurizing milk by heating it for 30 minutes at 142 F/ 61 C *versus* for 15 seconds at 160 F/71 C. Both these treatments do the same job in destroying bacteria they're aimed at.

And *No* to Processing in a Compost Heap

Who ever would have thought it? But there was a query based on the notion of holding containers of food at 140+ F/60+ C for a long time.

And *No* to Canning Pills

"Canning pills" went out with corset-covers. Old manuals might suggest that salicylic acid (read this "aspirin") be dropped in each canning jar before it was capped. Such things never helped then and would not help now. *No* preservative added can offset dirty handling or inadequate processing.

"Every-additive-at-once" preparations are used by some commercial canners, but none, thankfully, that *PFB* could discover is offered to home-canners by the salt industry.

THE ATMOSPHERIC STEAM CANNER

There came on the market in the late 1970's a canner devised to be a substitute for the not-always-deep-enough Boiling–Water Bath kettle that manufacturers of routine housewares had been so slow to produce in quantity at a reasonable price. (The question *PFB* gets asked most often is where to find affordable B–W Bath canners that live up to recognized standards for processing safely.)

This newcomer is referred to as an "atmospheric steam" canner, and it *must NEVER be confused with the traditional Pressure Canner*.

The atmospheric canner gets its name because the kettle is not sealed, and therefore the saturated steam inside it—not being under significantly more pressure per square inch than the air in the room outside—cannot get hotter than the boiling point of water in a utensil with an unsecured cover.

A previous edition of *PFB* described early research that led to the arrangement we're talking about. The anatomy of both models generally on sale in the U.S. is the same. It resembles a giant stove-top sterilizer for babies' nursing bottles. It comes in three sections: a shallow bottom to hold water that will boil hard to produce steam around the jars; a rack set just above the water to hold the jars of fruit or pickles; and a tall, domed lid that inverts to come down over the jars and rest on the lip around the edge of the water compartment. Small holes at the rim of the cover let steam escape near the base of the jars while they're being processed.

In mid-October 1987, The Pennsylvania State University's Center for Excellence researched the atmospheric canner's performance *using Boiling-Water Bath processing times*. The canner's disappointing results are summarized on pages 400–401, in the Appendix.

ABOUT JARS AND CANS

Either glass jars *specially made for home-canning* or metal cans may be used for putting food by.

Overall, canning jars are more versatile than cans are: (1) anything that is canned in cans may also be canned in glass; (2) Boiling–Water Bath and Pressure-processing are used for either jars or cans of food; but (3) there are certain foods that should be home-canned only in glass—the individual instructions later on will tell you which ones, but among them are jellies, preserves, pickles, relishes, a couple of fruits and vegetables, and seafoods.

Another advantage of jars is that they usually require one less step in the packing procedure than cans do; the step is called "exhausting," and will be described in detail in a minute.

Still another plus-mark for jars is that they're generally a lot easier to get, being sold in hardware and farm stores and supermarkets all over the country.

And of course jars show off their contents, and hence are a requisite for exhibiting your food at the fair. Jars of food must be stored in a cool, dry place that's also dark, otherwise the contents will fade in the light.

And last, jars are re-usable as long as they have no nicks or scratches, even minute ones, and as long as you can get the proper fittings for them.

Aside from having to be special-ordered from the distributor, cans require a machine for crimping their lids onto the sides to form a perfect seal. Also, certain foods need cans with special enameled linings.

Cans are not re-usable.

But cans of food require only dry, cool storage—not darkness, since the contents are not exposed to light. And cans stack easily. Further, when they're piping hot from the processing kettle they don't need to be handled so gingerly as glass does. (It's a mistake to be rough with them, though: a dented rim can mean a damaged seal.)

The Types of Jars for Home-Canning

The jars recommended for use to North American householders are all alike in three highly important respects:

First, the jars are manufactured specifically for canning food AT HOME. Generally speaking, this means that they are designed to withstand more cavalier treatment during filling, processing, and storage than their commercial counterparts undergo at the hands of industrial processors and shippers.

North America's largest makers of home-canning jars—which means in Canada and the U.S.A.—have told *PFB* that, though they also make commercial glass containers, their jars for salad dressing or whatever are "one-trip" containers, and therefore are not reliable for re-use in canning food at home.

Second, the canning jars come in ½-pint, 1-pint, 1½-pint, 1-quart, and ½-gallon sizes; of these, the ½-gallon does not have a wide-mouth version.

Neither *PFB* nor any established research publication gives instructions for processing ½-gallons, either for the B–W Bath or the Pressure Canner. Reason: dense and/or low-acid foods do not heat thoroughly enough in these out-size containers to be safe. With looser, and strong-acid foods, the outside would be cooked to death before the center was dealt with. Half-gallons are too tall for virtually all canners anyway.

Unless specified otherwise, ½-pints are processed like 1-pint jars, and 1½-pint jars are processed like 1-quart jars.

Third, the jars have types of closures (described below) for which fresh sealers are available at the start of each canning season. And our filling/processing/sealing instructions take these closures into account.

Modern Mason with Two-Piece Screwband Metal Lid

"Mason" goes back to the name of the originator of the screwtop canning jar, and denotes any jar with a neck threaded to take a closure that is screwed down. Today's modern mason is the highly refined home-canning jar being produced in the millions each year by manufacturers in the United States and Canada.

The jar mostly has standard and wide-mouth openings, and, depending on the maker, has a variety of shapes and sizes. The straight-sided and wide-mouth jars are especially handy for large pieces of food that you'd like to remove more or less intact.

The cartons will say which of these jars is O.K. for freezing.

The jars come packed with their closures. The lids also are sold separately, with or without screwbands.

Availability of jars, and currently their North American makers, are listed in the Appendix.

HOW THE TWO-PIECE METAL CLOSURE WORKS

The lid—called "dome" or "self-sealing" or "snap" by the individual makers, but of one basic design—is a flat metal disk with its edge flanged to seat accurately on the *rim* of the jar's mouth; the underside of the flange has a rubber-like sealing compound; the center surface next to the food is enameled, often white.

The lid is sterilized (more about this in a minute), placed on the clean-wiped rim of the jar of food, and then is held in place by the screwband, which is screwed down on the neck of the jar *firmly tight—AND IS NEVER TIGHTENED FURTHER*. "Firmly tight" means screwed down completely but without using full force, or without being yanked around as with a wrench.

The capped jar is processed, during which the "give" in the metal lid allows air in the contents to be forced out. As the jar cools, the pliant metal will be sucked down by the vacuum until the lid is slightly concave. You often hear the small *plink* as the lid snaps down, thus indicating that the jar is sealed.

STERILIZING JARS AND TWO-PIECE LIDS

Wash jars, screwbands, and lids in hot soapy water, rinse well in scalding water. Containers and closures that will be processed in a Boiling–Water Bath or Pressure Canner need not be sterilized further: but do let them stand filled/covered with the hot water until used, to protect them from dust and airborne spoilers.

Containers for food that will not be processed or "finished" in a B–W Bath at *212 F/100 C at the sea-level zone* must be sterilized. Wash jars and closures as above. Stand open jars upright in a big kettle, fill with hot water until the jars are submerged; bring the whole thing to boiling, and boil for 15 minutes. Remove the kettle from heat but let jars stay in the hot water.

Clean and scald the screwbands the way you do jars.

But scald, or boil, the metal lids *according to the manufacturer's directions.* This isn't a cop-out: the makers give different instructions here, and all of us consumers must assume that they know what's best for their own product. The result in every case is to sterilize, though; and all the lids are left in the hot water as a protection against dust, etc.

WHAT IS RE-USABLE

You should *NEVER use the lid itself again* for canning: the sealing compound on the lid will not seal right a second time around. And besides, it is ever so slightly warped now because you pried it off (and it may be punctured to boot). We scratch a big fat "X" on used lids on enameled or painted outside surface, then wash and toss them in a catch-all drawer to use sometime on a refrigerator-storage jar.

You of course can use the jars over and over again as long as they have perfectly smooth sealing rims, and have no cracks or scratches anywhere—including the minute scratches inside from digging out contents time and again.

You can use the screwbands again for canning, *unless* they're rusted, or bent, or the threads are marred—all signs that they're too tired to be safe in canning. (We often use "retired" screwbands for extra support under a springy wire rack in the bottom of our B–W Bath kettle, or to hold jars well apart in the canner.)

OLD-STYLE ZINC CAPS FOR MODERN MASONS

Not recommended for Pressure Canners, the one-piece threaded zinc cap with its porcelain liner was last catalogued by the Ball Corporation's 1979 *Blue Book,* but some can be found at flea markets or "turn-of-the-century" variety stores.

The zinc cap has a porcelain disk attached in the head where the metal would otherwise come in contact with food in the jar. It fits a modern standard-mouth mason, screwing down to rest on a separate and cushioning rubber ring that you have put around the neck of the jar *at the base of the threads* near the shoulder.

Jar rubbers are still gettable: a subcontractor currently is making "Good Luck" jar rubbers for Ball, and they're sold alongside separate disk lids.

Caps, particularly, must be uncorroded and with porcelain intact. Wash jars, caps, and rubbers in hot soapy water, and rinse well. Sterilize the clean caps by boiling for 15 minutes and letting them remain covered by the very hot water until each is used. But *do not boil the jar rubbers;* instead, pour briskly boiling water over them in a shallow pan and let them sit there, hot, until used.

Next, gently—but only slightly—stretch the rubber, now more pliable and still wet from being held in hot water, and ease it over the neck of the jar to sit flat around the narrow ledge of glass below the threads. Fill the container (how, in detail later), wipe any dribbles from the rubber, screw the cap down *tight* on the rubber ring—*and then give it a ¼-inch COUNTERturn* for a slight loosening that will allow air in the jar/contents to vent during processing.

As you remove the jars from the processing kettle, *retighten the cap slowly*—slowly, because a quick jerk is likely to shift the rubber, thus breaking the seal; and never tighten the cap again, especially after the jar has cooled. This gentle retightening when the jar is taken from the kettle is the action referred to later in the individual instructions as "complete the seals if necessary."

Only perfect caps and perfect jars are re-usable. Rubbers lose their sealing ability in just one use: *discard rubbers*.

Old-style Bailed Jars with Glass Lids

Not recommended for Pressure Canners because they're likely to be battle-weary, these haven't been manufactured since the early 1960's—but there are still thousands around. Sometimes called "lightning" or "ideal" type, they have a domed glass lid cushioned on a separate rubber ring that seats on a glass ledge a scant ¼-inch down on the neck of the jar. The lid is held in place during processing by the longer hoop of the two-part wire clamp.

As you remove the jars from the canner after processing, snap down the shorter spring-section of the clamp so it rests on the shoulder of the jar. Therefore, this is another case where "*complete the seals if necessary*" applies in the individual instructions.

FITTINGS FOR BAILED JARS

Discard any lids that have rough or chipped rims, or whose top-notch (which holds the longer hoop in place) is worn away.

Sometimes the wire bails are so rusted or old or tired that they have lost their gimp and can't hold the lid down tightly. *Please* don't go in for makeshift tightening by bending the wires or padding the lids. Retire the jars.

Rubbers are still sold in boxes of twelve, and come in standard and wide-mouth sizes. *Never re-use rubbers,* or use old, stale rubbers that have been hanging around for years. Stretch rubbers gently and only enough so they'll go over the neck of the jar.

To sterilize: boil jars and lids for 15 minutes, and let them wait, covered by the hot water, until used. But *don't boil rubbers:* wash well, then put them in a shallow pan, cover with boiling water, and let stand till used.

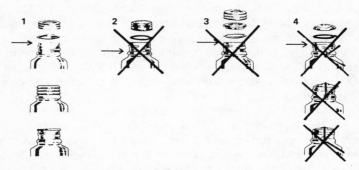

Jar closures: No. 1 (and down)—two-piece screwband lid seals modern masons on the rim; No. 2—zinc cap seals old-style masons at the shoulder; No. 3—glass lid and band; No. 4 (and down)—bailed lid. Nos. 2, 3, and 4 can still be found (used) but have not been made for years, thus their fittings (and safety) are suspect. (*Drawing by John Devaney*)

Masons with Three-Piece Glass Lids

These are like the modern mason with the two-piece screwband *metal* lid, and though no longer made in Canada, can be found off the beaten track.

The center of the glass lid protrudes down inside the jar; the seal occurs on the *rim* of the jar, where the edge of the lid is cushioned by a separate rubber ring applied, wet, to the lid.

The screwband is turned down tight, *then given* a ¼-inch counterturn (as with the one-piece zinc cap above) to allow air in the contents to vent during processing. *Retighten the band carefully as you remove each jar* from the processing kettle, and never retighten it further. (Another example of what you do to "complete seals if necessary.")

—But Don't Use These

Protect your family by using only the jars and fittings that are GRAS, as the Food and Drug Administration might put it (meaning "generally recognized as safe"). Don't be stampeded into using unreliable or makeshift closures (see Checklist for Safe Canning-Jar Closures, in a minute).

(1) Any jars imported from Europe or Asia for which you don't have full and explicit directions that live up to USDA or Agriculture Canada standards of safety. They're charming—as cannisters—but they're too bulky for safe processing of many dense, Raw-packed strong-acid foods; in addition, they are not O.K.'d as heat-tempered for Pressure-processing low-acid foods.

(2) Very old jars for which new fittings/closures are not gettable. There are a number of books about antique canning jars—like the ones for col-

lectors of old bottles—and you may have some real prizes (to keep or to sell, but not to can in). Look in the *Readers' Guide to Periodical Literature,* at every library, for articles in antiques magazines.

Jars In and Out of the Canner

Don't put cold jars of food into very hot water; don't fill a canner with jars of boiling-hot food and then slosh cold water into the canner. Don't clunk filled jars against each other—especially if they're filled with boiling-hot food.

After processing time is up, remove jars at once from the B–W Bath. However, let the Pressure Canner return to Zero naturally and rest for up to 5 minutes before you remove the lid and take out the jars. Often you have an "after-boil" bubbling in the jars as you remove them. Great: your seal has already started. Take care that you don't knock the jars against each other as you unload them. Complete seals if necessary. Then set them on a rack, a wooden surface, or one padded with cloth or newspaper, and be sure it's not in a draft of cold air. And *never* invert processed jars in the mistaken idea that you're helping the seal—quite the contrary!

Cooling. Jars must cool naturally: *don't* drape a towel over them with the idea of protecting them from air currents, because keeping them warm will invite flat sour. Let them sit undisturbed for 12 hours before you check the seals and perform the other chores described in Post-processing Jobs, later in this chapter.

Checklist for Safe Canning-Jar Closures

In the mid-seventies there was a scary shortage of canning-jar lids—scary, because it brought forth a number of fly-by-night producers in addition to some new and conscientious ones. In case such a crisis occurs again, here are the things to look for in a new product; and they apply to equipment other than jar lids.

Good Directions for Use?

First, *are* there any directions, either printed on the container or on a separate sheet/folder inside the package? (A *responsible* manufacturer—one who knows home-canning, and has food safety in mind—will want us to use his closure in the way that gets good, wholesome results which store well.)

Next, if there are directions, do they make sense? Although they may be terse, do they cite with helpful clarity the steps required for using that particular closure successfully?

Then, do the directions say how to handle the particular closure after the processed jar of food comes from the canner? Do they tell us explicitly *how this closure looks* when the cooled jar has a good seal? Or whether, if it's a conventional two-piece screwband type, the band may/may not be removed before storing?

And finally, do the directions say clearly that the steps they give cover only the procedure for using *that closure?* Do they therefore tell the home-canner to use an established guide for complete instructions on packing and processing specific foods? (Makers of canning-jar closures have an obligation to the people who use their products, and it takes only a few lines of print to give an address we can write to for a government bul-letin—or for their own home-canning booklet, if they've been in the busi-ness long enough to have one. See the Appendix.

Acceptable Characteristics for Its Type?

Is it well made and of durable material, able to withstand high heat for a maximum processing period without distorting? (A sleazy lid is unlikely either to behave well in a Pressure Canner at 240 F/116 C or to retain a seal after months on the jar.)

Does it have a ring of sealing compound that's evenly applied? And sufficiently wide to cover *amply* the jar's sealing rim?

Does the top have a degree of flexibility—meaning springiness—in the center? (On a well-sealed modern mason, all lids of this type will be slightly depressed in the center, which is pulled down by the strong vac-uum created as the food cools after correct processing.)

Does it have a tiny down-turning flange that lets it seat securely, and properly centered, on the jar's rim *before* the lid is held down by the band? (A piping-hot lid that slides back and forth on the jar requires some extra manipulation when we put on a piping-hot screwband. And the lid section of every two-piece closure *must* be set separately on the jar rim before the band is applied; even the most terse manufacturers' directions make a point of doing this. *Never* clap the lid inside the band and then screw the whole business on as if it were the top to a peanut-butter jar, because the ring of sealing compound on the separate lid may not be wide enough to compensate for such treatment—which could force the lid so badly off-center that it fails to hold a good seal.)

The Types of Cans for Home-Canning

The containers called "tin cans" are made of steel and merely coated with tin inside and outside. This tin coating is satisfactory for use with most foods. These are called *plain* cans.

But certain deeply colored acid foods will fade when they come in long contact with plain tin coating, so for them there is a can with an inner coating of special acid-resistant enamel that prevents such bleaching of the food. This is called an *R-enamel* can.

Then we have still other foods that discolor the inside tin coating—perhaps because some of them are high in sulfur, and act on the tin the way eggs tarnish silver; or because some are so low in acid as to nudge or straddle the Neutral line. Although there's no record of any damage to these foods from tin coating, there's a can for them that is lined with *C-enamel*.

A leading manufacturer of cans for home-canning says that we may use *plain* cans and *C-enamel* cans interchangeably—except for meats, which always take *plain* cans—with the sole detriment that sometimes the insides of plain cans will be discolored.

Note that although there are many sizes of canned foods on supermarket shelves, the cans used in putting food by at home are: No. 303—which holds about 16 ounces; and No. 401—which holds about 30 ounces. (The difference in measure between No. 303 and pints, and No. 401 and quarts accounts in part for the difference in processing time given for jars and cans.)

What Foods to Can in Which

Throughout the individual instructions we've included the type of can—plain, R-enamel, or C-enamel—to use if you use cans at all.

However, here's a rule-of-thumb to go by if you're canning a food we don't go into:

R-enamel. Think of "R" as standing for "red" and you'll have the general idea: beets, all red berries and their juices, cherries and grapes and their juices, plums, pumpkins and winter squash, rhubarb, and sauerkraut (which is very acid).

C-enamel. Think of "C" as standing for "corn" (which has no acid) and for "cauliflower" (whose typically strong flavor indicates sulfur) and you get: corn—and hominy, very low-acid Lima and other light-colored shell beans (and, combined with corn, succotash); cauliflower—and things with such related taste as plain cabbage, Brussels sprouts, broccoli, turnips, and rutabagas; plus onions, seafood, and tripe.

Plain. This is the catch-all, and may take these foods, as well as others: most fruits, tomatoes, meats, poultry, greens, peas, and green/snap/string/wax beans, and certain made dishes (like baked beans, etc.).

If you're canning mixed vegetables, use *C-enamel for preference* if one or more of the ingredients would go in C-enamel by itself.

Remember, though, that the heavens won't fall if you mix up *plain* and *C-enamel* for fruits and vegetables. Nor will "red" acid foods be bad if they're not canned in *R-enamel:* they just won't have their full color.

What's Re-usable?

Only the sealer apparatus. No damaged, rusted, dented, bent *new* can or lid may ever be used in the first place. No can or lid may ever be *re-used*.

For information on where to get cans and can sealers, see the Appendix.

Preparing Cans, Lids, and Sealer

If cans and lids are to be processed in a Boiling–Water Bath (212 F/100 C) or Pressure-processed, they need not be sterilized before filling.

Wash cans in clear hot water, scald; drain upside down on sterile cloths so air in the room cannot contaminate the inside of the can. To sterilize, wash, submerge in hot water and bring to boiling, and boil for 15 minutes; remove and drain as above.

The gaskets of some lids could be of a material that *must not be wetted* or it won't seal right; other gaskets are of a rubber-like composition that must be treated as carefully as the compound on nonboilable metal screwband lids. Therefore be sure to ask your supplier for explicit instructions for sterilizing the lids.

Adjust the sealer according to the manufacturer's directions, and test its efficiency by sealing a can that's partly filled with water, then submerge the can in boiling water: if any air bubbles rise from the can after a few

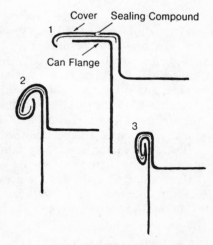

A can of food exhausted to 170 F/77 C (hence the dry cloth pad to protect fingers) will be sealed with the rolling seam shown in the drawing. (*USDA photograph; drawing by Norman Rogers*)

seconds, the lid is not seamed tightly enough. Adjust the chucks of your sealer until you have a perfect seam joining the lid to the can.

In and Out of the Canner

Unlike jars, cans do not vent during processing: they are sealed completely before they go into the B–W Bath or Pressure Canner. Therefore air in the food's tissues and in the canning liquid must be driven out *before* the lid is crimped on perfectly tight—otherwise you wouldn't end up with a vacuum when the cans are cooled.

So *all* cans of food require an extra intermediate step that may be bypassed with most jars (the exceptions being jars of meat, poultry, and seafood), and we'll deal with it now before we turn to the Raw and Hot packs in the general procedure for filling any container with food to be canned.

THE EXTRA STEP: "EXHAUSTING"

To drive out enough air to make the desired vacuum, you heat the food to a minimum of 170 F/77 C at the center of the filled can or jar. You achieve this temperature either by bringing loose food to boiling in a kettle and then ladling it into the container, or by heating the container of raw or cold food until a glass food thermometer thrust into the middle of the contents registers 170 F/77 C (or the 180 F/82 C preferred by the Ohio Extension Service, if you live in the Buckeye State).

Fill the cans according to the instructions for the specific food and put them, still open, on a rack in a large kettle; add hot water to about 2 inches *below* the tops of the cans. Cover the kettle and boil the whole business until your pencil-shaped food thermometer, stuck down in the center of the food until its bulb is halfway to the bottom of the can, registers 170 F/77 C. You can get the same result by boiling open cans in a Pressure Canner at Zero pounds pressure (as described for jars in "Canning Seafood," Chapter 11).

COOLING THE CANS

Cans *must be cooled quickly* after their processing time is up, lest their contents cook further (this is why, if you're using cans, you vent your Pressure Canner to hasten its return to Zero pounds).

Fill your sink or a washtub with very cold water and drop the processed cans into it. As each can is removed from the canner its ends should be slightly convex, bulging from the pressure of the hot food inside it; if the ends don't bulge, it means that the can was imperfectly sealed before it went into the processing kettle. The ends will flatten to look slightly concave when the contents have cooled and shrunk, indicating that the desired vacuum has formed inside.

Change the water when it warms, or add ice, to hasten cooling. Remove cans when they're still warm so the internal heat will hasten air-drying.

GENERAL STEPS IN CANNING

Let's see how to fill the jars/cans with food before they go into the Boiling–Water Bath or the Pressure Canner, and how to treat them afterward.

Raw Pack and Hot Pack

Many foods may be packed in their containers either *Raw* (in some manuals they used to be designated as "Cold") or *Hot*. The food is trimmed, cleaned, peeled, cut up, etc., in the same manner for both packs. The same amount of canning liquid is added to the container of solid food, regardless of whether it's Hot pack or Raw—roughly ½ to ¾ cup for pints or No. 303 cans, 1 to 1½ cups for quarts or No. 401 cans; also the liquid should always be very hot. The optional seasoning is added just before processing either pack. And with both packs the containers are handled identically after being removed from the canner and cooled.

Raw Pack

Today, Hot pack for low-acid (or borderline) foods is stipulated across-the-board, for safety's sake. Raw pack, though, may still be used for some fruits—mainly to protect their texture—and for many pickled foods.

Boiling syrup, juice, or water is added to raw foods that require added liquid for processing.

Jars of Raw-packed food must start their processing treatment in hot *but never boiling* water, otherwise they're likely to crack. Even when a jar has been exhausted, the water in the processing kettle should not be at a full boil.

Raw-packed foods usually take longer than Hot packs to process in a Boiling–Water Bath; this is especially true for the denser foods.

Hot Pack

An increasing number of foods are processed by Hot pack. Food that is precooked a little or almost fully is made more pliable, and so permits a closer pack. (Foods differ in the amount of preheating they need, though; spinach is merely wilted before it's packed Hot, but green string beans boil for 5 minutes.)

In a Boiling–Water Bath, Hot-packed food generally requires less processing time than Raw does, because it is thoroughly hot beforehand. However, there generally is no difference in the time required for Pressure Canning either pack: by the time you start counting—for example, the minute pressure reaches 10 pounds (240 F/116 C)—Raw-packed food has become as hot as if it had been packed Hot to begin with. (One of the interesting exceptions is summer squash, which needs *longer* Pressure-processing for Hot pack, because the precooked squash is much more dense in the container than the crisp raw pieces are.)

Leaving Headroom

In packing jars of food that will be processed in a Hot–Water Bath, Boiling–Water Bath, or Pressure Canner, there must be some leeway left between the lid and the top of the food or its liquid. This space—called *headroom* in the instructions that follow later—allows for expansion of solids or the bubbling-up of liquid during processing. Without it, some of the contents would be forced out with the air, thus leaving a deposit of food on sealing surfaces, and ruining the seal.

Too much headroom may cause food at the top to discolor—*and* could even prevent a seal, unless processing time was long enough to exhaust all the excess amount of air.

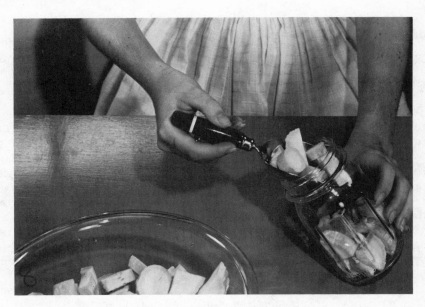

Pressure-processing will condense this Raw pack of summer squash a great deal, even after it's tamped down and given generous headroom. (*USDA photograph*)

Important Altitude Note: steam in the headroom expands more at high altitudes than it does in the sea-level zone (1000 feet/305 meters), so we suggest that you increase headroom by $\frac{1}{8}$ inch for every 1000 ft/305 m above the sea-level zone, but not to exceed $1\frac{3}{4}$ inches for quarts, 1 inch for pints, $\frac{3}{4}$ inch for $\frac{1}{2}$-pints.

Also, Pressure-processing has steam at a higher temperature than a B–W Bath does, and steam expands more when it's hotter.

All of which means that you should be prepared to make your own arithmetical adjustments if you live at, say 7000 ft/2134 m, and process at 15 psig.

The right headroom *in the sea-level zone* for each food and its processing liquid is specified in the individual instructions, with more headroom given to starchy foods—Lima/shell beans, green peas, corn, etc.—because they swell in the canner.

In the older three-piece closure for mason jars (glass lids + separate rubber ring + metal screwband to hold things in place), the center of the glass lid protrudes a bit down into the jar. With these glass lids, therefore, we recommend that you add from $\frac{1}{4}$ to $\frac{1}{2}$ inch *more* to the headroom given in the directions, in order to compensate for space lost in the neck of the jar.

Cans generally require no headroom between liquid and lid—air is driven out in the exhausting step.

Filling, Bubbling, and Capping

Whether it's in cans or jars, you pack most raw food firmly (except for the starchy vegetables that expand, mentioned above), and most hot foods rather loosely.

Prepare and fill only one container at a time—don't set them up in a row, standing open to airborne spoilers. Set the jar/can in a clean pie dish, or whatever, to catch spills and overflows.

As you pack Raw, shake the container or tunk it on the bottom to settle the pieces of food; use a slender rubber spatula to make room for chunks you're fitting in; don't compress the food so much that it will spring up again, though, and invade the headroom. Pour in your boiling liquid—syrup, water, juice—to the desired level, keeping track of the headroom you must leave.

Follow directions for headroom in Hot pack. Your wide-mouth funnel is most handy here for controlling dribbles of food that must be wiped away completely.

Removing Bubbles

In either pack in any container, run a slender plastic blade down between the food and the side of the jar/can at several points. There is likely to be more air trapped in the liquid between pieces of food in Raw pack than in Hot; but use your blade with both. Take care not to stir or to fold in more air. *Never use a metal blade:* it makes the minute scratches inside that cause jars to break.

Examples of headroom in canning in jars larger than 1 pint: left, 1 inch for chicken packed Raw and exhausted; right, $\frac{1}{4}$ inch for whole-kernel corn packed Hot in cans, with boiling precooking liquid added right up to the brim. (*USDA photographs*)

For insurance, use your blade again—but sparingly—when you "top up" a can with boiling liquid after exhausting it to 170 F/77 C.

Capping

With a fresh section of paper toweling, or a scrupulously clean cloth wrung out in boiling water, wipe the sealing rim of cans to remove any liquid, food tissue, or fatty substance, because any of these can interfere with the seal. Wrap a clean, damp cloth around the body of the hot can to make it comfortable to hold; set the can in the sealer. Wipe the lid (rinse it quickly in hot water if, in some way, it has become dusty—and this is the *first* time it may be wetted), place it on the sealing rim of the can, and crimp it in place according to the directions for making the seal complete. Set the can in the processing kettle, and deal with the next one.

Wipe the sealing edges of the jars—*rim* of the glass, if you're using two-piece screwband metal lids or the three-piece glass closure; *rubber ring* around the neck of the jar, if it's a bailed closure or takes a porcelain-lined cap. Remove every vestige of food or other substance that would prevent a perfect seal.

Set the metal lid of your modern masons on the rim of the jar, *sealing compound next to the glass;* screw down the band *firmly tight*. Put the jar in the canner.

Set a wet rubber ring on the rim of the jar that takes a three-piece glass closure; seat the glass lid on the rubber so the center of the lid protrudes down into the headroom; screw the metal band down tight, then give it a ¼-inch counterturn to allow the jar to vent during processing. (You'll retighten it after the jar is processed.)

Screw the porcelain-lined zinc cap down tight on the rubber-padded shoulder of the old-style mason jar, then give it a ¼-inch counterturn to allow for venting. (You'll retighten it after the jar is processed.)

Set the rounded glass lid of your bailed jar on the rubber ring, turning the cap so the notch in the top will catch the wire most securely; push the longer hoop up over the top of the lid until it's held firmly in the notch. (But *don't* snap the lower bail down on the shoulder of the jar: you'll do this after the jar is processed.)

Post-Processing Jobs

Your B–W Bath kettle is off the heat, your Pressure Canner has returned to Zero—but not until tomorrow will today's canning be finished.

As you remove *cans* to drop them in cold water, mark the ones with poor seals. Right now, hot from the canner, their ends *should* bulge from internal pressure sealed in; and in the cooling tub, tiny bubbles must *not* appear around the lids. Therefore flat ends or bubbles mean pinpoint gaps in the sealing seam.

On the other hand, your *jars* seal themselves as they cool. Of course you never tighten the modern "dome"/"snap"/"self-sealing" two-piece screwband metal closures. But when you used to "complete seals if necessary" by flipping down bails or screwing down caps or glass lids that have separate rubber rings, you were not actually making the seal: you were merely securing the lids firmly in place so the ensuing vacuum can complete the seal.

For this reason, *don't open the jar* to add liquid if you notice one whose liquid has partly boiled out during processing. Simply treat the jar normally and stand it up beside the others. After cooling undisturbed overnight, it might have a good, safe seal.

Two more things. Cans aren't shockproof just because they're already sealed and are made of metal, so don't bump them around or transport them until they have cooled and rested overnight. And don't invert or shake—or even tilt—jars before they're cold and sealed.

Checking Seals

The day after canning is the time for checking the seals of your containers, and preparing them for storage. And *this is your ONLY chance to salvage food that has failed to seal;* once it is stored, a bad seal means that you must destroy it.

CAN SEALS

By now the ends of your cans should be pulled slightly inward, proving that there is a vacuum inside. However, if the end of a can has not collapsed, press it hard: if it stays in, the seal is O.K.

Be right finicky as you check your cans, because at this early stage a poor seal will not have had time to become dramatic. Therefore the ends won't be bulging way out (certainly not even so much as they did right after the cans were processed yesterday); nor will the seams be leaking gassy, spoiled food (although traces of food at any seam are an obvious sign, since the cans got well rinsed by the water in which they cooled).

Springy ends and bits of food mean bad seals. Open the cans and refrigerate the contents, to serve or reprocess. Re-can and process anew *only* if failures in the batch warrant doing so.

JAR SEALS

First, don't be dumped if there's a haze of dried canning liquid on the outside of some jars. All your jars vented in the processing kettle, remember; and several may have lost liquid, which clouded the water in the canner and so left a slight deposit on the jars.

Food particles lodged around the base of the closure could mean trouble, though. You'll know for sure when you check your seals.

On your modern mason jars, the metal lids of the two-piece screwband closure will have snapped down, pulled in by the vacuum that means a good seal. If you find a lid that is not concave, press it: if it springs back, the seal is gone; if it stays down, its chances are better—but set this jar to one side for a tougher test you'll give questionable jars in a minute.

Test bailed jars and the old-style masons with porcelain-lined zinc caps by tilting them far enough so the food presses against the closure. If bubbles start at the lids and rise through the contents, the seal is no good. Moisture appearing at the sealing point is a bad sign too.

Test your flat glass lids by *gently* removing the screwbands that hold them down on their rubber rings. If the band is reluctant to turn, for heaven's sake don't force it—this could shift the lid enough to break the seal right there. A hot cloth held around the band is usually enough to make it expand and come free. Tilt the jar: any seepage or bubbles that mean a poor seal?

The really tough second test can be applied to all your jars except for old masons with the zinc cap. Take a jar you have doubts about—modern mason with its metal lid, any with a glass lid, any with a bailed closure—and remove the screwband or release the bail, including the longer hoop that holds the rounded glass top in place. Set the jar in a pan that's padded with a perfectly clean towel *and lift the jar by its sealed-on lid*. If the seal is weak, the weight of the jar will break it and the jar will drop and spill, leaving the metal disk or glass lid in your fingers. (The towel protects the fallen jar from cracking; collect the spilled contents to serve or reprocess.)

CHECKLIST FOR POOR SEALS

One poor seal out of a full canner batch is a disappointment, but not a worry. Nor is the food in only one container worth reprocessing, which means getting cranked up to start over from scratch with a fresh jar or can and a fresh closure, and repeating, with utmost care this time, every step in the whole canning procedure. Best to eat it right away.

But two poor seals?—not so good. And more than two seal failures in one canner batch will tell you that you're making at least one mistake in your canning technique. For your family's sake, see if it's one of these:

Short cuts in sanitation or preparing food "because it will sterilize anyway." Manufacturers' directions not followed in preparing closures.

Imperfect or makeshift containers/closures. Sealing edges of jars, permanent closures (glass tops/lids, porcelain-lined zinc caps) have nicks/cracks/warps. Jars are "one-trip" commercial containers, plus commercial lids from peanut butter, mayonnaise, pickle, etc., or babyfood jars. Rusty or bent bails on old jars (with extra padding to take up slack?). Re-used sealers (rubber rings or metal lids with sealing compound). Dented/bent cans or can lids.

Haphazard filling, exhausting, capping of containers. Packing too tightly or too loosely. Too much or too little headroom. Trapped air not removed. Cans not properly exhausted before capping. Sealing rims not wiped clean after filling. Screwbands of modern two-piece lids not "firmly tight" to hold lids snugly on and in place during processing; other jar lids not allowed necessary venting. Cans not properly exhausted before sealing. Can-sealer not tested/adjusted.

Processing kettle mismanaged. Water in B–W Bath not 2 inches over tops of containers; Pressure Canner not vented long enough to remove air (or, if used unlocked as a substitute B–W Bath kettle, not vented long or strongly enough before starting to time processing period). True boil not reached and maintained in B–W Bath; called-for pressure not reached (gauge unchecked?) or maintained; full processing time not used. Maker's instructions for mechanical operation not followed. Pressure allowed to fluctuate, so contents lost from jars and trapped between jar rim and lid—thus preventing good seal.

Processing method inadequate. Maverick or makeshift method used. Adjustments for altitude above sea-level zone (up to 1000 ft/305 m) not made. Incorrect venting or return-to-Zero of Pressure Canner (made to cool quickly? lid removed before Zero?).

Containers mishandled after processing. Modern two-piece screwband lids were fiddled with (notable offense being brisk tightening of the screwband, which usually loosens slightly during processing); seals not completed as needed for other closures.

Cans not cooled quickly enough; but jars left in B–W kettle after processing time is done OR were wrapped to slow down natural cooling.

Jars inverted to cool: wrong and unnecessary but maybe prompted by misconceived notion of how post-World War II closures work.

Jars/cans roughly stacked for storage or in other ways knocked around, thus hurting seals.

Cleaning and Labeling

Twelve hours after they're processed, remove the screwbands and wipe the jars carefully with a clean, damp cloth, paying special attention to the area around the seals.

Never force a reluctant screwband. Hold a hot cloth around it to make it loosen by expanding; if this doesn't work, mark the jar for special watching while it is stored, and turn to the next jar.

The modern metal closure doesn't need a screwband to ensure the seal, of course—but the band does protect the seal in case you plan to transport the jars or stack them. *Don't replace the screwbands if you take them off,* because you could twist the lid just enough so the torque will break the seal.

At county and state fairs, canned foods are entered with all screwbands off, thereby proving faith in the seal.

The glass lids of three-piece closures are especially vulnerable to jostling and bumping, so weigh carefully the possibility of broken seals before removing bands here.

Label each container with the name of the food, the date it was canned, and any special treatment you gave the food (the last information will be mighty handy if the product is notably good or notably bad when you come to serve it).

Storing

Storage for all canned food must be cool and dry and—if the food is in jars—dark. Even when home-canned foods are adequately processed, they will lose Vitamin C, carotene, thiamine, riboflavin, and niacin at temperatures above 50 F/10 C; and light hastens oxidation of fats and oils, destroys fat-soluble and light-sensitive vitamins, and fades the color of the food.

"Cool" means 32–50 F/Zero–10 C. Containers must not freeze, lest the food expand and break the seal. Canned food that is held in storage too warm can still spoil, because certain thermoduric bacteria can reactivate and grow at room temperature or higher. Therefore keep canned food away from heating pipes or cozy nooks behind furnaces, etc.

Damp or humid storage can corrode or rust the metal of cans and closures, and thereby endanger the seals.

Berries and fruits are especially likely to bleach in jars exposed to light, but other foods, too, can become pale and unappetizing if stored in the light. "Cool" and "dry" have priority over "dark," so you may have to protect jars of food from light by wrapping them individually in paper, or putting them in the cartons the jars came in (be sure to put cardboard dividers between the jars).

Put any containers with suspect seals in a special place so you will use them first or be able to keep an eye on them easily.

Arrange your food on the storage shelves so "last in" is "last out"—this way you'll keep a good rotation.

And do check your canned food periodically for signs of spoilage that have developed during storage.

Signs of Spoilage and What to Do

Before a container is opened, you can see signs of spoilage that indicate the food is unfit or actually dangerous to eat:

- Seeping seams, bulging ends on cans.
- Seepage around the seal, even though it seems firmly seated.
- Mold around the seal or visible in the contents.
- Gassiness (small bubbles) in the contents.
- Cloudy or yeasty liquid.
- Shriveled or spongy-looking food.
- Food an unnatural color (often very dark).

When the container is opened, these are additional signs of spoilage:

- Spurting liquid, pressure from inside as the container is opened.
- Fermentation (gassiness).
- Food slimy, or with too soft a texture.
- Musty or disagreeable or downright nasty odor.
- Mold, even a fleck, on the underside of the lid or in the contents.

If any such signs are evident in unopened or opened containers DESTROY THE CONTENTS SO THEY CANNOT BE EATEN BY PEOPLE OR ANIMALS. Burn the food if you can—it is sometimes not safe from animals if you bury it. If you can't burn or bury the contents, dump them into a large kettle, adding water to prevent scorching, and boil the spoiled food hard for 15 minutes to destroy toxins, and then flush down the toilet whatever will not clog the plumbing. Soak all metal cans, lids, jars, closures in the very strong disinfecting solution of 1 part household chlorine bleach to 4 parts tepid water for at least 15 minutes. Discard the sterilized metal cans and closures, and all sealers. Sterilized jars and glass lids may be used again if they are perfect and undamaged. In a fresh 1:4 solution of chlorine bleach and water, wash all utensils, cloths, and surfaces that might have come in contact with the spoiled food.

Before tasting any canned low-acid food: you must be unshakably certain that your Pressure Canner was operated correctly—pressure gauge accurate and/or deadweight gauge signaling properly—and that requirements for times and corrections for altitude were followed.

Unless you are sure that these safeguards were observed, a margin of protection is added by boiling the canned low-acid food hard for 15 minutes to destroy any hidden toxins (corn and greens require 20 minutes), and stirring to distribute the heat. If the food foams unduly or smells bad during boiling, destroy it completely so it cannot be eaten by people or animals.

Note: at least two groups whose work *PFB* respects have studied additional ways to destroy toxins that cause illness (the toxins here have not, obviously, given such advance warnings as bad smell, foaminess, etc.). One is a team made up of Oregon State University's noted microbiologist Dr. Margy Woodburn and researchers at the University of Wisconsin; the other was led by Mrs. Betty Wendstadt of National Presto Industries. Their addresses, so you can inquire about recent published material, are given in the Appendix, "Where to Find Things": Mrs. Wendstadt's under the listings for Chapter 6, "Canning Methods," and Dr. Woodburn's under the listings for Chapter 11, "Canning Seafood."

Community Canning Kitchens

The concept of community canning kitchens was nothing new in the midseventies, when they seemed to spring up by the dozen in every state: they actually went back several generations, and especially during the Great Depression they were a godsend. With varying degrees of sophistication and amounts of equipment, they offered householders without adequate means at home a place to bring their food for canning. Sometimes the facility was at a church or school, and the neighbors pooled their money to buy the canners and the containers. As years went by— and up to the present—the kitchens became fully equipped with retorts, blanchers, mechanized can-sealers if cans are used, and the cooling arrangements for jars. In addition, highly trained personnel operate the machinery, with the householders generally giving a hand with food preparation and the less technical jobs.

It is the general practice now for the kitchen to furnish containers and charge such-and-such per container and processing the food.

After the seventies, most of the newer canning kitchens began closing, except for the ones that were taken over cooperatively to produce specialty foods for sale. There still remain well-established kitchens in farming areas. Your county's Cooperative Extension Service can give you information about ones in your region.

Equipment for Canning Kitchens

Community kitchens using glass jars were originated by the Ball Corporation and in 1980 were donated to the World Ministry of Brethren Churches, and recently have been taken over by Joel M. Jackson, the man who had been instrumental in developing them in the first place. Information about them can be got from Mr. Jackson, Director of Food Preservation Systems, 1604 Old New Windsor Road, Old New Windsor, Maryland 21776 (telephone 301/635-2765).

Information about everything involved in canning in metal cans can be got from Dixie Canner Equipment Company, Inc., P.O. Box 1348, Athens, Georgia 30601 (telephone 404/549-1914). Direct your queries to Charlotte Blume, or to William Stapleton, the owner and developer of the equipment. Dixie Canner does not sell the cans, but can steer you to a number of companies that do: its prime function is designing canning facilities ranging from a sealer that can be clamped on a kitchen table and be used by one determined person, to what are assembly-line facilities. Dixie published the *Nutrition Education and Food Conservation* (NEFCO) *Canning Book* for use with metal cans, and it is a good source of information.

See the reference for this chapter in the Appendix for more on suppliers of equipment and containers.

7
CANNING FRUITS

Prediction: there will be a resurgence of interest in canning fruit, now that it has been made inescapably plain that the average diet in the United States contains far too much sugar. The old idea that canned fruit must wallow in Heavy Syrup in order to be acceptable is a concept long discredited. Indeed, fruit may be canned without any added sweetness, because the amount of sweetening put into canned fruits is merely there to help hold color or to enhance flavor: it has no virtue to speak of as a preservative. So what does one use in place of a syrup, especially since water alone may be too blah for words?—fruit juice. It takes only a little more thought and a bit more time to extract it from fruit set apart for the purpose. The result is delightful.

As with all food that is being put by, fruit and berries must be in fine condition before they are canned. Therefore use only firm just-ripe fruits and berries. They must be fresh—strictly fresh. They must be processed within hours of being brought into the house, or they must be refrigerated overnight.

If they are merely a shade overripe they have lost some of their natural acid content—and they'll also be more likely to float to the top of the jar.

If they are extremely overripe they have lost a critical amount of acid, and, aside from the fact that their flavor and texture are disappointing, they might not be able to discourage growth of spoilers.

Never use "drops"—drops being fruit that has fallen from the tree or been shaken from the bush. These are on the verge of rotting, and they can all but ruin a good batch of food.

The individual instructions that follow use the Boiling–Water Bath at 212 F/100 C at the sea-level zone, with, to protect the flavor of juice alone, the *Hot*–Water Bath at 190 F/88 C at sea level. To people living at high altitudes we recommend another look at Correcting for Altitude in Chapter 3; and we've inserted just ahead of the specifics for each fruit a reminder about increasing headroom above 1000 ft/305 m, and especially for Pressure-processing.

Note: procedures for canning Tomatoes—which are classed botanically as fruit—are given separately in their own Chapter 8.

No Short Cuts

You simply must follow carefully all the right steps for selecting, preparing, and processing fruits: there have been cases reported of botulism poisoning from home-canned apricots, blackberries, figs, huckleberry juice, peaches, and pears—even though these are all in the traditional strong-acid grouping. Records of these cases indicate that there was careless handling somewhere along the line. (See Dealing with the Spoilers in Chapter 2.)

In case you find a tempting old recipe that relies on the "open-kettle" procedure now proven to be inadequate, here is how to translate it for the B–W Bath. Prepare and cook the fruit according to the old recipe. Pack the boiling-hot fruit and its juice in clean jars, leaving correct headroom for container size and altitude. Adjust lids, and process in a B–W Bath for the time given for Hot pack in the directions for that sort of food, later in this chapter.

CONVERSIONS FOR CANNING FRUITS

Do look at the conversions for metrics, (with workable roundings-off) and for altitude—both in Chapter 3—and apply them.

Liquids for Canning Fruits

Sugar in the relatively small amounts used in canning fruits does *not* act as a preservative: it merely helps to retain the texture and color of canned fruit. That sugar added cannot act as a preservative is borne out by the disquieting appearance of spoilage in opened jars of fruit that have remained too long in the refrigerator.

Commercial canners in the past have leaned toward Heavy Syrup, except in the so-called diet pack that declares that the fruit is either canned in its own juices or that water only has been added to the pack. Formerly, these ostensible diet packs were found only in the special sections of supermarkets devoted to low-sodium, low- or non-sugar products. Now, though, fruit canned in its own juices or with Very Light Syrup is mingled with the rest of the fruits on store shelves.

Heavy Syrup has not been used very much by home-canners for some time. The instructions for individual fruits recommend which syrups to use—but remember, please, that if you prefer to can without any sweetening, it is perfectly safe to do so. The recommendation of syrups are merely for flavor and texture, and this fact cannot be emphasized too much.

Boiling water is the easiest liquid to add: it is poured into the jars after the fruit has been packed; then, as with all packs involving added liquid, a plastic blade or a very thin spatula is moved around the inside of the jar to release any bubbles of trapped air.

Altitude note: at sea level the usual headroom for fruits is a scant ½ inch for ½-pint jars (not used very often unless they are for special diets, which are discussed later). Standard headroom for fruit in pints or quarts in a B–W Bath at sea level is ½ inch. At high altitudes—especially above 3,000 feet, and *always at 5,000 or more feet*—extra headroom must be added because steam in the jars as the atmospheric pressure gets *lower,* expands as the altitude gets higher. Therefore a housewife in Santa Fe would can a 1-pint jar of fruit with up to 1 inch of headroom (and most likely in a Pressure Canner).

How Much Liquid

Roughly estimate up to ½ to ¾ cup of syrup/liquid for each pint or No. 303 can, and increase the allowance proportionally for larger containers; this means that quarts would take from 1 to 1½ cups of added liquid, with the same amount estimated for No. 401 cans. The middle-sized 1½-pint jars—and tapered wide-mouth—would take an in-between amount.

It's a good idea to do your fruit juice/syrups before you actually start preparing your fruit for canning; of course the syrup can be made several days in advance and refrigerated. If you take a quick look at the yield chart for fresh fruits to can you'll find that the *pounds* needed per 1-quart jar are the most sensible measurement to go by. (Not many people buy grapes or apricots by the bushel any more; cherries and the smaller fruits are bought by the pound and, unless you have your own fruit trees or berry plants, you're watching the cost as you go along.) However, a whole bushel of peaches will produce 18 to 24 quarts, depending on whether the fruit is packed as halves or quarters or slices.

The rules-of-thumb are: for Thin Syrup use 4 cups of water or other liquid and 2 cups of sugar to make 5 cups.

For Medium, use 4 cups of water or other liquid and 3 cups of sugar to make 5½ cups.

For Heavy, use 4 cups of water or other liquid and 4¾ cups of sugar to get 6½ cups.

To make any of these, mix the sugar with the water/liquid in the proportions given, heat them together until the sugar is dissolved, skimming if necessary. Skimming is sometimes needed because of perfectly edible substances in the sugar that can make a small amount of opaque froth as the syrup is heated.

Using Juice for Canning Liquid

The best source for estimating the yield of juice from fruit is in Chapter 18, "Jellies, Jams, and Other Sweet Things"; or the leaflet that comes with any commercial pectin to be used for making jelly and jam at home. Naturally these juices that are the base for jellies are likely to overpower fruit if they are used full strength as canning liquid; therefore, if you make up a batch of clear, strained juice as if it were for jelly, you will want to dilute it with an equal amount of water, or to suit the family's taste. Try mixing maverick juices for unusual and delicious flavors. However, ground spices usually change flavor too much or cloud the liquid. Compatible juices are apple (not cider) and pineapple; both can be bought in large containers, to be diluted, etc.

Either buy extra fruit from which you'll make the juice, or set aside a portion of a large amount, like a bushel, that you already have. Pick over the fruit carefully (reserving the perfect examples for canning and using the homelier ones for making juice). Wash these and cut away all blemishes, bruises, gouges; certainly any areas where spoilage has started must be cut out drastically. It is not necessary to peel the fruit to be used in the canning liquid, but just make sure that your fruit is impeccably clean, and has been cut in small pieces after all blemishes have been removed. Put the prepared fruit in an enameled or stainless-steel kettle with an equal amount of water. Bring to simmering and, when the fruit is softened, crush it with a potato masher or pastry blender or some such thing. Then bring it to a boil and boil it for a couple of minutes. The fruit pulp need not be put through a real jelly bag (again, see Chapter 18) if you are in a hurry. Still, it is nice to have a reasonably clear liquid surrounding your nice canned fruit, so flop your pulp around in a sieve to get all the juice you can and then—because the juice is much less concentrated than it would be for making jelly—pour it through a standard coffee filter into a series of clean, hot quart jars. Our favorite for fast filtering is a large circle that folds into quarters, but any filter paper will do as liners for smaller sieves that hold the cone in place to let the hot juice drip through very quickly. Cap with the sterile lid and store in the refrigerator, not longer than 3 or 4 days.

The Sweeteners to Use

Chapter 5, "Common Ingredients and How to Use Them," offers details of the relative sweetening powers of natural ingredients—light corn syrup, mild-flavored honey, maple syrup, brown sugar, and sorghum—and also comments on the virtues of non-nutritive sweeteners in special diets.

Meanwhile, *fructose,* which is fruit sugar, could be substituted for standard sucrose (table sugar)—*if* you can afford to do so. Commercial canners use the "HFS" (high fructose syrup) that is not always handy for the general public. If you are drawn to fructose, it would be better economy to add it when the fruit is served.

Steps in Canning All Fruits

After making your syrup, collect your utensils and containers. It's vital to have scrupulously clean utensils, cloths, and work surfaces, including cutting-boards and counters, and a good supply of fresh water of drinking quality.

Jars/cans and their sealers must be perfect and perfectly clean. They need not be sterilized, since the adequate Boiling–Water Bath will sterilize the inside of containers during processing. They must, however, be clean and hot.

Work with only one canner batch at a time. Wash the fruit thoroughly in fresh drinking water, but don't let it soak. Lift it from the water to allow sediment to settle at the bottom of the wash water. Be extra gentle with berries. Remove stems, hulls, pits, skins, cores as described in instructions for individual fruits. Cut away all soft or bruised spots and any places where skin is broken: such blemishes can spoil your batch.

To Prevent Canned Fruit from Discoloring

Cut apples, apricots, nectarines, peaches, and pears discolor in air. Either coat the cut pieces well as you go along with a solution of 1 teaspoon crystalline ascorbic acid (Vitamin C, the best anti-oxidant) to each 1 cup water; OR drop the pieces in a solution of 2 tablespoons salt and 2 tablespoons vinegar for each 1 gallon of cold water—but not for longer than 20 minutes, lest nutrients leach out too much—then rinse and drain the pieces well before packing Raw or Hot. Optional: to prevent their darkening while in the containers, add $\frac{1}{4}$ teaspoon Vitamin C to each 1 quart during packing—IF they haven't been treated with ascorbic acid as they were being prepared. (See more anti-oxidant treatments in Chapter 5.)

And remember: fruit canned with too much headroom or too little liquid will tend to darken at the top of the container.

Packing, Processing, and All the Rest

We refer you to the blow-by-blow General Steps in Canning in the previous chapter instead of offering a quick paraphrase here. Everything in that detailed account applies to packing, processing, checking, and storing jars or cans of fruit.

Each step is essential to successful canning, but if we had to assign a No. 1 priority as most critical for food safety (assuming that the containers are sound and the food is appetizing), it would be *adequate processing*—meaning full heat for the full time in a Pressure Canner for low-acid foods, or in a B–W Bath for strong-acid ones.

CANNED FRUIT TROUBLES AND WHAT TO DO

The only time you can tinker SAFELY with a container of canned food is during the 24 hours after it comes from the canner and before it is stored away. And then only if you find a faulty seal during this lull. Repack and reprocess the fruit from scratch according to the original instructions, cutting no corners. There of course will be a loss in quality, especially in texture, from doing it over; and if there's only one poor seal it's probably simpler to eat the fruit right away or refrigerate it for a day or so, then serve it.

Examples: failing to exhaust the contents of a can to a minimum of 170 F/77 C, or leaving insufficient headroom in jars (the latter can cause bits of food to be forced out during the processing period, with resulting poor seals). Heat the food—exhausting if in cans, or as Hot pack—in clean containers, *with fresh, new lids/sealers,* and reprocess for the full time. However, several poor seals warn you that something was dangerously wrong with your packing or processing, and the failure is likely to affect the whole batch.

Failing to keep at least 2 inches of boiling water over the tops of the containers, and not keeping the water in the canner at a full boil from beginning to end of the processing period—both are fairly common causes of loss of liquid in the jars, and poor seals. Repack in clean containers with fresh, new sealers, and reprocess for the full time.

But, if after the containers have been stored away, you find any of the following, DESTROY THE CONTENTS SO THAT THEY CANNOT BE EATEN BY PEOPLE OR ANIMALS. Then give a 15-minute soak for containers and closures in a 1:4 household bleach solution, discarding all closures except *sound* all-glass lids for bail-type jars, and discard all cans, metal caps, rubber rings, and sealers. If sound, the jars may be used again.

- Broken seals, bulging lids on cans.
- Seepage around the seal, even though it seems firmly seated.
- Mold, even a fleck, in the contents or around the seal or on the underside of the lid.
- Gassiness (small bubbles) in the contents.
- Spurting liquid, pressure from inside as the container is opened.
- Spongy or slimy food.
- Cloudy or yeasty liquid.
- Off-odor, disagreeable smell, mustiness.

Not Prize-Winning, but Edible

If the containers and contents offer none of the signs of spoilage noted above, and if the storage has been properly cool, canned fruit with the following less-than-perfect characteristics is still O.K. to eat:

Floating fruit. The fruit was overripe, or it was packed too loosely, or the syrup was too heavy.

Darker fruit at the top of the container. Too much headroom above the liquid.

Bleached-looking berries. With no signs of spoilage present, this could mean that jars were exposed to light in the storage area; next time, wrap the jars in paper or stash them in closed cartons if the storage isn't dark.

CANNING EACH FRUIT

Altitude and headroom reminders: do look at the Correcting for Altitude section in Chapter 3 for required adjustments you must make in times (or even in the processing procedure), and jot down your total adjusted processing times in the spaces we provide in the individual instructions that follow (you will also find these spaces in other chapters' instructions).

And remember that, if you have elected to use Pressure-processing according to the makers' most recent instructions (see the Appendix for sources) instead of the Boiling–Water Bath, you should increase slightly the amount of headroom, because the air (gas) inside the jar expands more at the greater heat of the Pressure Canner. If you remain with the B–W Bath and merely add processing *time,* you still should increase headroom for altitudes of more than, say, 3000 ft/914 m—¾ inch for pints is not too much in this case.

Apples

Even with root-cellaring and drying (see both), you'll want some apples put by as sauce, dessert slices, or pie timber. And of these, probably the handiest thing is to can applesauce and slices done in syrup, and to freeze the slices you'll use for pies (Chapter 14). There's a handy rule for Apple Pie Filling in Chapter 12 and for the method/uses of Boiled Cider in Chapter 23. How to do Apple Cider is described later in this chapter, under the Canning Juices section.

GENERAL HANDLING

Boiling–Water Bath. Use Hot Pack only. Use jars or plain cans. Process with Thin Syrup, plain water, or with natural juice as desired.

Because apples oxidize in the air, work quickly with only one canner batch at a time. Wash, peel, and core apples (save peels and cores for jelly, as described in Applesauce, in a minute); treat prepared pieces with either of the anti-discoloration solutions described in Chapter 5. Drain, and carry on with the specific handling.

SLICES (HOT PACK ONLY)

Rinse drained, prepared pieces. Cover with hot Thin Syrup or water; boil gently for 5 minutes. Lift out and drain, saving cooking syrup or water. Pack Hot.

In jars. Fill clean, hot jars, leaving ½ inch of headroom. Add boiling-hot canning liquid of your choice, leaving ½ inch of headroom; adjust lids. Process in a Boiling–Water Bath (212 F/100 C)—pints or quarts for 20 minutes. Remove jars; complete seals if necessary.

> • **Adjustment for my altitude** _____ •

In plain cans. Fill, leaving only ¼ inch of headroom. Add boiling-hot Thin Syrup or water to the top of the can. Exhaust to 170 F/77 C (approx. 10 minutes); seal. Process in a B–W Bath (212 F/100 C)—20 minutes for either No. 303 or 401 cans. Remove cans; cool quickly.

> • **Adjustment for my altitude** _____ •

APPLESAUCE (HOT PACK ONLY)

Prepare by your favorite rule and according to how you'll use it—chunky or strained smooth; sweetened or not; with spices (cinnamon, nutmeg, whatever) or plain. Because of complete precooking and being packed so hot, processing time is relatively short and is designed to ensure destruction of spoilers and a good seal.

Pare crisp, red apples, cut in quarters or eighths, and remove core parts; drop pieces in anti-discoloration solution. (*Don't throw away the peels and cores: save them to boil up for a beautiful juice for jelly.*) Put about 1 inch of water in a large enameled or stainless-steel kettle, fill with well-rinsed apple pieces to within 2 inches of the top. Bring to a boil, stirring now and then to prevent sticking, and cook until apples are tender. Leave as is for chunky sauce, or put it through a sieve or food mill for smoothness. Sweeten to taste if you like; bring it briefly to boiling to dissolve any sweetening. Pack very hot.

In jars. Fill clean, hot jars with piping-hot sauce, leaving ½ inch of headroom; adjust lids. Process in a Boiling–Water Bath (212 F/100 C)— 20 minutes for either pints or quarts. Remove jars; complete seals if necessary.

> • **Adjustment for my altitude** _____ •

In plain cans. Pack to the top with hot sauce. Exhaust to 170 F/77 C (approx. 10 minutes); seal. Process in B–W Bath (212 F/100 C)—20 minutes for either No. 303 or No. 401 cans. Remove cans; cool quickly.

> • **Adjustment for my altitude** _____ •

BAKED APPLES (HOT PACK ONLY)

Sometimes people can baked apples. Prepare them in a favorite way and bake until *half done;* pack Hot in wide-mouth jars or plain cans as for Apple Slices, adding hot Thin Syrup. Adjust jar lids or exhaust and seal cans. Process in a Boiling–Water Bath (212 F/100 C)—20 minutes for either pint or quart jars, 10 minutes for either No. 303 or 401 cans (reaching 170 F/100 C by exhausting shortens processing time). Complete seals if necessary for jars; cool cans quickly.

    • **Adjustment for my altitude** _____ •

Apricots

Can these exactly as you would Peaches, but leave the skins on if you like. Some varieties tend to break up when they're heated, so handle them very gently. Hot pack preferred.

Berries

See the major subsection following Fruit for Special Diets section.

Cherries, Sour (for Pie)

Because these are used primarily as pie timber, they may be canned in water—but they have better flavor in Thin Syrup. Either way, you'll add the extra sweetening at the time you thicken the juice for pies.

GENERAL HANDLING

Boiling–Water Bath. Use Raw or Hot pack. Use jars or R-enamel cans. Prepare Thin Syrup for Raw pack; heat in their own juice with sugar for Hot pack.

Wash, stem, and pit cherries. (Use a small sterilized hairpin or the looped end of a paper clip if you don't have a pitting gadget.) Shake fruit down in the containers for a firm pack.

RAW PACK

In jars. Jog cherries down several times during packing; leave ½ inch of headroom. Add boiling syrup, leaving ½ inch of headroom; adjust lids. Process in a Boiling–Water Bath (212 F/100 C)—pints or quarts for 25 minutes. Remove jars; complete seals if necessary.

    • **Adjustment for my altitude** _____ •

In R-enamel cans. Make a firm pack, leaving only ¼ inch of headroom. Add boiling syrup to top. Exhaust to 170 F/77 C (approx. 10 minutes); seal. Process in a B–W Bath (212 F/100 C)—No. 303 cans for 20 minutes, No. 401 cans for 25 minutes. Remove cans; cool quickly.

    • **Adjustment for my altitude** _____ •

Measure pitted cherries and put them in a covered kettle with ½ cup of sugar for every 1 quart of fruit. There should be enough juice to keep the cherries from sticking. Set on lowest burner. Cover the kettle, and bring fruit very slowly to a boil to bring out the juice. Be prepared to add a little boiling water to each jar if you haven't enough juice to go around.

In jars. Fill with hot fruit and juice, leaving ½ inch of headroom; adjust lids. Process in a Boiling–Water Bath (212 F/100 C)—pints for 10 minutes, quarts for 15 minutes. Remove jars; complete seals if necessary.

• **Adjustment for my altitude** _____ •

In R-enamel cans. Fill to the top with hot fruit and juice. Exhaust to 170 F/77 C (approx. 10 minutes); seal. Process in a B–W Bath (212 F/100 C)—No. 303 cans for 15 minutes, No. 401 for 20 minutes. Remove cans; cool quickly.

• **Adjustment for my altitude** _____ •

Cherries, Sweet

GENERAL HANDLING

Boiling–Water Bath. Use Raw or Hot pack. Use jars or cans (plain cans for light varieties like Royal Ann; R-enamel for dark red or "black" types such as Bing).

If you're going to serve these as is, or combined with other fruits in a compôte, you don't pit them (they'll hold their shape better unpitted); but do prick each cherry with a needle to keep it from bursting while it's processed. Use Medium or Heavy Syrup for Raw pack; for Hot pack add more sugar than for Sour Cherries.

Wash cherries, checking for blemishes, and discard any that float (they may be wormy); remove stems. Shake down for a firm pack.

RAW PACK

In jars. Fill firmly, leaving ½ inch of headroom. Add boiling syrup, leaving ½ inch of headroom; adjust lids. Process in a Boiling–Water Bath (212 F/100 C)—pints and quarts for 25 minutes. Remove jars; complete seals if necessary.

• **Adjustment for my altitude** _____ •

In plain or R-enamel cans. Make a firm pack, leaving only ¼ inch of headroom. Fill to top with boiling syrup. Exhaust to 170 F/77 C (approx. 10 minutes); seal. Process in a B–W Bath (212 F/100 C)—No. 303 cans for 25 minutes, No. 401 cans for 25 minutes. Remove cans; cool quickly.

• **Adjustment for my altitude** _____ •

PREFERRED: HOT PACK

Measure washed and pricked cherries into a covered kettle, adding ¾ cup of sugar for every 1 quart of fruit. Because there is not much juice in the pan, add a little water to keep fruit from sticking as it heats. Cover and bring very slowly to a boil, shaking the pan gently a few times (instead of stirring, which breaks the fruit). Heat some Medium or Heavy Syrup to have on hand in case there's not enough juice to go around when you fill the containers.

In jars. Proceed and process 15 minutes for pints, 20 minutes for quarts.

   • **Adjustment for my altitude** _____ •

In plain or R-enamel cans. Proceed and process 25 minutes for No. 303 and No. 401 cans.

   • **Adjustment for my altitude** _____ •

Dried Fruits

FEASIBILITY

Any dried fruit may be freshened and canned. But why do it, when they keep so well as is (see "Drying," Chapter 21)—unless you foresee a particular need for a few servings of them stewed up ready for the table?

GENERAL HANDLING

Boiling–Water Bath only. Use Hot pack. Use jars or plain cans.

If you're in a hurry, cover with water, bring to a boil, and simmer until the fruit is plumped. Drain, saving the cooking water for processing, and proceed with Hot pack.

HOT PACK ONLY

In jars. Fill with hot fruit, sweeten; and add hot cooking water as for Raw pack. Process in a B–W Bath (212 F/100 C)—pints for 15 minutes, quarts for 20 minutes. Remove jars; complete seals if necessary.

   • **Adjustment for my altitude** _____ •

In plain cans. Fill with hot fruit, sweeten, and add hot cooking water as for Raw pack. Process in a B–W Bath (212 F/100 C)—No. 303 cans for 15 minutes, No. 401 for 20 minutes. Remove cans; cool quickly.

   • **Adjustment for my altitude** _____ •

Figs

The green-colored Kadota variety makes a particularly attractive product, but whatever kind you use should be tree-ripened yet still firm.

If you come into a trove of frozen fruit, here's how to handle a really big chunk if there's no room in your freezer.

GENERAL HANDLING

Defrost unopened box slowly; drain off all juice to make the sweet syrup desired. Sweeten as liked. Bring juice to boiling; add fruit and simmer 3 minutes. Proceed with Hot pack and processing for the particular fruit as listed here. Note that *figs require added acid.*

Some casual old instructions would have you soften (or even remove) fig skins by treating the fruit with a strong soda solution, even lye—*but don't do it.* Any such alkaline will counteract some of the acidity upon which we rely to make the stipulated Boiling–Water Bath efficient. Use long Boiling–Water Bath. Use Hot pack only. Use jars or plain cans. Prepare Thin Syrup.

Wash ripe, firm figs; do not peel or remove stems. Cover with boiling water and let simmer for 5 minutes. Drain and pack hot, not too tightly.

HOT PACK ONLY

In jars. Fill with hot figs, leaving ½ inch of headroom for pints, 1 inch of headroom for quarts. Add 1 tablespoon bottled lemon juice to pints, 2 tablespoons lemon juice to quarts (an optional very thin slice of fresh lemon may also be added to each jar for looks). Add boiling syrup, retaining headroom. Adjust lids. Process in a Boiling–Water Bath (212 F/ 100 C)—pints for 45 minutes, quarts for 50 minutes. Remove jars; complete seals if necessary.

     • **Adjustment for my altitude** _____ •

In plain cans. Fill with hot fruit, leaving only ¼ inch of headroom. Top off with boiling syrup and 2 teaspoons lemon juice to No. 303 cans, 4 teaspoons lemon juice to No. 401 cans (an optional very thin slice of fresh lemon may also be added to each can for looks). Exhaust to 170 F/77 C (approx. 10 minutes); seal. Process in a B–W Bath (212 F/100 C)—No. 303 cans for 45 minutes, No. 401 cans for 50 minutes. Remove cans; cool quickly.

     • **Adjustment for my altitude** _____ •

Grapes

Tight-skinned seedless grapes are the ones to can if you can any—for fruit cocktail, compôtes, gelatin desserts, and salads (but grapes for juice may be any sort you have plenty of).

GENERAL HANDLING

Boiling–Water Bath. Use Raw or Hot pack. Use jars or cans (plain, or R-enamel cans if it's a dark grape).

Sort, wash, and stem.

YIELDS IN CANNED FRUIT

Since the legal weight of a bushel of fruits differs among states, the weights given below are average; the yields are approximate.

FRUITS	FRESH	QUARTS CANNED
Apples	1 bu (48 lbs)	16–20
	2½–3 lbs	1
Apple juice	1 bu (48 lbs)	10
Applesauce	1 bu (48 lbs)	15–18
	2½–3½ lbs	1
Apricots	1 bu (50 lbs)	20–24
	2–2½ lbs	1
Berries (excluding strawberries)	24-qt crate	12–18
	5–8 cups	1
Cherries, as picked	1 bu (56 lbs)	22–32
	2–2½ lbs	1
Figs	2–2½ lbs	1
Grapes	28-lb lug	7–8
	4 lbs	1
Grapefruit	4–6 fruit	1
Nectarines	18-lb flat	6–9
	2–3 lbs	1
Peaches	1 bu (48 lbs)	18–24
	2–2½ lbs	1
Pears	1 bu (50 lbs)	20–25
	2–2½ lbs	1
Pineapple	2 average	1
	5 lbs	2
Plums and Prunes	1 bu (56 lbs)	24–30
	2–2½ lbs	1
Rhubarb	15 lbs	7–11
	2 lbs	1
Strawberries	24-qt crate	12–16
	6–8 cups	1

RAW PACK

In jars. Fill tightly but without crushing grapes, leaving ½ inch of headroom. Add boiling Medium Syrup, leaving ½ inch of headroom; adjust lids. Process in a Boiling–Water Bath (212 F/100 C)—pints for 15 minutes, quarts for 20 minutes. Remove jars; complete seals if necessary.

• **Adjustment for my altitude** _____ •

In cans (plain or R-enamel). Fill, leaving ¼ inch of headroom. Add boiling Medium Syrup to top. Exhaust to 170 F/77 C (approx. 10 minutes);

seal. Process in a B–W Bath (212 F / 100 C)—No. 303 cans for 20 minutes, No. 401 for 25 minutes. Remove cans; cool quickly.

• **Adjustment for my altitude** _____ •

PREFERRED: HOT PACK

Prepare as for Raw pack. Bring to a boil in Medium Syrup. Drain, reserving syrup, and pack.

In jars. Pack with hot grapes, leaving ½ inch of headroom. Add boiling syrup, leaving ½ inch of headroom; adjust lids. Process in a Boiling–Water Bath (212 F / 100 C)—pints for 15 minutes, quarts for 20 minutes. Remove jars; complete seals if necessary.

• **Adjustment for my altitude** _____ •

In cans (plain or R-enamel). Fill with hot grapes, leaving ¼ inch of headroom. Add boiling syrup to the top. Exhaust to 170 F/77 C (approx. 10 minutes); seal. Process in a B–W Bath (212 F / 100 C)—No. 303 cans for 15 minutes, No. 401 for 20 minutes. Remove cans; cool quickly.

• **Adjustment for my altitude** _____ •

Grapefruit (or Orange) Sections

FEASIBILITY

Only if you have a good supply of tree-ripened fruits is canning worthwhile—but canning makes an infinitely handier product than freezing does. Don't overlook Mixed Fruit; and don't forget marmalades and conserves.

GENERAL HANDLING

Boiling–Water Bath only. Use Raw pack only. Use jars only (cans could give a metallic taste to home-canned citrus).

Wash fruit and pare, removing the white membrane as you go. Slip a very sharp thin-bladed knife between the dividing skin and pulp of each section, and lift out the section without breaking. Remove any seeds from individual sections. Prepare Thin Syrup.

RAW PACK ONLY

In jars only. Fill hot jars with sections, leaving ½ inch of headroom. Add boiling Thin Syrup, leaving ½ inch of headroom; adjust lids. Process in a Boiling–Water Bath (212 F / 100 C)—10 minutes for either pints or quarts. Remove jars; complete seals if necessary.

• **Adjustment for my altitude** _____ •

Juices

See grouped handling at the end of Fruits section.

Peaches

FEASIBILITY

The benefits and pleasures of canning are exemplified in peaches: home-canned peaches are full of flavor, are versatile, and are considered by many cooks to be better than frozen ones.

GENERAL HANDLING

Boiling–Water Bath only. Use Raw or Hot pack. Use jars or plain cans.

Unusually good tip from the NEFCO (Nutrition Education and Food Conservation) canning book, a publication of Dixie Canner Equipment Co. Inc., of Athens, Georgia, and edited ably by Dr. William Hurst, food scientist at the University of Georgia: halve and pit Freestone peaches before they are peeled; doing so blanches the cut surfaces and prevents discoloration of the pit area, etc.

Wash, halve, and pit; blanch. From the cavity, scrape away dark fibers with a melon-baller or small spoon. Hold peeled fruit in an anti-discoloration solution as for Apples; rinse and drain before packing. Use Thin or Medium Syrup, or a combination of juice and water/syrup.

For Peach Melba, pack as halves, Raw for best texture, in wide-mouth jars. Light corn syrup is good here: look at Liquids for Canning Fruits, earlier in this Chapter.

Skins easily come off scalded peaches, and cut fruit is held in an anti-darkening solution. (*USDA photograph*)

RAW PACK

In jars only. Pack halves or slices attractively, leaving ½ inch of headroom. Add boiling syrup, leaving ½ inch of headroom; adjust lids. Process in a Boiling–Water Bath (212 F/100 C)—pints for 25 minutes, quarts for 30 minutes. Remove jars; complete seals if necessary.

• **Adjustment for my altitude** _____ •

PREFERRED: HOT PACK

Simmer prepared peaches in hot syrup for 2 minutes. Drain, reserving syrup.

In jars. Fill with hot peaches, leaving ½ inch of headroom. Add boiling syrup, leaving ½ inch of headroom; adjust lids. Process in a Boiling–Water Bath (212 F/100 C)—pints for 20 minutes, quarts for 25 minutes. Remove jars; complete seals if necessary.

• **Adjustment for my altitude** _____ •

In plain cans. Fill with hot peaches, leaving only ¼ inch of headroom. Add boiling syrup to the top. Exhaust to 180 F/82 C (approx. 13 minutes); seal. Process in a B–W Bath (212 F/100 C)—No. 303 cans for 15 minutes, No. 401 for 25 minutes. Remove cans; cool quickly.

• **Adjustment for my altitude** _____ •

Boiling syrup must cover the beautifully packed peaches with headroom to spare. (*USDA photograph*)

Peaches, Brandied

GENERAL HANDLING

Boiling–Water Bath only. Use Hot pack only. Use jars only (because they look so pretty: which is part of their fun).

The peaches should be small to medium in size, firm-ripe, and with attractive color; blemish-free of course. Wash. Using a coarse-textured towel, rub off all their fuzz. Weigh them.

For every 1 pound of peaches, make a Heavy Syrup of 1 cup sugar to 1 cup water. Bring syrup to boiling and, when sugar is dissolved, add the whole peaches and simmer them for 5 minutes. Drain; save the syrup and keep it hot.

HOT PACK ONLY, IN JARS ONLY

Without crushing, fit peaches in hot jars, leaving ½ inch of headroom. Pour 2 tablespoons of brandy over the peaches in each 1-pint jar, using proportionately more brandy for quarts. Fill jars with hot syrup, leaving ½ inch of headroom; adjust lids. Process in a Boiling–Water Bath (212 F/ 100 C)—pints for 20 minutes, quarts for 25 minutes. Remove jars; complete seals if necessary.

  • **Adjustment for my altitude** _____ •

Pears

Bartlett pears are ideal for canning, to serve alone or as a salad or in a compôte. They will be too soft for successful canning, though, if they've ripened on the tree: use ones that were picked green (but full grown) and allowed to ripen in cool storage, between 60 and 65 F (16 and 18 C).

Very firm-fleshed varieties like Seckel and Kieffer are generally spiced or pickled; they make a satisfactory product if ripened in storage and simmered in water till nearly tender before packing with syrup. The so-called winter pears—such as Anjou and Bosc—are usually eaten fresh; they are cold-stored much the way apples are, but they are not likely to keep as long.

GENERAL HANDLING

Use a Boiling–Water Bath. Hot pack preferred. Use jars or plain cans.

Wash; cut lengthwise in halves or quarters. Remove stems, core (a melon-ball scoop is handy for this); pare. Treat pieces against oxidation with either of the solutions described for apples.

Make Thin or Medium Syrup.

Small pears may be canned whole: pare them, but leave the stems on. It takes about 9 small whole pears to fill a pint jar or a No. 303 can.

HOT PACK ONLY

Quality suffers when pears are packed Raw. Simmer fruit in syrup for 2 minutes; drain, reserving hot syrup.

In jars. Fill with hot pears, leaving ½ inch of headroom. Add boiling syrup, leaving ½ inch of headroom; adjust lids. Process in a Boiling–Water Bath (212 F/100 C)—pints for 20 minutes, quarts for 25 minutes. Remove jars; complete seals if necessary.

> • **Adjustment for my altitude** _____ •

In plain cans. Fill with hot pears, leaving only ¼ inch of headroom. Add boiling syrup to the top. Exhaust to 170 F/77 C (approx. 10 minutes); seal. Process in a B–W Bath (212 F/77 C)—No. 303 cans for 15 minutes, No. 401 cans for 25 minutes. Remove cans; cool quickly.

> • **Adjustment for my altitude** _____ •

MINT VARIATION

Prepare as above; the pears may be cut up or left whole. To Medium Syrup, add enough natural peppermint extract and green food coloring to give the desired taste and color.

Simmer the pears in this syrup for 5 to 10 minutes, depending on size and firmness of fruit, before packing Hot and processing in a B–W Bath as above.

Pears, Spiced

Seckel, Kieffer, and similar hard varieties are best for spicing. Bartletts or other soft pears may be used if they are underripe.

GENERAL HANDLING

Use a Boiling–Water Bath. Use Hot pack only. Use jars only (like Brandied Peaches, these are very attractive to look at; and you could take a ribbon at the fair!).

Wash, peel, and core 6 pounds of pears. Gently boil them covered in 3 cups of water until they start to soften.

Make a very heavy syrup of 4 cups sugar and 2 cups white vinegar. Tie in a small cloth bag 3 or 4 three-inch sticks of cinnamon, ¼ cup whole cloves, and 4 teaspoons cracked ginger. Simmer the spice bag in the syrup for 5 minutes.

Add the pears and the water in which they were partially cooked to the spiced syrup, and simmer for 4 minutes. Drain pears, saving the hot syrup and discarding the spice bag.

HOT PACK ONLY, IN JARS ONLY

Pack hot pears attractively in clean hot jars. Add spiced syrup, leaving ½ inch of headroom; adjust lids. Process in a Boiling–Water Bath (212 F/ 100 C)—pints for 20 minutes, quarts for 25 minutes. Remove jars; complete seals if necessary.

> • **Adjustment for my altitude** _____ •

Pineapple

Fresh pineapple is as easy to can as any fruit—and may be packed in any plain or minted and colored syrup.

GENERAL HANDLING

Use a Boiling–Water Bath. Use Hot pack only. Use jars or plain cans.

Scrub firm, ripe pineapples. Cut a thin slice from each end. Cut like a jelly roll in ½-inch slices, or in 8 lengthwise wedges. Remove the skin, the "eyes," and the tough-fiber core from each piece. Leave in slices or wedges, or cut small or chop: let future use guide your hand.

Simmer pineapple gently in Light or Medium Syrup for about 5 minutes. Drain; save the hot syrup for packing.

HOT PACK ONLY

In jars. Fill with fruit, leaving ½ inch headroom. Add hot syrup, leaving ½ inch of headroom; adjust lids. Process in a Boiling–Water Bath (212 F/ 100 C)—pints for 15 minutes, quarts for 20 minutes. Remove jars; complete seals if necessary.

- **Adjustment for my altitude** _____ •

In plain cans. Fill with fruit, leaving ¼ inch of headroom. Add hot syrup to the top of the cans. Exhaust to 170 F/77 C (approx. 10 minutes); seal. Process in a B–W Bath (212 F/100 C)—No. 303 cans for 20 minutes, No. 401 for 25 minutes. Remove cans; cool quickly.

- **Adjustment for my altitude** _____ •

Plums (and Italian Prunes)

GENERAL HANDLING

Use a Boiling–Water Bath. Use Raw or Hot pack. Use jars or cans—R-enamel for red plums, plain for greenish-yellow varieties.

Firm, meaty plums (such as the Greengage) hold their shape better for canning whole than the more juicy types do. Freestone plums and prunes are easily halved and pitted for the tighter pack.

Choose moderately ripe fruit. Wash. If fruit is kept whole, the skins should be pricked several times with a large needle to prevent the fruit from bursting. Halve and pit the freestone varieties. Prepare Medium or Heavy Syrup, and have it hot.

RAW PACK

In jars. Fill with raw fruit, leaving ½ inch of headroom. Add boiling syrup, leaving ½ inch of headroom; adjust lids. Process in a Boiling–Water Bath (212 F/100 C)—pints for 20 minutes, quarts for 25 minutes. Remove jars; complete seals if necessary.

- **Adjustment for my altitude** _____ •

In cans (R-enamel for red fruit, plain for light-colored). Pack raw fruit, leaving ¼ inch of headroom. Add boiling syrup to the top of the can. Exhaust to 170 F/77 C (approx. 10 minutes); seal. Process in a B–W Bath (212 F/100 C)—No. 303 cans for 15 minutes, No. 401 for 20 minutes. Remove cans; cool quickly.

· Adjustment for my altitude _____ ·

PREFERRED: HOT PACK

Heat prepared plums to boiling in syrup. If they're halved and are very juicy, heat them slowly to bring out the juice; measure the juice, and for each 1 cup juice add ¾ cup sugar—give or take a little, according to your taste—to make a Medium Syrup. Reheat to boiling for just long enough to dissolve the sugar. Drain fruit, saving the hot syrup. Have some hot plain Medium Syrup on hand for eking out sweetened juice.

In jars. Pack hot fruit, leaving ½ inch of headroom. Add boiling syrup, leaving ½ inch of headroom; adjust lids. Process in a Boiling–Water Bath (212 F/100 C)—pints for 20 minutes, quarts for 25 minutes. Remove jars; complete seals if necessary.

· Adjustment for my altitude _____ ·

In cans (R-enamel for red fruit, plain for light-colored). Pack hot fruit, leaving ¼ inch of headroom. Fill to top with boiling syrup. Exhaust to 170 F/77 C (approx. 10 minutes); seal. Process in a B–W Bath (212 F/100 C)—No. 303 cans for 15 minutes, No. 401 for 20 minutes. Remove cans; cool quickly.

· Adjustment for my altitude _____ ·

Rhubarb (or Pie Plant)

Never eat rhubarb LEAVES: they are high in oxalic acid, which is poisonous.

Safe to eat are the tart, red stalks of this plant, which are excellent pie timber, make a tangy dessert sauce, are a favorite ingredient in old-time preserves. However, your best use would be to can sweetened sauce by the method given below, and to freeze the raw pieces for pies. Rhubarb juice makes a good hot-weather drink.

GENERAL HANDLING AS SAUCE

Use a Boiling–Water Bath. Use Hot pack. Use jars or R-enamel cans.

For best results, can it the same day you cut it. If the stalks are young enough, they need not be peeled (their red color makes an attractive product). *Discard leaves,* trim away both ends of the stalks, and wash; cut stalks in ½-inch pieces. Measure. Put rhubarb in an enameled kettle (because of the tartness), mixing in ½ cup of sugar for each 1 quart (4 cups) of raw fruit. Let it stand, covered, at room temperature for about 4 hours to draw out the juice. Bring slowly to a boil; let boil no more than

1 minute (or the pieces will break up). (*Alternative:* bake sugared rhubarb in a heavy, covered pan in a slow oven, (approx. 275 F/135 C, for 1 hour.)

HOT PACK ONLY

In jars. Fill with hot fruit and its juice, leaving ½ inch of headroom; adjust lids. Process in a Boiling–Water Bath (212 F/100 C)—15 minutes for either pints or quarts. Remove jars; complete seals if necessary.

• **Adjustment for my altitude** _____ •

In R-enamel cans. Pack hot fruit and juice to the top of the cans. Exhaust to 170 F/77 C (approx. 10 minutes); seal. Process in a B–W Bath (212 F/100 C)—15 minutes for either No. 303 or No. 401 cans. Remove cans; cook quickly.

FRUIT FOR SPECIAL DIETS

For canning large pieces of fruit without sugar or other sweetener, follow individual instructions given earlier, but in addition to omitting sugar *use Hot pack only,* and eke out the natural liquid with extra unsweetened boiling juice (not water) if necessary to fill containers.

Pint jars (½-pints for infants or the person with a small appetite) are usually the best size to use unless you're canning sugarless fruit for several people in the family. Processing time is the same for pint and ½-pint jars.

Fruit Purées

Infants and those on low-residue diets require fruit whose natural fiber has been reduced to tiny particles in a sieve, food mill, blender, or food processor with the steel blade in place (in the last instance, the purée will be runniest, and this looseness can be reduced by longer precooking). These purées are generally processed without sweetening—certainly the strained fruits for babies are unsweetened.

Any favorite fruit may be canned as a purée except for figs. Apples, apricots, peaches, and pears are the most popular for purées.

Use a Boiling–Water Bath. Use Hot pack only. Use standard ½-pint canning jars with appropriate closures (*not* commercial babyfood jars, whose sealers are not re-usable).

Apple Purée

Follow directions for Applesauce earlier, but omit sweetening; sieve, pack, and process as for Apricot Purée, below.

Apricot Purée (or Peach or Pear)

Use perfectly sound, ripe fruit. Wash, drain; pit and slice. In a large kettle crush a 1-inch layer of fruit to start the juice, then add the rest of the prepared fruit; if there seems not to be enough juice to keep the fruit from sticking or scorching, add no more than ¼ cup water for every firmly packed 1 cup of fruit. Simmer over medium heat, stirring as needed, until the fruit is soft—about 20 minutes. Push the cooked fruit through a sieve or food mill; or whirl briefly in a blender at a high setting, or in a processor, and strain. Measure, and add 1 tablespoon fresh lemon juice for each 2 cups of pulp. Reheat to a 200 F/93 C simmer. Pack hot.

HOT PACK ONLY, IN JARS ONLY

Pour hot purée into clean, scalded ½-pint jars, leaving ½ inch of headroom. Adjust lids. Process in a Boiling–Water Bath (212 F/100 C) for 15 minutes. Remove jars, complete seals if necessary.

• **Adjustment for my altitude** _____ •

CANNING BERRIES

Don't get so taken up with making jellies and jams in berry-time that you forget to can some too: they may be done for serving solo and in compôtes and salads; or, with slightly different handling, for use in cobblers, pies, and puddings.

For purposes of general handling, berries—except for strawberries, which are a law unto themselves—are divided in two categories: *soft* (raspberries, blackberries, boysenberries, dewberries, loganberries, and youngberries), and *firm* (blueberries, cranberries, currants, elderberries, gooseberries, and huckleberries). The texture usually determines which pack to use; but some of the firm ones may be dealt with in more than one way, and such variations are described separately below for the specific berries.

It goes without saying that you'll want to use only perfect berries that are ripe without being at all mushy. Pick them over carefully, wash them gently, and drain; stem or hull them as necessary. Work with only a couple of quarts at a time because all berries, particularly the soft ones, break down quickly by being handled.

General Procedure for Most Berries

All berries are acid, so a Boiling–Water Bath for the prescribed length of time is the best process for them.

Use Raw pack generally for *soft* berries, because they break down so much in precooking.

A Hot pack in general makes a better product of most *firm* berries.

Use jars or cans—R-enamel cans for all red berries, but plain cans for gooseberries.

All may be canned either with sugar or without—but just a little sweetening helps hold the flavor even of berries you intend to doll up later for desserts. Thin and Medium Syrups are used more often than Heavy, with Medium usually considered as giving a better table-ready product than Thin.

Raw Pack (Soft Berries)

In jars. Fill clean, hot jars, shaking to settle the berries for a firm pack; leave ½ inch of headroom. Add boiling Thin or Medium Syrup, leaving ½ inch of headroom; adjust lids. Process in a Boiling–Water Bath (212 F/ 100 C)—pints for 15 minutes, quarts for 20 minutes. Remove jars; complete seals if necessary.

• **Adjustment for my altitude** _____ •

In R-enamel cans. Fill, shaking for a firm pack; leave only ¼ inch of headroom. Add boiling Thin or Medium Syrup to the top of the can. Exhaust to 190–200 F/88–93 C (approx. 4–5 minutes); seal. Process in a B–W Bath (212 F/100 C)—10 minutes for both No. 303 and No. 401 cans. Remove cans; cool quickly.

• **Adjustment for my altitude** _____ •

Standard Hot Pack (Most Firm Berries)

Measure berries into a kettle, and add ½ cup of sugar for each 1 quart of berries. On lowest burner, bring very slowly to a boil, shaking the pan to prevent berries from sticking (rather than stirring, which breaks them down). Remove from heat and let them stand, covered, for several hours. *This plumps up the berries and keeps them from floating to the top of the container when they're processed.* For packing, reheat them slowly. As insurance, have some hot Thin or Medium Syrup on hand in case you run short of juice when filling the containers.

In jars. Fill with hot berries and juice, leaving ½ inch of headroom. Proceed and process as for Raw pack.

• **Adjustment for my altitude** _____ •

In R-enamel or plain cans. Fill to the top with hot berries and juice, leaving no headroom. Proceed and process as for Raw pack.

Unsweetened Hot Pack (Most Firm Berries)

This is often used for sugar-restricted diets; it is also another way of canning berries intended for pies.

Pour just enough cold water in a kettle to cover the bottom. Add the berries and place over very low heat. Bring to a simmer until they are hot throughout, shaking the pot—not stirring—to keep them from sticking.

Pack hot fruit and its juice, leaving headroom as above; remove any air bubbles by running a plastic blade around the inner side of the container. Process as for Raw pack.

Specific Berries (Except Strawberries)

Blackberries

Raw pack. Usually considered soft, so for over-all versatility use Raw pack under the General Procedure above. With boiling water or Thin or Medium Syrup—but Medium Syrup if you want them table-ready. In jars or R-enamel cans.

Blueberries

Though in the firm category, they actually break down too much in the standard Hot pack (but they make a lovely sauce for ice cream, etc., if you want to can them by Hot pack with a good deal of extra sweetening). Old-timers dried them to use like currants in fruit cake.

Raw pack. With boiling water or syrup (Medium recommended). In jars or R-enamel cans. Proceed and process under General Procedure above.

Raw pack variation. If you want to hold them as much like their original texture and taste as possible when canned (to use like fresh berries in cakes, muffins, pies), you must blanch them. Put no more than 3 quarts of berries in a single layer of cheesecloth about 20 inches square. Gather and hold the cloth by the corners, and dunk the bundle to cover the berries in boiling water until juice spots show on the cloth—*about 30 seconds.* Dip the bundle immediately in cold water to cool the berries. Drain them.

Fill jars, leaving ½ inch of headroom. Add no water or sweetening; adjust lids. Process as for standard Raw pack under General Procedure above.

Boysenberries

Soft; in Raw pack as under General Procedure. Use jars or R-enamel cans.

Cranberries

These hold so well fresh in proper cold storage (see "Root-Cellaring," Chapter 22) or in the refrigerator, and they also freeze, so they probably make the most sense canned if they're done as whole or jellied sauce.

Use jars only.

For about 6 pints Whole Sauce. Boil together 4 cups sugar and 2 cups water for 5 minutes. Add 8 cups (about 2 pounds) of washed, stemmed cranberries, and boil without stirring until the skins burst. Pour boiling hot into clean *hot jars,* leaving ½ inch of headroom, and run a plastic blade or spatula around the inner side of the jar to remove trapped air. After filling with ½ inch of headroom, adjust lids, and process pints in a B–W Bath (212 F/100 C) for 10 minutes. Remove jars; complete seals if necessary.

• **Adjustment for my altitude** _____ •

For 4 pints Jellied Sauce. Boil 2 pounds of washed, stemmed berries with 1 quart of water until the skins burst. Push berries and juice through a food mill or strainer. Add 4 cups sugar to the resulting purée, return to heat, and boil almost to the jelly stage (see Testing for Doneness in Jelly section in Chapter 18). Pour into hot, straight-sided jars (so it will slip out easily); cap, and process in a B–W Bath for 10 minutes. Remove jars; cool upright.

Currants

Currants are a novelty in certain sections of the United States where the bushes were uprooted because they harbored a fungus destructive to the white pine. If you are fortunate enough to have some, by all means make jelly with them. Turn extra ones into dessert sauce; dry some; and they freeze.

Classed as firm berries, they can be packed Hot as under General Procedure; in jars or R-enamel cans.

Dewberries

Soft; in Raw pack as under General Procedure. Use jars or R-enamel cans.

Elderberries

Best use is for jelly or wine.

If you do can them, use a Hot pack under General Procedure—and add 1 tablespoon lemon juice for each 1 quart of berries, to improve their flavor. In jars or R-enamel cans.

Gooseberries

Another scarce fruit in many sections of the country, like the currant. They make such heavenly old-fashioned pies, tarts, and preserves!

Although they're firm, they do well in Raw pack with very sweet syrup. Or they may be done with Hot pack (where they hold their shape less well).

Wash; pick them over, pinch off stem ends and tails. Some cooks prick each berry with a sterile needle to promote a better blending of sweetening and juice (but we can't imagine a quicker way to drive ourselves up the wall; we'll leave it to osmosis).

Raw pack. Heavy Syrup is recommended for these very tart berries. And they'll probably pack better if you put ½ cup of hot syrup in the bottom of the container before you start filling. Use jars or plain cans. Process as in General Procedure above.

Hot pack. Follow the steps given earlier for a standard Hot pack—but you may want to increase the sugar to ¾ cup for each 1 quart of berries. Process.

Huckleberries

Being cousins of the blueberry, these firm berries are handled like Blueberries.

Loganberries

They're soft, so pack Raw as under General Procedure.

Raspberries

Tenderest of the soft berries, these really do better if they're frozen. Put by some as jam or jelly, of course; try canning some sauce or juice. And if you have them in your garden, please don't forget to take a basket of fresh-picked raspberries to some older person who doesn't have a way to get them any more.

If you can them, use Raw pack and Medium Syrup. Use jars or R-enamel cans. Follow the General Procedure above.

Youngberries

Another softie. Raw pack, as under General Procedure.

Strawberries

The most popular berry in the United States, these nevertheless are often a disappointment when canned, because they will fade and float if they are handled in the standard way recommended for most other soft berries. Here's how to have a blue-ribbon product—and no short cuts, please.

GENERAL HANDLING

Use a Boiling–Water Bath. Use Hot pack only (even though they're soft). Use jars or R-enamel cans.

Wash and hull perfect berries that are red-ripe, firm, and without white or hollow centers. Measure berries. Using ½ to 1 cup sugar for each 4 cups of berries, spread the berries and sugar in shallow pans in thin alternating layers. Cover with waxed paper or foil if necessary as a protection against insects, and let stand at room temperature for 2 to 4 hours. Then turn into a kettle and simmer for 5 minutes in their own juice. Have some boiling Thin Syrup on hand if there's not enough juice for packing.

HOT PACK ONLY

In jars. Fill, leaving ½ inch of headroom (adding a bit of hot syrup if needed); adjust lids. Process in a Boiling–Water Bath (212 F/100 C)— pints for 10 minutes, quarts for 15 minutes. Remove jars; complete seals if necessary.

> • **Adjustment for my altitude** ——————— •

In R-enamel cans. Fill to the top with hot berries and juice. Exhaust to 190–200 F/88–93 C (approx. 10 minutes); seal. Process in a B–W Bath (212 F/100 C)—No. 303 cans for 15 minutes, No. 401 for 20 minutes. Remove cans; cool quickly.

> • **Adjustment for my altitude** ——————— •

CANNING FRUIT JUICES

Beverage juices will have better flavor if they are presweetened at least partially. Use sugar, or the sweetener of your choice. (But for sugar-restricted diets, *use only the non-nutritive artificial sweetener approved by your doctor;* postpone adding the sweetener until serving time—saccharin-based substitutes may leave an unpleasant aftertaste in cooking.)

Juices intended for jelly are not sweetened until you are making your jelly.

General Procedure for Most Juices

Because boiling temperature (212 F/100 C at sea level) can impair the fresh flavor of almost all fruit juices, these are usually processed by the *Hot*–Water Bath (given as 190 F/88 C), which is pasteurization. Be sure there is at least 1 to 2 inches of definitely simmering water above the tops of the containers throughout the processing time, as for the B–W Bath.

The various nectars are processed in a Boiling–Water Bath (212 F/100 C) because of their greater density.

Use Hot pack only. Use jars or R-enamel cans.

Choose firm-ripe, blemish-free fruit or berries; wash carefully, lifting the fruit from the water to let any sediment settle, and to avoid bruising. Then stem, hull, pit, core, slice—whatever is needed for preparing the particular fruit. *Simmer* the fruit until soft; strain through a jelly bag to extract clear juice (see Equipment for Jellies, Jams, Etc. in Chapter 18).

Containers for Canned Juices

Modern glass home-canning jars with fresh sealers are your best choice for juices or nectars. One manufacturer recommends R-enamel cans for all juices, even though the fruits from which they are made do not lose color in plain cans. Juices in cans, which are light-proof, need not be stored in the dark as do glass jars.

Not suitable for re-use in heat-processing at home are soft-drink bottles; nor would crimped-on caps be satisfactory.

Apple Cider (Beverage)

Get cider fresh from the mill and process it without delay (though it can be held in a refrigerator in sterilized covered containers for 12 hours, if necessary). To prepare, strain it through a clean, dampened jelly bag, and in a large kettle bring it to a good simmer at 200 F/93 C, *but do not boil.* Pack Hot.

Hot pack only, in jars. Pour strained fresh cider into hot jars, leaving ½ inch of headroom; adjust lids. Process in a Hot–Water Bath at 190 F/ 88 C, for 30 minutes for either pints or quarts. Remove jars; complete seals if necessary.

> • **Adjustment for my altitude** _____ •

Hot pack only, in cans (R-enamel suggested). Fill to the top with strained fresh cider leaving no headroom. Exhaust to 170 F/77 C (approx. 10 minutes); seal. Process in a H–W Bath at 190 F/88 C, for 30 minutes for either No. 303 or No. 401 cans. Remove cans; cool quickly.

> • **Adjustment for my altitude** _____ •

Apple Juice (for Jelly Later)

Add some underripe apples to the batch for more pectin. Wash and cut up apples, discarding stem and blossom ends. *Do not peel or core* (you may even use the peels left over from making Applesauce). Barely cover with cold water and bring to a boil over moderate heat, and simmer until apples are quite soft—about 30 minutes. Strain hot through a dampened jelly bag.

Reheat to 200 F/93 C and pack *hot;* process as for Apple Cider, above.

Apricot Nectar

Nectars—most often made from apricots, peaches, and pears—are simply juices thickened with *finely* sieved pulp of the fruit; usually they are "let down" with ice water when served. For sweetening, which is optional, honey or corn syrup may be substituted (see earlier in this chapter for proportions); artificial non-nutritive sweetening, if wanted, should be added at serving time to avoid a flavor change caused by heat-processing.

Use a Boiling–Water Bath. Use Hot pack only. Use ½-pint or pint jars or No. 303 cans (R-enamel suggested).

Wash, drain; pit and slice. Measure, and treat with an anti-oxidant if desired (see General Handling for All Fruits, earlier in this chapter). In a large enameled kettle add 1 cup boiling water to each 4 cups of prepared fruit, bring to simmering, and cook gently until fruit is soft. Put through a fine sieve or food mill. Measure again, and to each 2 cups of fruit juice-plus-pulp add 1 tablespoon lemon juice and about ½ cup sugar, or sweetening to taste. Reheat and simmer until sugar is dissolved. Pack Hot.

Hot pack only, in jars. Pour hot nectar into ½-pint or pint jars, leaving ½ inch of headroom; adjust lids. Process in a Boiling–Water Bath (212 F/ 100 C)—15 minutes for either ½-pints or pints. Remove jars; complete seals if necessary.

• **Adjustment for my altitude** _____ •

Hot pack only, in No. 303 cans (R-enamel suggested). Fill No. 303 cans to the top with simmering nectar, leaving no headroom; seal. Process in a Boiling–Water Bath (212 F/100 C)—for 15 minutes. Remove cans; cool quickly.

• **Adjustment for my altitude** _____ •

Berry Juices

Crush and simmer berries in their own juice until soft; strain through a jelly bag—allow several hours for draining. If you twist the bag for a greater yield, the juice should be strained again through clean cloth to make it clear.

Measure, and to each 4 quarts of strained juice add 4 tablespoons lemon juice, plus sugar to taste—usually 1 to 2 cups. (If the juice is for jelly later, omit lemon juice and sugar at this time.) Reheat juice to a 200 F/ 93 C simmer. Pack.

Hot pack only, in jars. Pour simmering juice into hot scalded jars, leaving ½ inch of headroom; adjust lids. Process in a Hot–Water Bath at 190 F/88 C, for 10 minutes for either pints or quarts. Remove jars; complete seals if necessary.

• **Adjustment for my altitude** _____ •

Hot pack only, in R-enamel cans. Fill cans to the top with hot juice, leaving no headroom; seal (at simmering stage it will already be more than 170 F/77 C, so exhausting is not necessary). Process in a H–W Bath at 190 F/88 C, for 10 minutes for either No. 303 or No. 401 cans. Remove cans; cool quickly.

• **Adjustment for my altitude** _____ •

Cranberry Juice

Boiling–Water Bath only. Hot pack only. Use jars only.

Pick over the berries and wash. Measure, and add an equal amount of water. Bring to boiling in an enameled kettle and cook until berries burst. Strain through a jelly bag (squeezing the bag adds to the yield: re-strain if you want beautifully clear juice). Add sugar to taste, and bring just to boiling. Pack hot.

If you're canning this juice only for special-diet reasons, omit sugar. Add the artificial non-nutritive sweetener *prescribed by your doctor* just before serving.

Hot pack only, in jars only. Pour boiling juice into clean hot jars, leaving ½ inch of headroom; adjust lids. Process in a B–W Bath (212 F/100 C)— 10 minutes for either pints or quarts. Remove jars; complete seals if necessary.

• **Adjustment for my altitude** _____ •

Currant Juice

Prepare and process as for Berry Juices, above.

Grape Juice

Hot–Water Bath only. Use Hot pack only. Use jars or R-enamel cans.

The extra intermediate step of refrigerating the juice will prevent crystals of tartaric acid (harmless, but not beautiful) in the finished product. It's easier to work with not more than 1 gallon of grapes at a time.

Select firm-ripe grapes; wash, stem. Crush and measure into an enameled or stainless-steel kettle; add 1 cup water for each 4 quarts of crushed grapes. Cook gently *without boiling* until fruit is very soft—about 10 minutes. Strain through a jelly bag, squeezing it for a greater yield.

Refrigerate the juice for 24 hours. Then strain again for perfect clearness, being mighty careful to hold back the sediment of tartaric acid crystals in the bottom of the container.

Add ½ cup sugar for each 1 quart of juice (or omit sweetening), and heat to a 200 F/93 C simmer.

Hot pack only, in jars. Pour simmering juice into hot scalded jars, leaving ½ inch of headroom; adjust lids. Process in a Hot–Water Bath at 190 F/88 C, for 10 minutes for either pints or quarts. Remove jars; complete seals if necessary.

• **Adjustment for my altitude** _____ •

Hot pack only, in R-enamel cans. Fill cans to the top with simmering juice, leaving no headroom; seal (the step of exhausting to 170 F/77 C is not necessary when juice is simmering-hot). Process in a H–W Bath at 190 F/88 C—30 minutes for either No. 303 or No. 401 cans. Remove cans; cool quickly.

• **Adjustment for my altitude** _____ •

Peach Nectar

Prepare and process as for Apricot Nectar, above.

Pear Nectar

Prepare and process as for Apricot Nectar, above.

Plum Juice (and Fresh Prune)

Hot–Water Bath only. Use Hot pack only. Use jars or R-enamel cans.

Choose firm-ripe plums with attractive red skins. Wash; stem; cut in small pieces. Measure. Put in an enameled or stainless-steel kettle, add 1 cup water for each 1 cup prepared fruit. Bring slowly to simmering, and cook gently until fruit is soft—about 15 minutes. Strain through a jelly bag. Add ¼ cup sugar to each 2 cups juice, or to taste. Reheat just to a 200 F/93 C simmer.

Pack and process as for Berry Juices, above.

Rhubarb Juice

This makes good sense if you have extra rhubarb, because it can be used for a delicious quencher, and was the main ingredient of a hill-country wedding punch in olden days. And rhubarb is said to be good for our teeth.

Boiling–Water Bath only. Use Hot pack only. Use jars or R-enamel cans.

Wash and trim fresh young red rhubarb, but *do not peel*. Cover the bottom of the kettle with ½ inch of water, add rhubarb cut in ½-inch pieces. Bring to simmering, and cook gently until soft—about 10 minutes. Strain through a jelly bag. Reheat juice, adding ¼ cup sugar to each 4 cups of juice to hold the flavor, and simmer at 200 F/93 C until sugar is dissolved.

Hot pack, in jars. Pour simmering juice into hot scalded jars, leaving ½ inch of headroom; adjust lids. Process in a B–W Bath (212 F/100 C)—10 minutes for either pints or quarts. Remove jars; complete seals if necessary.

> • **Adjustment for my altitude** _____ •

Hot pack, in R-enamel cans. Fill cans to the top with simmering juice; seal (exhausting is not necessary if juice is simmering-hot). Process in a B–W Bath (212 F/100 C)—10 minutes for either No. 303 or No. 401 cans. Remove cans; cool quickly.

> • **Adjustment for my altitude** _____ •

8
CANNING TOMATOES

Tomatoes are far and away the most popular food for canning at home.

One reason is their versatility. Canned as "cut-up plain" in the specific how-to coming in a minute, they may be heated with a favorite herb and turned into a basic *salsa di pomidoro* for pasta; with another vegetable and chicken or rabbit, they become a hunter's stew; added to a glorious fisherman's catch, they help build a bouillabaise. When we can them puréed they are a mainstay for soups; canned whole and titivated with mild onion and tangy dressing, they are served cold as a winter salad at the side of roast beef. They can be reduced to a paste and put by in small jars to help give blush to a sauce. Juiced alone or in combination they become the healthiest of appetizers.

Another attribute is their abundance. They are grown at home in every likely corner of North America, outdoors or in solar rooms; professional truck gardeners sell them from uncountable roadside stands. We still have relatives of the great classic canning tomatoes, so juicy that they were peeled, quartered, and pressed into jars to create their own splendid canning liquid. However, more seedsmen each year are offering the small, blunt, oblong varieties with few seeds, reluctant juice, and plenty of firm flesh. It is these so-called pasta tomatoes that are now described in the catalogs as "canners": ideal for the commercial processor, they require extra attention and time from the householder. As we saw in the section How Heat Penetrates a Container of Food, in Chapter 6, "The Canning Methods," heat is carried by the swirl of liquid to the innermost part of the pack: this is convection, and it requires less time to be effective than does heat led by conduction to the center through a dense pack, one with less liquid. Know your tomato before you can it!

The average bushel of tomatoes weighs about 53 pounds/25 kilos, and will yield from 15 to 20 quarts when simply cut up and canned; it takes

2½ to 3 pounds of fresh tomatoes to do 1 quart. The yield in jars will be smaller, though, if the fruit is an ultramodern hybrid that is heavy on flesh and light on juice. Since no water may be added to this pack—doing so will reduce the acidity of the pack to a dangerous degree—some of the tomatoes must be juiced beforehand to provide extra canning liquid for filling the jars. Some householders even have on hand an emergency out-size can or two of commercially canned tomato juice at the height of the tomato season.

And be picky if you're buying from a produce stand. Sometimes truck gardeners advertise poor cull fruit as "canning tomatoes."

Tomatoes' third great virtue is still the relative ease with which they're canned.

Adding Acid—Why and Which

Nowadays home-grown tomatoes often nudge the 4.6 *pH* cut-off, beyond which they are considered *low*-acid, and must be Pressure-processed accordingly. To ensure that they are satisfactorily within Boiling–Water Bath range at sea level, the 1988 *Guide* from the Center for Excellence stipulates adding any one of three acid substances: powdered citric acid, OR bottled lemon juice (which always is a guaranteed 5 percent acidity), OR distilled white vinegar (which you use twice as much of as lemon juice, because vinegar is so much more volatile). Any of these acids may be added first at the bottom of the pack, or at the top; the swirl of the contents during processing will distribute it, along with the optional salt. Traditionally ½ teaspoon canning salt was added at the top of a packed pint, 1 teaspoon atop a quart; the salt is merely a flavor-enhancer, though, and often is omitted by today's home-canners.

Word-Play and Some History

Tomatoes' quiet revolution coincided with tragedy in 1974, and the result was alarm about canning tomatoes in the accepted way.

Seedsmen had established their new strains of table tomatoes—fat, meaty, sweet varieties that were grown to be served sliced to garnish a juicy cut of steak. Less juice, fewer seeds, above all a new sweetness: these were the selling points, and almost at once the public began equating sweetness with less acid. This casual reaction was not wholly justified because hybridizers often were breeding for special sweetness, rather than for notably less acidity. The relationship will be underscored in Chapter 19 dealing with pickles and other tart things: too much tartness must never be offset by reducing the vinegar that makes the product safe, but rather by increasing the sweetness with additional sugar.

The matter would have resolved itself without undue fanfare if the Center for Disease Control of the U.S. Public Health Service had not dropped a bomb: it reported that early in 1974 there were two deaths from botulism poisoning traced directly to home-canned tomatoes and tomato juice. Consumer activists immediately queried if the "new, low-acid" tomatoes were responsible, and even suggested in print that, from then on, tomatoes would need to be Pressure-processed at 15 psig instead of in a Boiling–Water Bath.

Meanwhile the public health officers discovered what allowed the spores of *C. botulinum* to make the toxin that killed the victims. Common bacteria or molds grew in the food in the jars and thereby reduced the acidity because the natural acid in the tomatoes was metabolized by the micro-organisms as they grew and developed. It was established after compassionate, but thorough, investigation that these bacteria or molds survived either because the tomatoes were canned by the discredited open-kettle method, or entered under the lid of a jar that wasn't adequately sealed. Inadequate processing is virtually always the cause of food poisoning that develops during shelf life.

. . . But the *Care* Never Varies

No matter which strain of tomato you can in which form—whole, stewed, or puréed; as juice, chili sauce, or ketchup—any sloppiness, any cutting of corners will result in tomato products that are disappointing or even nasty or possibly downright dangerous.

The federal, state, and non-government experts whom *PFB* has consulted since the "great tomato revolution" of 1974 agree 100 percent that clean, careful handling, and due respect for the *Whys* of good packing and processing, are the primary safeguards in canning tomatoes of all varieties.

Selecting the fruit. Use only *firm*-ripe, unblemished tomatoes, ones that have not quite reached the table-ready stage wanted for slicing and serving raw.

Discard any that have rotten spots or mold. (The regulations governing commercial canning regard *just one decomposed tomato per 100 sound fruit* as reason enough to condemn the entire lot as unfit for human consumption.)

Discard any that have open lesions.

Washing. Wash the fruit carefully in fresh water of drinking quality. If many are spattered with field dirt, or have not been staked or mulched in your own garden, add a little mild detergent and 4 teaspoons of 5 percent chlorine bleach to each 1 gallon of wash water; rinse well in fresh water. (This thoroughness cuts down bacterial load.)

Peeling and cutting. You will be working with clean equipment and cutting surfaces—just as you do when handling any food you're putting by. Peel tomatoes by dipping a few at a time in briskly boiling water, then dunking immediately in cold, clean water: the skins will strip off.

Without cutting into the seed cavity, ream out the stem end and core (the point of an apple-corer does a good job). Cut off the blossom end. Cut off any green shoulders to ensure a product of uniform tenderness and flavor. Cut out any bruises, no matter how small.

Cut/chop as individual instructions say to.

Packing. Pack in clean, scalded containers. All tomato products are packed Hot, with the contents of cans exhausted to a minimum of 170 F/ 77 C if the tomatoes have cooled after precooking. Add acid and optional salt. ADDING ACID IS NOT A CRUTCH. Increasing acidity does not allow you to short-cut any step in safe canning procedures.

Processing. Whether you process in a Boiling–Water Bath or in a Pressure Canner, *time the processing accurately:* from return to the full boil in a B–W Bath; or after 10 pounds is reached, following a 7-to-10-minute strong flow of steam from the vent (depending on the size of the canner) to ensure adequate pressure inside.

Removing and cooling containers. Follow instructions given toward the end of Chapter 6. Complete the seals on bail-type jar lids as you take them from the canner. NEVER RE-TIGHTEN TWO-PIECE SCREW-BAND LIDS. AT ANY TIME. EVER.

Remember that *hastening* the cooling of jars can cause them to break; *retarding* the natural cooling of jars can cause thermophilic spoilage (like "flat sour") to develop in the contents. But cool *cans* quickly.

Check, clean, label and store containers according to Post-Processing Jobs in Chapter 6.

CONVERSIONS FOR CANNING TOMATOES

Do look at the conversions for metrics, with their workable roundings-off, and for altitude—both in Chapter 3—and apply them.

The Choice Between Processing Methods

Until the extensive research on lower-acid tomatoes is completed and the reports are correlated, there can be no consensus that says tomatoes must be Pressure-processed to ensure a safe home-canned food. In the meantime, each householder must make a judgment call.

If you have reason to think your tomatoes are not quite within the acid range for the B–W Bath (a *pH* rating more than 4.6, or even the 4.7 cutoff some experts sanction)—because of the way they were grown, or if you're stuck with fruit that's past its ideal condition—then just let informed good sense choose between:

Either (1) increasing the acid content of the pack yourself (how, is told below), and use a proper B–W Bath (212 F / 100 C) for the time specified.

Or (2) processing at 5 or 10 pounds pressure (228 F / 109 C, or 240 F / 116 C) for the length of time given for the individual tomato products.

Or (3) processing at 15 pounds pressure (250 F / 121 C) because you live at high altitude that requires this adjustment (see Correcting for Altitude in Chapter 3).

Or (4) processing at 15 pounds pressure for barely a moment in a procedure researched at the University of Minnesota and reported by Isabel D. Wolf and Edmund A. Zottola. *PFB* is edgy about describing this method from instructions given in the University of Minnesota's Food Science and Nutrition *Fact Sheet No. 13, Home Canning Tomatoes* (1976, with a 1980 updated notation in the Minnesota *Extension Bulletin* 413; see the Appendix under the subheading for this chapter). Success with this method from a research project relies on the abilities of the householder and an impeccably accurate and faultlessly operated Pressure Canner. The Appendix tells where to get your own copy.

And there's always *freezing* (in Chapter 17, "Freezing Convenience Foods,") in a space-saving form like sauce.

Some Random Notes

PFB has seen some odd and hair-raising things over the years where tomatoes are concerned.

County fairs always have rows of jars of tomatoes—whole fruit, cut-up plain, or done with special touches that just could be hazardous to your health. One interesting exhibit featured quart jars of routine cut tomatoes: seals intact, good-looking liquid, no dark gurry showing at the sealing rim (jars are automatically regarded as unacceptable if they are displayed with their screwbands still on), no extrusions, no tiny mold spots. But the seeds had tiny tails—*they had sprouted*, indicating that the processing had not been adequate to prevent germination, and that the storage was warm enough to encourage sprouting (which halted only because acidity prevented further growth). It was recommended that the contents be destroyed.

Sometimes we hear of tomatoes canned whole in a B–W Bath with boiling water used as the processing liquid. *PFB* regards this practice as courting grave trouble. Of course, if the tomatoes were treated like a low-acid vegetable and processed in a Pressure Canner at 10 psig, we would relax.

Is it feasible to buy respectable commercially canned tomato juice in economically large containers? Yes, if the tomatoes are the tasty plum type with very little juice of their own, and the batches are small (for small families).

The 1988 *Guide* has a pint jar of crushed whole tomatoes, with the usual acid added but with no added liquid, processed in a Boiling–Water Bath for 85 minutes. Reason for the long bath (or alternative Pressure-processing): the pack of modern meaty "canners" is so dense as to require such heroic processing.

Pressure-Processing for Tomatoes

It is easy to cite Pressure-processing as the alternative method for canning tomatoes safely at home. However, a sound and well-tested timetable— a timetable of the reasoned sort we all rely on now for canning other foods—will take a good while to establish.

Meanwhile, for the householder who feels secure only with tomatoes done in a Pressure Canner, we offer the following stopgap. First, though, five things:

1. Pressure-processing is NOT A CRUTCH. It DOES NOT MEAN THAT YOU CAN SHORT-CUT ANY STEP of good canning procedure—careful selection, sanitation, correct packing, maintaining pressure, accurate timing, ensuring seals, proper storage.

2. Nutrients in some degree and of course texture to a greater extent will suffer more than they do in a Boiling–Water Bath. The tender flesh of tomatoes will disintegrate more (unless a firming additive is included: see calcium hydroxide under Firming Agents in Chapter 5), and an excessive amount of juice is likely to separate from the tissues.

3. Independent food scientists around the country agree that *5 pounds pressure is TOO LOW* to get the result desired from Pressure-processing plain tomatoes.

4. The processing times given below are for *cut-up plain tomatoes ONLY.* The times are not long enough to deal safely with a mixture of tomatoes and lower-acid vegetables like onions, celery, green peppers, or whatever.

5. The processing vessel used is a conventional Pressure *Canner— NOT a pressure saucepan,* even though it might hold several pint jars. The much smaller size of the saucepan plays hob with any pressure timetable (for why, see "Canning Methods," Chapter 6). And anyway, such a little saucepan-size batch of cut-up plain tomatoes would be better converted to Plain Sauce and done in the proper Boiling–Water Bath.

These points made, for Pressure-canning cut-up plain tomatoes: use Hot pack only. Use jars or plain cans (if necessary, exhausting the contents to a minimum of 170 F/77 C after packing). Vent the heated canner 7 minutes for medium-size kettles, 10 minutes for large ones. Time the processing after internal pressure of the canner has reached *10 pounds* (240 F/116 C)—15 minutes for pint jars, 20 minutes for quarts, 15 minutes for No. 303 cans, 20 minutes for No. 401 cans. Complete jar seals if necessary, cool naturally; remove cans, cool quickly in cold water.

Before Tasting Canned Food with Any Low-Acid Ingredients: you must be unshakably certain that your Pressure Canner was operated correctly—pressure gauge accurate and dead-weight gauge signaling properly—and that the Boiling–Water Bath kept the necessary water at a full boil around and over the containers of food; and that requirements for times and corrections for altitude were followed.

Unless you are sure that these safeguards were observed, a margin of protection is added by boiling the canned food hard for 15 minutes to destroy any hidden toxins and stirring to distribute the heat. If the food foams unduly or smells bad during boiling, destroy it completely so it cannot be eaten by people or animals.

Canned Tomato Troubles and What to Do

New developments bring new strictness, so canned tomatoes in your storeroom should be examined for the same signs that mean vegetables are unfit, or dangerous, to eat:

- Broken seals.
- Bulging cans.
- Seepage around the seal.
- Mold, the tiniest spot, around the seal or on the underside of the lid or in the contents.
- Gassiness in the contents, spurting liquid from pressure inside any container when it is opened.
- Cloudy or yeasty liquid.
- Unnatural color.
- Unnatural or unpleasant odor.

If any of these indications is present in the smallest degree, play safe and *do not even taste the tomatoes before boiling them for 15 minutes to destroy hidden toxins. Then, during boiling, if they foam or smell bad, destroy them so they cannot be eaten by people or animals.* Wash the containers and sealers in hot soapy water, then cover with fresh water and boil hard for 15 minutes; salvage only sterilized jars—destroy cans and all closures, etc.

Basic Tomato Products

For more tomato products see Chapter 12, "Canning Convenience Foods," and relishes in Chapter 19.

Unless you can pick with finicky selectiveness from a well-managed small garden of your own, there is bound to be a range of ripeness, size, and condition in any good-size batch of tomatoes you're getting ready to can. So pick them over carefully, of course discarding any rotten or banged-up ones, and let size and degree of acceptable ripeness dictate the form you'll can them in.

Perfect just *firm-ripe*—better underripe than overripe—uniform and small enough to slip easily down into the jar—these are the ones for canning whole to use in salads. Misshapen or overly large fruit are cut to stewing size—quarters, eighths, or chunks—and are canned plain or with added vegetables for flavor; they also go for sauce or juice.

Regardless of the form or whether they're to be processed in a Boiling–Water Bath or a Pressure Canner, prepare the tomatoes according to the general directions given earlier in this chapter under . . . But the Care Never Varies and Added Acid + B–W Bath.

Cut-up Plain Tomatoes

Boiling–Water Bath preferred (but for Pressure-canning see times, etc., in the general introduction for this chapter). Use Hot pack only. Use jars or plain cans.

Select, wash, peel, according to general handling earlier; cut in quarters or eighths, saving all juice possible. In a large enameled kettle bring cut tomatoes to a boil in their own juice, and cook gently for 5 minutes, stirring so they don't stick. Pack.

Note: individual food scientists around the country do not agree that 5 pounds pressure is enough to get the result desired from Pressure-processing Plain Tomatoes, especially at altitudes above 3000 feet/914 meters. Nor do they agree across-the-board on B–W Bath timing in every case: check with your regional Extension Service for special handling in your area.

HOT PACK ONLY

B–W Bath, in jars. Fill clean scalded jars with boiling-hot tomatoes and their juice, leaving ½ inch of headroom. Add ¼ teaspoon crystalline citric acid (or 1 tablespoon lemon juice) to pints, ½ teaspoon citric acid (or 2 tablespoons lemon juice) to quarts. (Optional: add ½ teaspoon salt to pints, 1 teaspoon salt to quarts.) Adjust lids. Process in a Boiling–Water Bath (212 F/100 C)—35 minutes for pints, 45 minutes for quarts. Remove jars. Complete seals if necessary.

　　• **Adjustment for my altitude** _____ •

B–W Bath, in plain cans. Fill to the rim with boiling-hot tomatoes and juice. Add ¼ teaspoon citric acid (or 1 tablespoon lemon juice) to No. 303 cans, ½ teaspoon citric acid (or 2 tablespoons lemon juice) to No. 401 cans. Fill to the top with boiling juice. (Optional: add ½ teaspoon salt to No. 303 cans, 1 teaspoon salt to No. 401 cans.) If the tomatoes have cooled unavoidably, exhaust to 170 F/77 C (approx. 10 minutes); seal. Process in a B–W Bath (212 F/100 C)—35 minutes for No. 303 cans, 45 minutes for No. 401 cans. Remove cans. Cool quickly.

　　• **Adjustment for my altitude** _____ •

Plain Tomato Sauce (Purée)

This is a handy way indeed to can tomatoes, and it makes a better base for red Italian-style pasta sauces than does Tomato Paste (which many Americans tend to use too much of in such sauces anyway). The texture should fall about halfway between juice and paste.

Do not add onions or celery, etc., now *or you must* Pressure-process the sauce.

HANDLING

Use a Boiling–Water Bath. Use Hot pack only. Use ½-pint or pint jars only.

Prepare and sieve the fruits as for Tomato Juice (below). In a large enameled kettle bring the juice to boiling, and boil gently until thickened but not so stiff as Tomato Paste—about 1 hour or a little longer. Stir often so it doesn't stick.

Hot pack only, in ½-pint or pint jars. Pour into clean scalded jars, leaving ¼ inch of headroom in ½-pints, ½ inch in pints. Add ⅛ teaspoon citric acid (or 1½ teaspoons lemon juice) to ½-pints. (Optional: add ½ teaspoon salt to ½-pints.) Add ¼ teaspoon citric acid (or 1 tablespoon lemon juice) to pints. (Optional: add ½ teaspoon salt to pints.) Adjust lids. Process in a B–W Bath (212 F/100 C)—30 minutes for either ½-pints or pints. Remove jars; complete seals if necessary.

• **Adjustment for my altitude** _____ •

Tomato Paste

If you're canning many tomatoes, you'll surely want a few little jars of paste put by too. Work with small batches, because it scorches easily during the last half of cooking. And forgo onions, garlic, celery, etc., because such flavors may not be wanted in delicate sauces you merely want to color with the paste.

HANDLING

Use a Boiling–Water Bath. Use Hot pack only. Use ½-pint jars only.

Carefully wash, peel, trim, and chop the tomatoes saving all the juice possible (4 to 4½ pounds of tomatoes will make about four ½-pint jars of paste). In an enameled kettle bring the chopped tomatoes to a boil, then reduce heat and simmer for 1 hour, stirring to prevent sticking. Remove from heat and put the cooked pulp and juice through a fine sieve or food mill. Measure, return to the kettle, and for every 2 cups of sieved tomatoes add ¼ teaspoon citric acid (or 1 tablespoon lemon juice). (Optional: add ½ teaspoon salt.) Reheat, and continue cooking very slowly, stirring frequently, until the paste holds its shape on the spoon—about 2 hours more.

Hot pack only, in ½-pint jars. Ladle hot paste into clean hot jars, leaving ¼ inch of headroom. Adjust lids, and process in a B–W Bath (212 F/100 C) for 35 minutes. Remove jars; complete seals if necessary.

• **Adjustment for my altitude** _____ •

Tomato Juice

Time out for an interesting point: preparing juice from *uncooked* tomatoes—in a blender at high speed, or in a food processor—gives a thin product that separates (enzyme action is the reason).

Canned tomato juice is noted for encouraging growth of the highly heat-resistant bacillus that causes *flat-sour* spoilage, a sneaky and nasty-tasting condition indeed. However, even though the organism is very hard to destroy, it can be avoided quite easily: just follow carefully all the requirements for handling food in a sanitary way.

This care extends from every piece of sterilized equipment to the tomatoes themselves. Choose only firm-ripe, red, perfect tomatoes—no injured ones, none with soft spots or broken skins. Wash them thoroughly. With a stainless-steel knife cut away stem and blossom ends and cores. Cut the fruit in eighths and put it in an enameled kettle (which won't react with the acid of the tomatoes), and simmer it, stirring often, until soft. Put the tomatoes through a fine sieve or food mill: the finer the pulp, the less likely that the juice will separate during storage. Measure the juice into the kettle, and for each 4 cups of juice add ½ teaspoon citric acid (or see Chapter 5). (Optional: add ½ teaspoon salt to pints, 1 teaspoon salt to quarts.) Reheat all to simmering. Pack hot.

Hot pack only, in jars. Fill clean, hot jars with the very hot juice, leaving ½ inch of headroom; adjust lids. Process in a Boiling–Water Bath—35 minutes for either pints or quarts. Remove jars, complete seals if necessary.

• **Adjustment for my altitude** _____ •

Hot pack only, in plain cans. Fill to the top with boiling juice, leaving no headroom; seal (juice already 170 F/77 C or over doesn't need further exhausting). Process in a B–W Bath (212 F/100 C) for 45 minutes for either No. 303 or No. 401 cans. Remove cans; cool quickly.

• **Adjustment for my altitude** _____ •

9
CANNING VEGETABLES

Anyone who has the enterprise to can vegetables at home surely has the good sense to want to can them safely. *Because all fresh natural* (as opposed to pickled) *vegetables are low-acid, they MUST be processed in a regular Pressure Canner.* And no short cuts, no scamping.

The outbreaks of food-borne botulism have fallen off from their peak in the mid-seventies, when hundreds of thousands of Americans were discovering canning, but still at least 90 percent of the traceable sources of this dreaded type of poisoning are laid to *home*-canned low-acid food. And inadequate processing is the cause named in virtually every case investigated by public-health teams.

The spores of *C. botulinum*—and it's the spores that make the toxin, remember—can survive 5 hours or more of boiling at 212 F/100 C, even though the vegetative form of the bacterium is much more fragile. Therefore anything less than adequate Pressure-processing is a monstrous gamble. People who count on getting away with processing natural vegetables in a Boiling–Water Bath are playing for stakes too high.

CONVERSIONS FOR CANNING VEGETABLES

Do look at the conversions for metrics, with their workable roundings-off, and for altitude—both in Chapter 3—and apply them.

YIELDS IN CANNED VEGETABLES

Since the legal weight of a bushel of vegetables differs among states, the weights given below are average; the yields are approximate.

VEGETABLES	FRESH	QUARTS CANNED
Asparagus	1 bu (45 lbs)	11
	3–4 lbs	1
Beans, Lima, in pods	1 bu (32 lbs)	6–8
	4–5 lbs	1
Beans, snap/green/wax	1 bu (30 lbs)	15–20
	1½–2 lbs	1
Beets, without tops	1 bu (52 lbs)	17–20
	2½–3 lbs	1
Broccoli	25-lb crate	10–12
	2–3 lbs	1
Brussels sprouts	4 qts	1–1½
	1 lb	½
Carrots, without tops	1 bu (50 lbs)	16–20
	2½–3 lbs	1
Cauliflower	2 medium heads	1½
	1 bu (12 lbs)	4–6
Corn, in husks	1 bu (35 lbs)	8–9
	4–5 lbs	1
Eggplant	2 average	1
Kale	1 bu (18 lbs)	6–9
	2–3 lbs	1
Okra	1 bu (26 lbs)	17
	1½ lbs	1
Peas, green (pods)	1 bu (30 lbs)	6–8
	2–2½ lbs	½
Potatoes, white	1 bu (50 lbs)	20
	2½–3 lbs	1
Potatoes, sweet (and yams)	1 bu (50 lbs)	20
	2½–3 lbs	1
Pumpkin	50 lbs	15
	3 lbs	1
Spinach (most greens)	1 bu (18 lbs)	6–9
	2–3 lbs	1
Squash, summer	1 bu (40 lbs)	16–20
	2–2½ lbs	1
Squash, winter (chunks)	3 lbs	1½

Equipment for Canning Vegetables

All the utensils and standard kitchen furnishings that you used for fruits will do, with two exceptions:

(1) Use only a *standard Pressure Canner* for processing. In Chapter 6, "Canning Methods," we went on record as not liking pressure *saucepans* much for canning: here we say a flat "don't" when the food is low-acid *and* higher altitude is a factor. These pots simply are too small to trust for psig's and times needed to make food safe. THEY CAN RUN DRY.

(2) You will need a blanching basket or some other means of holding prepared vegetables loosely in boiling water for the partial precooking in Hot pack. Blanching may be done in a microwave oven or in steam, but it is handier to precook in boiling water, which you can then use as the canning liquid—thereby saving nutrients.

Canned Vegetable Troubles and What to Do

If you find in your vegetables any of the following, *destroy the contents so that they cannot be eaten by people or animals*. Wash containers and closures vigorously as described in Chapter 6. Throw away the cans, lids, sealing disks, and rubber rings; if sound, the jars may be used again.

- Broken seal.
- Bulging cans.
- Seepage around the seal.
- Mold, even a speck, in the contents or around the seal or on the underside of the lid.
- Gassiness (small bubbles) in the contents.
- Spurting liquid, pressure from inside the container as it is opened.
- Cloudy or yeasty liquid.
- Unnatural or unpleasant odor.

Before tasting any canned low-acid food: you must be unshakably certain that your Pressure Canner was operated correctly—pressure gauge accurate and dead-weight gauge signaling properly—and that requirements for times and corrections for altitude were followed.

Unless you are sure that these safeguards were observed, a margin of protection is added by boiling the canned low-acid food hard for 15 minutes to destroy any hidden toxins (corn and greens require 20 minutes), and stirring to distribute the heat. If the food foams unduly or smells bad during boiling, destroy it completely so it cannot be eaten by people or animals.

Some Special Considerations

Salt added to vegetables in canning is merely a seasoning and therefore *of course is optional.* Pure canning salt is ideal, but the amounts called for are so small that the fillers, etc., in your regular table salt won't cloud the canning liquid. However, salt substitutes should *not* be added to the pack before processing, lest the finished product have an unwanted aftertaste: add your salt substitute to the heated vegetable just before serving.

Starchy vegetables swell during processing, especially so if they are packed Raw, so they need double the headroom usually supplied to nonstarchy foods. Be particularly careful about shell beans of all kinds, green peas, and whole-kernel corn (hominy).

Add extra headroom at altitudes above 2000 feet/610 meters (see Correcting for Altitude in Chapter 3), because the lower atmospheric pressure allows the steam inside the jars to expand more. Also, the greater temperatures in Pressure-processing mean that the steam will expand more than in boiling not under pressure.

Some precooking water can be bitter. This of course depends on the hardness of the water to begin with, and to some extent on the growing conditions of the vegetable. However, water in which asparagus, some greens, and members of the turnip family are precooked for Hot pack can be bitter; taste the water, and if it is too strong or has a bitterness, substitute boiling water as the canning liquid.

Asparagus

Asparagus keeps more spring flavor if you freeze it; but it cans easily—whole or cut up.

GENERAL HANDLING

Only Pressure Canning for asparagus: it has even less acid than string beans. Hot pack preferred. Use jars or plain cans.

Wash; remove large scales that may have sand behind them; break off tough ends; wash again. If you're canning it whole, sort spears for length and thickness, because you'll pack them upright; otherwise cut spears in 1-inch pieces.

PREFERRED: HOT PACK

Whole spears—stand upright in a wire blanching basket and dunk it for 3 minutes in boiling water up to *but not covering* the tips; drain and pack upright (tight but not squdged). Cut-up—cover clean 1-inch pieces with boiling water for 2 to 3 minutes; drain and pack.

The added processing liquid can be the boiling-hot precooking water—if it's free of grit—instead of fresh boiling water.

In jars. Pressure-process at 10 psig (240 F/116 C)—pints for 30 minutes; quarts for 40 minutes.

• **Adjustment for my altitude** _____ •

In plain cans. Pack as for jars, leaving only ¼ inch of headroom; add ½ teaspoon salt to No. 303 cans, 1 teaspoon to No. 401 (optional). Fill to top with boiling water. Exhaust to 190 F/83 C (about 7 minutes); seal and pressure process at 10 psig (240 F/116 C)—No. 303 cans for 25 minutes, and No. 401 for 30.

• **Adjustment for my altitude** _____ •

Beans, "Butter"

See Beans, Lima (fresh).

Beans—Green / Italian / Snap / String / Wax

Perhaps next to tomatoes, these beans are the most popular vegetable canned by North American householders. They also are established as being the single most likely source of botulism poisoning among home-canned foods.

But please don't forgo canning them: just process them in a Pressure Canner for the time stipulated, making any needed adjustment for altitude.

They freeze well, especially the young ones whose seeds have barely begun to form bumps; freeze some of these, and can the more mature ones. Look up these beans in Chapter 21, "Drying": old-timers called them Leather Britches.

BEFORE WE START—A WORD ABOUT MUSHINESS

People have written to *PFB* often about how to avoid having mushy canned green beans, and the answer first and always is *never cut down on the Pressure-processing time*. The warning holds true regardless of the part of North America the query comes from, and regardless of the varieties, growing conditions, and hardness or softness of the water for canning.

Variety is the least important factor in the result, actually. If you grow your own you might plant an old-fashioned pole bean. But do ask your County Agricultural Agent for varieties that do well for canning in your area; especially look at some of the tender new hybrids—these are likely to be more satisfactory when frozen.

The hardness or softness of the water in your area of course has a bearing on the texture of the finished product. Rainwater, or water chemically treated to be very soft, can make the beans slough or get soft very quickly when brought to a full, rolling boil for serving at the table. This is the reason why commercial canners sometimes give their beans a meticulously controlled low-temperature blanch for a few minutes before proceeding to Hot-pack and Pressure-process them (a pre-treatment that sets the calcium pectate in the beans' outer tissues). You can achieve much the same result by Raw-packing your beans.

Perhaps the most sensible solution: simply choose beans for canning that are a little more mature than you'd use immediately for the table or for freezing. *PFB* was given this tip by several plant scientists, who added that signs of bumpiness indicate that the bean seeds are starting to develop and fill the pods, and the tissues therefore will be more likely to stay firm in canning.

GENERAL HANDLING

Pressure Canning only. Use Raw or preferred Hot pack (Hot makes them supple, permits a more solid pack; and the procedure described below as a blanch is a paraphrase of the commercial canners' pre-treatment.) Use jars or plain cans.

Wash, trim ends, unzip strings as needed. Sort for size and maturity: some may be packed whole and upright like asparagus; others may be frenched or cut (slanted) in 1-inch pieces.

Blanch for Hot pack (optional for Raw pack). PFB prefers hot-water to steam blanching as more reliable at high altitudes. In a large kettle of water at 170–180 F/77–82 C, place a blanching basket or loosely tied cheesecloth bag of prepared beans and slow-cook for 5 minutes after water returns to temperature; lift out, remove beans, and pack them. Repeat for rest of the batch. Save blanching water and bring to boil to use for canning liquid.

PREFERRED: HOT PACK

In jars. Fill with hot beans, whole beans upright—use a wide-mouth funnel for cut ones—leaving ½ inch of headroom. (Optional: add ½ teaspoon salt to pints, 1 teaspoon salt to quarts.) Add boiling blanching liquid, leaving ½ inch of headroom; remove trapped air with slender plastic spatula. Adjust lids; process at 10 pounds (240 F/116 C)—pints for 20 minutes, quarts for 25. Remove jars; complete seals if necessary.

• **Adjustment for my altitude** _____ •

In plain cans. Fill loosely with hot beans, leaving only ¼ inch of headroom. (Optional: add ½ teaspoon salt to No. 303 cans, 1 teaspoon to 401.) Fill to top with boiling blanch water. Exhaust to 170 F/77 C (up to 10 minutes); seal. Process at 10 pounds (240 F/116 C)—No. 303 cans for 25 minutes, No. 401 for 30 minutes. Remove cans; cool quickly.

RAW PACK

The hot-water blanch is recommended. If not blanched, beans must be packed deliberately tight.

In jars. Leave ½ inch of headroom. (Add optional salt, as for Hot pack above.) Add boiling liquid, leaving ½ inch of headroom; adjust lids, and process as for Hot pack, above.

In plain cans. Fill cans as for Hot pack; add optional salt and fill with boiling liquid. Exhaust to 170 F/77 C (approx. 10 minutes if beans are *not* blanched); seal. Process at 10 pounds (240 F/116 C)—No. 303 cans for 25 minutes, 401 for 30 minutes. Remove cans; cool quickly.

• **Adjustment for my altitude** _____ •

Beans, Fresh Lima (Shell Beans)

Pressure Canning only. Use Hot pack. Use jars or C-enamel cans.

Deal with one variety at a time: different-sized types require different amounts of headroom and perhaps processing.

Shell the beans and wash them before packing. They must be packed loosely because, like all starchy legumes, they swell.

HOT PACK ONLY

In jars. Boil beans 1 timed minute in water to cover; fill jars loosely with drained hot beans leaving 1 inch of headroom for pints or quarts. Cover with hot cooking water; adjust lids. Pressure-process at 10 pounds (240 F/116 C)—pints for 40 minutes, quarts for 50 minutes. Remove jars; complete seals if necessary.

• **Adjustment for my altitude** _____ •

In plain cans. Fill with hot beans, leaving ¾ inch of headroom for either No. 303 or No. 401 cans; *don't press or shake down.* (Optional: add ½ teaspoon salt to No. 303 cans, 1 teaspoon salt to No. 401.) Fill cans to top with boiling water. Exhaust to 180 F/82 C (about 10 minutes); seal. Pressure-process at 10 pounds (240 F/116 C)—40 minutes for either No. 303 or No. 401. Remove cans; cool quickly.

• **Adjustment for my altitude** _____ •

Beets

Beets keep well in a root cellar (see Chapter 22). Between canning and freezing, can them: they can beautifully. Use canned beets plain, titivated as a relish, etc.

GENERAL HANDLING

Only Pressure Canning for beets: they rank with home-canned string beans as carriers of *C. Botulinum* toxin. Because they're firm-fleshed, use Hot pack only. Use jars or R-enamel cans.

Sort for size; leave on tap root and 2 inches of stem (otherwise they bleed out their juice before they get in the containers). Wash carefully. Cover with boiling water and boil until skins slip off easily (15 to 25 minutes, depending on size/age). Drop them in cold water for just long enough to be able to slip off skins; skin, trim away roots, stems, any blemishes. Leave tiny beets whole; cut larger ones in slices or dice. Now they are ready to pickle or can.

HOT PACK ONLY

In jars. Fill with hot beets, leaving $\frac{1}{2}$ inch of headroom. (Optional: add $\frac{1}{2}$ teaspoon salt to pints, 1 teaspoon salt to quarts.) Add fresh boiling water, leaving $\frac{1}{2}$ inch of headroom; adjust lids. Pressure-process at 10 pounds (240 F/116 C)—pints for 30 minutes, quarts for 35 minutes. Remove jars; complete seals if necessary.

> • Adjustment for my altitude _____ •

In R-enamel cans. Fill with hot beets, leaving only $\frac{1}{4}$ inch of headroom (Optional: add $\frac{1}{2}$ teaspoon salt to No. 303 cans, 1 teaspoon salt to No. 401.) Fill to top with boiling water. Exhaust to 170 F/77 C (about 10 minutes); seal. Pressure-process at 10 pounds (240 F/116 C)—35 minutes for either No. 303 or No. 401 cans. Remove cans; cool quickly.

> • Adjustment for my altitude _____ •

Beets, Pickled

Boiling–Water Bath (vinegar makes them so acid that a B–W Bath is quite O.K.). Use Hot pack only. Use jars only.

Scrub; leave the tap root and a bit of stem to help prevent "bleeding." Boil until tender—how long, depends on size—in unsalted water (if wanted, salt may be added later). Dunk in cold water to handle; trim, strip off skins, slice. While beets are cooking, make a Pickling Syrup of equal parts of vinegar and sugar, adding 25 percent more sugar if you like a less-sharp pickle: remember that you counteract acidity *by increasing sweetness, NOT by lowering acidity*—especially for a low-acid food like beets. Figure on $\frac{1}{2}$ to $\frac{3}{4}$ cup syrup for each pint jar; larger containers of course need more, so allow for this; and leftover syrup can be refrigerated until used in any number of ways.

Fill clean, hot jars with hot beet slices, leaving $\frac{1}{2}$ inch of headroom for pints, 1 inch for quarts. (Optional: add $\frac{1}{2}$ teaspoon salt to pints, 1 teaspoon salt to quarts.) Add boiling Pickling Syrup, leaving $\frac{1}{2}$ inch headroom in pints, 1 inch in quarts. Adjust lids. Process in a Boiling–Water Bath (212 F/100 C)—30 minutes for either pints or quarts. Remove jars; complete seals if necessary.

> • Adjustment for my altitude _____ •

Broccoli, Brussels Sprouts, Cabbage

These discolor when canned and grow even stronger in flavor. So instead, freeze broccoli or Brussels sprouts, ferment cabbage as Sauerkraut (see how to can later); or use any or all in mixed pickles as liked.

Carrots

Like beets, carrots can be harvested late in the season to can when weather is cooler. Don't bother with overlarge, woody ones.

GENERAL HANDLING

Only Pressure Canning. Raw or Hot pack. In jars or C-enamel cans.

Sort for size. Wash, scrubbing well, and scrape. (An energetic scrub with your stiffest brush often will do for the very small ones; or parboil them just enough to loosen the skins, dunk them in cold water, slip off their skins, then use Hot pack.) Slice, dice, cut in strips—whatever.

RAW PACK

In jars. Fill tightly, leaving 1 inch of headroom. (Optional: add ½ teaspoon salt to pints, 1 teaspoon salt to quarts.) Add boiling water, leaving ½ inch of headroom (water comes above the carrots); adjust lids. Pressure-process at 10 pounds (240 F/116 C)—pints for 25 minutes, quarts for 30 minutes. Remove jars; complete seal if necessary.

• **Adjustment for my altitude** _____ •

In C-enamel. Fill tightly, leaving ½ inch of headroom. (Optional: add ½ teaspoon salt to No. 303 cans, 1 teaspoon salt to No. 401.) Add boiling water to top. Exhaust to 170 F/77 C (approx. 10 minutes); seal. Pressure-process at 10 pounds (240 F/116 C)—No. 303 for 30 minutes, No. 401 for 35 minutes. Remove cans; cool quickly.

• **Adjustment for my altitude** _____ •

PREFERRED: HOT PACK

Cover clean, scraped, cut, or whole carrots with boiling water and bring again to a full boil; drain, but save the water to put in the jars for processing.

In jars. Pack hot carrots, leaving just ½ inch of headroom. Proceed as for Raw pack, using the cooking water for the added processing liquid.

• **Adjustment for my altitude** _____ •

In C-enamel cans. Pack, leaving only ¼ inch of headroom. Proceed as for Raw pack, using cooking water as the processing liquid, and reducing Pressure-processing time—No. 303 cans for 30 minutes, No. 401 for 35 minutes.

• **Adjustment for my altitude** _____ •

Celery

Given for reference, as a key factor in combinations (e.g., with other vegetables or tomato-based sauces).

HOT PACK ONLY, IN JARS ONLY

Cut up stalks. Boil 3 minutes; drain, and save liquid. Fill jars with hot celery, leaving 1 inch of headroom. (Optional: add ½ teaspoon salt to pints, 1 teaspoon salt to quarts.) Add boiling-hot cooking water, leaving 1 inch of headroom; adjust lids. Pressure-process at 10 pounds (240 F/116 C)— pints for 30 minutes, quarts for 35 minutes. Remove jars; complete seals if necessary.

> • **Adjustment for my altitude** _____ •

Corn, Cream Style

Canning is the better, and certainly handier, way of putting by cream-style corn. Its density demands that it be home-canned *only in pint jars or No. 303 cans:* an extremely low-acid vegetable, it would be pressure-cooked to death for the much longer time needed to process the interior of larger containers.

GENERAL HANDLING

Pressure Canning only. Use Raw or Hot pack. Use pint jars or No. 303 C-enamel cans.

Get it ready by husking, de-silking, and washing the ears. Slice the corn from the cob *halfway through the kernels,* then scrape the milky juice that's left on the cob in with the cut corn (this is where the "cream" comes in).

RAW PACK

In pints jars only. Fill with corn-cream mixture, leaving 1½ inches of headroom (more space than usual is needed for expansion). (Optional: add ½ teaspoon salt.) Add boiling water, leaving ½ inch of headroom (water will be well over the top of the corn); adjust lids. Pressure-process at 10 pounds (240 F/116 C) for 95 minutes. Remove jars; complete seals if necessary.

> • **Adjustment for my altitude** _____ •

In No. 303 C-enamel cans only. Fill without shaking or pressing down, leaving ½ inch of headroom. (Optional: add ½ teaspoon salt.) Fill to top with boiling water. Exhaust to 185 F/85 C (approx. 15 minutes); seal. Pressure-process at 10 pounds (240 F/116 C) for 95 minutes. Remove cans; cool quickly.

> • **Adjustment for my altitude** _____ •

PREFERRED: HOT PACK

Prepare as for Raw pack. To each 4 cups of corn-cream mixture, add 2 cups boiling water. Heat to boiling, stirring, over medium heat (it scorches easily).

In pint jars only. Fill with boiling corn and liquid, leaving 1 inch of headroom. (Optional: add ½ teaspoon salt.) Adjust lids. Pressure-process at 10 pounds (240 F/116 C) for 85 minutes. Remove jars; complete seals if necessary.

· **Adjustment for my altitude** _____ ·

In No. 303 C-enamel cans only. Fill to the top with boiling corn and liquid. (Optional: add ½ teaspoon salt.) Exhaust to 185 F/85 C (approx. 15 minutes); seal. Pressure-process at 10 pounds (240 F/116 C) for 95 minutes. Remove cans; cool quickly.

· **Adjustment for my altitude** _____ ·

Corn, Whole Kernel

Pressure Canning only. Use Raw or Hot pack. Use jars or C-enamel cans. Less dense than cream-style corn, it may be canned in quarts and No. 401 cans just as well as in pints and No. 303 cans.

Husk, de-silk, and wash fresh-picked ears. Boil ears 3 minutes to set the milk. Cut from the cob at about ⅔ the depth of the kernels (this is deeper than for cream-style).

RAW PACK

In jars. Fill, leaving 1 inch of headroom—and don't shake or press down. (Optional: add ½ teaspoon salt to pints, 1 teaspoon salt to quarts.) Add boiling water, leaving ½ inch of headroom (water will come well over top of corn); adjust lids. Pressure-process at 10 pounds (240 F/116 C)—pints for 55 minutes, quarts for 85 minutes. Remove jars; complete seals if necessary.

· **Adjustment for my altitude** _____ ·

In C-enamel cans. Fill, leaving ½ inch of headroom—and don't shake or press down. (Optional: add ½ teaspoon salt to No. 303 cans, 1 teaspoon salt to No. 401 cans.) Add boiling water to top. Exhaust to 170 F/77 C (approx. 10 minutes); seal. Pressure-process at 10 pounds (240 F/116 C)—55 minutes for No. 303 or 65 minutes for No. 401 cans. Remove cans; cool quickly.

· **Adjustment for my altitude** _____ ·

PREFERRED: HOT PACK

Prepare as for Raw pack. To each 4 cups of kernels, add 2 cups boiling water. Bring to boiling over medium heat, stirring so it won't scorch. Drain, saving the hot liquid.

In jars. Fill with kernels, leaving 1 inch of headroom. (Optional: add ½ teaspoon salt to pints, 1 teaspoon salt to quarts.) Add boiling-hot cooking liquid, leaving 1 inch of headroom; adjust lids. Pressure-process at 10 pounds (240 F/116 C)—pints for 55 minutes, quarts for 85 minutes. Remove jars; complete seals if necessary.

· **Adjustment for my altitude** _____ ·

In C-enamel cans. Fill with hot kernels, leaving ½ inch of headroom. (Optional: add ½ teaspoon salt to No. 303 cans, 1 teaspoon salt to No. 401.) Add boiling-hot cooking liquid to top (water will come well over top of corn). Exhaust to 170 F/77 C (about 10 minutes); seal. Pressure-process at 10 pounds (240 F/116 C)—55 minutes for No. 303 and 85 minutes for No. 401. Remove cans; cool quickly.

　　• **Adjustment for my altitude** _____ •

Eggplant

This loses its looks when it's canned, but does well in casseroles.

GENERAL HANDLING

Pressure Canning only. Use Hot pack, in jars only.

　　Wash, pare, and slice or cube eggplant. Soak for 20 to 30 minutes in 1 quart water with 1 tablespoon of salt; drain. In fresh water, with 1 tablespoon of lemon juice, simmer for 5 minutes. Drain. Fill clean hot jars, leaving 1 inch of headroom. Add fresh boiling water but no salt; adjust lids. Pressure-process at 10 pounds (240 F/116 C)—pints for 30 minutes, quarts for 40 minutes. Remove jars; complete seals if necessary.

　　• **Adjustment for my altitude** _____ •

Greens—Spinach, etc., and Wild

All garden greens—spinach, chard, turnip or beet tops—can nicely; so do wild ones like dandelions and milkweed (fiddleheads and cowslips are usually such treats that they're eaten as they come in). Greens freeze well, and with more garden freshness.

GENERAL HANDLING

Pressure Canning only. Use Hot pack only (to make greens solid enough in the container). Use jars or plain cans.

　　Using spinach as the example *for garden greens:* remove bits of grass, poor leaves, etc., from just-picked leaves; cut out tough stems and coarse midribs. Wash thoroughly, lifting from the water to let any sediment settle. Put about 1 pound at a time of clean, wet leaves in a large cheesecloth, tie the top, and steam until well wilted. Accumulate and pack.

HOT PACK ONLY

In jars. Fill with greens, leaving ½ inch of headroom. (Optional: add ½ teaspoon salt to pints, 1 teaspoon salt to quarts.) Add boiling water, leaving ½ inch of headroom; adjust lids. Pressure-process at 10 pounds (240 F/116 C)—pints for 70 minutes, quarts for 90 minutes. Remove jars; complete seals if necessary.

　　• **Adjustment for my altitude** _____ •

In No. 303 plain cans only. Fill with greens, leaving only ¼ inch of headroom. (Optional: add ½ teaspoon salt to No. 303 cans. Cover to top with boiling water. Exhaust to 170 F/77 C (approx. 10 minutes); seal. Pressure-process at 10 pounds (240 F/116 C—No. 303 for 70 minutes. Remove cans; cool quickly.

　　• **Adjustment for my altitude** _____ •

Jerusalem Artichokes

Prepare and process like Carrots.

Mixed Vegetables (in General)

Pressure Canning only. Hot pack only. Use jars or plain or C-enamel cans (for which type of can, see the list in Chapter 6).

Rule-of-thumb for processing: choose the time required by the single ingredient requiring the longest processing (e.g., stewed tomatoes and celery, do as celery).

Wash, trim vegetables, peeling if necessary. Cut to uniform size. Cover with boiling water, boil 5 minutes. Drain; save the cooking water for processing.

HOT PACK ONLY

In jars. Fill with hot mixed vegetables, leaving ½ inch of headroom. (Optional: add ½ teaspoon salt to pints, 1 teaspoon salt to quarts.) Add boiling cooking water, leaving ½ inch of headroom; adjust lids. Pressure-process at 10 pounds (240 F/116 C)—pints for 60 minutes, quarts for 85 minutes. Remove jars; complete seals if necessary.

　　• **Adjustment for my altitude** _____ •

In plain or C-enamel cans. Fill with hot mixed vegetables, leaving ½ inch of headroom. (Optional: add ½ teaspoon salt to No. 303 cans, 1 teaspoon salt to No. 401.) Fill to the top with boiling water (fresh, or the cooking water). Exhaust to 170 F/77 C (approx. 10 minutes); seal. Pressure-process at 10 pounds (240 F/116 C)—No. 303 cans for 60 minutes, No. 401 for 85 minutes. Remove cans; cool quickly.

　　• **Adjustment for my altitude** _____ •

Mixed Corn and Beans (Succotash)

Pressure Canning only. Use Hot pack only. Use jars or C-enamel cans.

Boil freshly picked ears of corn for 5 minutes; cut kernels from cobs (as for Whole-Kernel Corn, *without* scraping in the milk). Prepare fresh lima beans or green/snap beans, and boil by themselves for 3 minutes. Measure and mix hot corn with ½ to an equal amount of beans.

HOT PACK ONLY

In jars. Fill with hot corn-and-bean mixture, leaving 1 inch of headroom. (Optional: add ½ teaspoon salt to pints, 1 teaspoon salt to quarts.) Add boiling water, leaving 1 inch of headroom; adjust lids. Pressure-process at 10 pounds (240 F/116 C)—pints for 60 minutes, quarts for 85 minutes. Remove jars; complete seals if necessary.

• **Adjustment for my altitude** _____ •

In C-enamel cans. Fill with hot mixture, leaving ½ inch of headroom. (Optional: add ½ teaspoon salt to No. 303 cans, 1 teaspoon salt to No. 401.) Fill to the top with boiling water. Exhaust to 170 F/77 C (approx. 10 minutes); seal. Pressure-process at 10 pounds (240 F/116 C)—No. 303 cans for 70 minutes, No. 401 for 95 minutes. Remove cans; cool quickly.

• **Adjustment for my altitude** _____ •

Mushrooms

Canned mushrooms have a bad track record as carriers of *C. botulinum* toxin, so use only fresh edible mushrooms—preferably those grown in presterilized soil.

GENERAL HANDLING

Pressure Canning only. Use Hot pack only. Use ½-pint or pint jars, plain No. 303 cans.

Soak them in cold water for 10 minutes to loosen field dirt, then wash well. Trim blemishes from caps and stems. Leave small buttons whole; cut larger ones in button-size pieces. In a covered saucepan simmer them gently for 15 minutes.

HOT PACK ONLY

In jars. Fill with hot mushrooms, leaving ½ inch of headroom. (Optional: add ¼ teaspoon salt to ½-pints, ½ teaspoon salt to pints.) To prevent color change, add ¼ teaspoon of crystalline ascorbic acid to ½-pint jars, ½ teaspoon to pints. Add boiling water, leaving ½ inch of headroom; adjust lids. Pressure-process at 10 pounds (240 F/116 C)—45 minutes for either ½-pints or pints. Remove jars; complete seals if necessary.

• **Adjustment for my altitude** _____ •

In plain No. 303 cans. Fill with hot mushrooms, leaving ½ inch of headroom. (Optional: add ½ teaspoon salt.) Add ½ teaspoon crystalline ascorbic acid. Fill to top with boiling water. Exhaust to 170 F/77 C (approx. 10 minutes); seal. Pressure-process at 10 pounds (240 F/116 C) for 45 minutes. Remove cans; cool quickly.

• **Adjustment for my altitude** _____ •

Okra (Gumbo)

Freezes well: but if you plan to use it cut up in soups and stews, it's probably handier canned.

GENERAL HANDLING

Pressure Canning only. Use Hot pack only. Use jars or plain cans.

Wash tender young pods; trim stems but don't cut off caps. Cover with boiling water and boil 1 minute; drain. Leave whole with cap, or cut in 1-inch pieces, discarding cap.

HOT PACK ONLY

In jars. Fill with hot okra, leaving ½ inch of headroom. (Optional: add ½ teaspoon salt to pints, 1 teaspoon salt to quarts.) Add boiling water, leaving ½ inch of headroom; adjust lids. Pressure-process at 10 pounds (240 F/116 C)—pints for 25 minutes, quarts for 40 minutes. Remove jars; complete seals if necessary.

> • **Adjustment for my altitude** _____ •

In plain cans. Fill with hot okra, leaving only ¼ inch of headroom. (Optional: add ½ teaspoon salt to No. 303 cans, 1 teaspoon salt to No. 401.) Fill to top with boiling water. Exhaust to 170 F/77 C (approx. 10 minutes); seal. Pressure-process at 10 pounds (240 F/116 C)—No. 303 cans for 30 minutes, No. 401 for 45 minutes. Remove cans; cool quickly.

> • **Adjustment for my altitude** _____ •

Onions, White

Onions that are properly cured and stored (see "Drying," Chapter 21) carry over so well that many cooks don't bother to can them—on top of which home-canned onions are apt to be dark in color and soft in texture. They are often canned in combinations; take into account that their processing time at 240 F/116 C is 25 minutes for pints, and 30 minutes for quarts.

Parsnips

This is probably the only vegetable that actually improves by wintering over in frozen ground—so why take the shine off it as the "first of spring" treat?

But if you can't keep them in a garden or a root cellar: wash, trim, scrape, and cut them in pieces, then proceed as for Carrots.

Peas, Black-eyed (Cowpeas, Black-eyed Beans)

Pressure Canning only. Use Hot pack; jars or C-enamel cans.

Shell and wash before packing. Take care *not* to shake or press down the peas when you pack the containers: they swell in the containers.

PREFERRED: HOT PACK

After shelling and washing, cover with boiling water, bring to a full, high boil. Drain, saving the blanching water for processing.

In jars. Pack hot, leaving 1¼ inches of headroom in pints, 1½ inches of headroom in quarts. (Optional: add ½ teaspoon salt to pints, 1 teaspoon salt to quarts.) Add boiling water (or blanching liquid), leaving ½ inch of headroom in either size of jar; adjust lids. Pressure-process at 10 pounds (240 F/116 C)—pints for 45 minutes, quarts for 55 minutes. Remove jars; complete seals if necessary.

• **Adjustment for my altitude** _____ •

In C-enamel cans. Pack hot, leaving ½ inch of headroom for either size of can. (Optional: add ½ teaspoon salt to No. 303 cans, 1 teaspoon salt to No. 401.) Add boiling water, leaving ¼ inch of headroom. Exhaust to 170 F/77 C (approx. 10 minutes); seal. Pressure-process at 10 pounds (240 F/116 C)—No. 303 cans for 40 minutes, No. 401 for 45 minutes. Remove cans; cool quickly.

• **Adjustment for my altitude** _____ •

Peas, Green

Shell and wash. *Don't shake or press down in packing:* peas swell.

Edible-pod and *Snow* varieties. Unstring young edible-pod ones, proceed to pack Hot and process like Beans, Green, earlier. If too mature, these peas may be shelled and treated as in Preferred: Hot Pack, below. Snow peas are much better frozen, but if you must can them, do them like Beans, Green, with Hot pack.

RAW PACK

In jars. Fill, leaving 1 inch of headroom. (Optional: add ½ teaspoon salt to pints, 1 teaspoon salt to quarts.) Add boiling water to within 1½ inches of the top of the jar (water will come well below the top of the peas); adjust lids. Pressure-process at 10 pounds (240 F/116 C)—40 minutes for both pints and quarts. Remove jars; complete seal if necessary.

• **Adjustment for my altitude** _____ •

In plain cans. Fill, leaving only ¼ inch of headroom. (Optional: add ½ teaspoon salt to No. 303 cans, 1 teaspoon salt to No. 401.) Fill to top with boiling water. Exhaust to 170 F/77 C (approx. 10 minutes); seal. Pressure-process at 10 pounds (240 F/116 C)—No. 303 cans for 40 minutes, No. 401 for 50 minutes. Remove cans; cool quickly.

• **Adjustment for my altitude** _____ •

PREFERRED: HOT PACK

Cover peas with boiling water, return to a full boil; save the water.

In jars. Fill loosely with hot drained peas, leaving 1 inch of headroom.

(Optional: add ½ teaspoon salt to pints, 1 teaspoon salt to quarts.) Add boiling water or blanching liquid, leaving 1 inch of headroom; adjust lids. Pressure-process at 10 pounds (240 F/116 C)—40 minutes for either pints or quarts. Remove jars; complete seals if necessary.

• **Adjustment for my altitude** _____ •

In plain cans. Fill loosely with hot peas, leaving only ¼ inch of headroom. (Optional: add ½ teaspoon salt to No. 303 cans, 1 teaspoon salt to No. 401.) Fill to brim with boiling water or blanching liquid. Exhaust to 170 F/77 C (approx. 10 minutes); seal. Pressure-process at 10 pounds (240 F/116 C)—No. 303 cans for 40 minutes, No. 401 for 50 minutes. Remove cans; cool quickly.

• **Adjustment for my altitude** _____ •

Peppers, Green (Bell, Sweet)

Pressure Canning only. Hot pack only. Use pint jars only.

Wash, remove stems, cores, and seeds. Cut in large pieces or leave whole. Put in boiling water and boil 3 minutes. Drain and pack. (If you like them peeled, take them from the boiling water, dunk in cold water to cool just enough for handling, and strip off the skins; pack.)

Acid is added to green peppers in order to can them safely.

HOT PACK ONLY

In pints only. Pack hot peppers flat, leaving 1 inch of headroom. Add 1 tablespoon bottled lemon juice, optional ½ teaspoon salt; pour in boiling water, leaving ½ inch of headroom, so that water covers the peppers. Adjust lids. Pressure-process at 10 pounds (240 F/116 C)—pints for 35 minutes.

• **Adjustment for my altitude** _____ •

Pimientos

Pressure Canning only. Use Hot pack only. Use ½-pint or pint jars.

Warning note: it used to be permissible to can tightly packed or densely textured food without added liquid: *no longer*. Pimientos are among the foods that now require added liquid.

Wash, cover with boiling water, and simmer until skins can be peeled off—4 to 5 minutes. Dunk in cold water so they can be handled, trim stems, blossom ends, and skin them like Green Peppers, above; pack hot, adding acid for safety as indicated.

HOT PACK ONLY, IN JARS ONLY

Pack flat in clean, hot ½-pint or pint jars, leaving ¾ inch of headroom. Add 1½ teaspoons bottled lemon juice to ½-pints, 1 tablespoon to pints. (Optional: add ¼ teaspoon salt to ½-pint jars, ½ teaspoon salt to pints.) Pour in just enough boiling water to cover pimientos, leaving ½ inch of headroom in both ½-pints and pints. Adjust lids. Pressure-process at 10 pounds (240 F/116 C)—35 minutes for either ½-pints or pints. Remove jars; complete seals if necessary.

· **Adjustment for my altitude** _____ ·

Potatoes, Sweet (and Yams)

Updated research says that these, too, must be canned with some liquid added, to let heat penetrate safely. For glazing: remove pieces whole from the container, pat off moisture, and finish drying them on foil in a slow oven for a few minutes, turning once; then glaze.

GENERAL HANDLING

Pressure Canning only. Hot pack only; Wet pack only; in wide-mouth jars.

Sort for size, wash; boil or steam until only half cooked, and the skins come off easily—20 minutes or so. Dunk in cold water so they can be handled, slip off skins, cut away blemishes. Cut large potatoes in pieces lengthwise.

HOT PACK, WET ONLY

In jars. Pack loosely; upright if you'll glaze them later. Leave 1 inch of headroom in both pints and quarts. (Optional: add ½ teaspoon salt to pints, 1 teaspoon salt to quarts.) Add boiling water or Medium Syrup (see Liquids for Canning Fruits in Chapter 7), leaving 1 inch of headroom; adjust lids. Pressure-process at 10 pounds (240 F/116 C)—pints 65 minutes, quarts 90 minutes. Remove jars; complete seals if necessary.

· **Adjustment for my altitude** _____ ·

Potatoes, White ("Irish")

These potatoes don't home-freeze at all well (unless they're partially precooked in a combination dish)—so cold-store them (see "Root-Cellaring," Chapter 22). But it's possible for them to be too immature to store without spoiling—so you can them. Delicate tiny, new potatoes can well, and are good served hot with parsley butter or creamed.

GENERAL HANDLING

Pressure Canning only. Use Hot Pack. Use jars or plain cans. If around 1 to 1½ inches in diameter, they may be canned whole; dice the larger ones.

Wash and scrape just-dug new potatoes, removing all blemishes. (If you're dicing them, prevent darkening during preparation by dropping the dice in a solution of 1 teaspoon salt for each 1 quart of cold water.) Drain before packing by either method.

Cover clean, scraped *whole* potatoes with boiling water, boil 10 minutes; drain. Drain anti-discoloration solution off *diced* potatoes; cover them with boiling fresh water, boil 2 minutes; drain.

In jars. With whole or diced potatoes, leave ½ inch of headroom. (Optional: add ½ teaspoon salt to pints, 1 teaspoon salt to quarts.) Add boiling water, leaving ½ inch of headroom; adjust lids. Pressure-process at 10 pounds (240 F/116 C)—*whole,* pints for 30 minutes, quarts for 40 minutes; *diced,* pints for 35 minutes, quarts for 40 minutes. Remove jars; complete seals if necessary.

> • **Adjustment for my altitude** _____ •

In plain cans. Fill with whole or diced potatoes, leaving only ¼ inch of headroom. (Optional: add ½ teaspoon salt to No. 303 cans, 1 teaspoon salt to No. 401.) Fill to top with boiling water. Exhaust to 170 F/77 C (approx. 10 minutes); seal. Pressure-process at 10 pounds (240 F/116 C)—*whole* or *diced*—No. 303 cans for 30 minutes, No. 401 for 50 minutes. Remove cans; cool quickly.

> • **Adjustment for my altitude** _____ •

Pumpkin (and Winter Squash)

Pressure Canning only. Hot pack only. Use jars or R-enamel cans. *Cube it only:* heat transfer fails in strained foods of this type.

Dry-fleshed pumpkin (or winter squash) is best for canning: test it with your thumbnail—it's dry enough if your nail won't cut the surface skin easily.

Wash, cut in manageable hunks, pare, and remove seeds; cut in 1-inch cubes. Cover with water and boil 2 minutes. Drain, reserving hot liquid, and pack hot.

In jars. Fill with hot cubes, leaving ½ inch of headroom. (Optional: add ½ teaspoon salt to pints, 1 teaspoon salt to quarts.) Add boiling cooking water, leaving ½ inch of headroom; adjust lids. Pressure-process at 10 pounds (240 F/116 C)—pints for 55 minutes, quarts for 90 minutes. Remove jars; complete seals if necessary.

> • **Adjustment for my altitude** _____ •

In R-enamel cans. Fill, leaving only ¼ inch of headroom. (Optional: add ½ teaspoon salt to No. 303 cans, 1 teaspoon salt to No. 401.) Fill to brim with boiling cooking water. Exhaust to 170 F/77 C (approx. 10 minutes); seal. Pressure-process at 10 pounds (240 F/116 C)—No. 303 cans for 75 minutes, No. 401 for 115 minutes. Remove cans; cool quickly.

> • **Adjustment for my altitude** _____ •

Rutabagas (and White Turnips)

Root-cellaring is best, but rutabagas are good to ferment like Sauerkraut (see how in "Curing," Chapter 20)—in which case they are canned like Sauerkraut, below.

Salsify (Oyster Plant)

Like parsnips and horseradish, this delicately flavored, old-fashioned vegetable is able to winter in the ground, and it may be root-cellared. With a little extra attention so it won't discolor, it cans well.

GENERAL HANDLING

Pressure Canning only. Use Hot pack only. Use jars or C-enamel cans.

Its milky juice turns rather rusty when it hits the air, and you may prevent discoloration of the vegetable by either one of two ways: (1) scrub roots well, scrape as for carrots, and slice, dropping each slice immediately in a solution of 2 tablespoons vinegar and 2 tablespoons salt for each 1 gallon of cold water; rinse well, cover quickly with boiling water, boil for 2 minutes; pack hot. Or (2) scrub roots; in a solution of 1 tablespoon vinegar to each 1 quart of water, boil whole until skins come off easily— 10 to 15 minutes; rinse well in cold water, skin, leave whole or slice; pack Hot.

HOT PACK ONLY

In jars. Fill, leaving 1 inch of headroom. Add fresh boiling water, leaving 1 inch of headroom; adjust lids. Pressure-process at 10 pounds (240 F/ 116 C)—pints for 35 minutes, quarts for 40 minutes. Remove jars; complete seals if necessary.

• **Adjustment for my altitude** _____ •

In C-enamel cans. Fill, leaving ½ inch of headroom; add fresh boiling water to brim. Exhaust to 170 F/77 C (approx. 10 minutes); seal. Pressure-process at 10 pounds (240 F/116 C)—35 minutes for both No. 303 and No. 401 cans. Remove cans; cool quickly.

• **Adjustment for my altitude** _____ •

Sauerkraut (Fermented Cabbage)

By all means can your sauerkraut—unless you are able to guarantee cool enough storage for the crock after it is fermented.

GENERAL HANDLING

Pasteurizing Hot–Water Bath processing is adequate for a food as acid as this. Hot pack only. Use jars or R-enamel cans.

HOT PACK ONLY

Heat to simmering—170–180 F (77–80 C)—*do not boil,* and pack as directed below. If you don't have enough sauerkraut juice, eke it out with a brine made of 1½ tablespoons salt for each 1 quart of water.

In jars. Fill clean, hot jars, leaving ½ inch of headroom. Add hot juice (or hot brine, above), leaving ½ inch of headroom; adjust lids. Process in a simmering Hot (180–190 F)–Water Bath—pints for 15 minutes, quarts for 25. Remove jars; complete seals if necessary.

• **Adjustment for my altitude** _____ •

In R-enamel cans. Pack hot, leaving only ½ inch of headroom. Fill to the brim with hot juice or brine (again, as above). Exhaust to 170 F/77 C if needed; seal. Process in a simmering bath—25 minutes for either No. 303 or No. 401 cans. Remove cans; cool quickly.

• **Adjustment for my altitude** _____ •

Soybeans

Although not particularly interesting by themselves, soybeans are a splendid *natural* high-protein addition—aside from being an economical "stretcher"—for ground meats, stews, chowders, casseroles, etc. They are good keepers when dried and they also freeze well.

HOT PACK ONLY

Prepare and process like Lima Beans except *increase the processing time* at 10 pounds (240 F/116 C) to *55 minutes* for pint jars, *65 minutes* for quart jars.

Spinach

See Greens.

Squash—Chayote, Summer, Zucchini

Pressure Canning only. Use Hot pack. Use jars or plain cans.

Wash, trim ends, but do not peel. Cut in ½-inch slices; halve or quarter the slices to make the pieces uniform.

HOT PACK ONLY

Cover with boiling water, bring to a boil. Drain, saving the hot cooking liquid for processing.

In jars. Pack hot squash loosely, leaving ½ inch of headroom. Top with cooking water with ½ inch headroom; add optional salt. Pressure-process pints for 30 minutes, quarts for 40 minutes. Remove jars; complete seals if necessary.

• **Adjustment for my altitude** _____ •

In plain cans. Pack hot squash loosely, leaving ¼ inch of headroom; add optional salt, and top with cooking water. Exhaust; pressure-processing No. 303 cans for 30 minutes and No. 401 cans for 35 minutes. Remove cans; cool quickly.

• **Adjustment for my altitude** _____ •

Squash, Winter

See Pumpkin.

Tomatoes

See Chapter 8.

Turnips, White

See Rutabagas.

10
CANNING MEATS

Beef, pork, lamb, and chicken are the most popular meats for canning. Domestic rabbits and small-game animals are canned like Poultry. Choose only good meat for canning, and handle it quickly and with total cleanliness, because bacteria grow at a frightening rate in meats and poultry if given half a chance. Any meat picks up bacteria, so don't keep it waiting at room temperature until you can handle it. If you have a large amount to do, temporarily store the part you're not working on in the refrigerator or a meat cooler (32–38 F/Zero–3 C). Can first thing *tomorrow* what you could not can today.

Process all canned animals and birds only in a Pressure Canner at 10 or 15 pounds (240 F/116 C or 250 F/121 C) to destroy bacteria—including the *spores of C. botulinum*. The 15 psig processing is specially recommended to offset the effects of high altitude. By all means see the rules-of-thumb in the Canning at High Altitude subsection in Chapter 3.

Equipment and Its Care

A Pressure Canner is essential. And you'll need a pencil-shaped glass food thermometer *for exhausting jars* as well as cans.

Containers may be jars—we recommend the modern straight-sided ones, because they're easier to get the meat out of—or plain cans (meat sometimes makes the coating flake off the interiors of enameled cans: harmless, but unattractive).

Use good-sized sharp knives, including a 3- to 4-inch boning knife.

Wooden cutting-boards and surfaces can be very handsome and they are easier on knives than harder surfaces are, but *PFB* uses a heavy-duty skidproof acrylic panel whose side supports of metal can be extended to

146

straddle a sink. Easy to work on, much easier to keep reasonably free of the sort of bacteria that are likely to be harbored in the pores of wood.

If you plan to do large quantities of meat, you should have a high-sided roasting pan and a very big kettle (as large as your Boiling–Water Bath kettle).

All tools and utensils must be scrubbed in hot soapy water and rinsed well with fresh boiling water before each use.

To control bacteria, cutting-boards and wooden working surfaces must be scrubbed hard in hot soapy water *both before and after* you handle meat on them, and must be disinfected with a solution of ¼ cup chlorine bleach to 4 cups of lukewarm water; leave it on for several minutes, then rinse off with fresh water. Or leave on the surface for 15 minutes a solution of 2 tablespoons chlorine bleach in 4 cups of water; rinse with boiling water.

Signs of Spoilage

- Broken seal.
- Bulging cans.
- Seepage around the seal.
- Mold, even a speck, in the contents or around the seal or on the underside of the lid.
- Gassiness (small bubbles) in the contents.
- Spurting liquid, pressure from inside the container as it is opened.
- Cloudy or yeasty liquid.
- Unnatural or unpleasant odor.

The fats in meat, plus its high protein content, make it more susceptible to spoilage than vegetables are. Take this warning to heart:

Before tasting home-canned meat and poultry, and their broths: you must be unshakably certain that your Pressure Canner was operated correctly—pressure gauge accurate and deadweight/weighted gauge signaling properly—and that requirements for times and Correcting for Altitude in Chapter 3 were followed.

Unless you are sure that these safeguards were observed, a margin of protection is added by boiling the food hard for 20 minutes to destroy possible hidden toxins, stirring to distribute the heat and adding water if necessary. If the food foams unduly or smells bad during boiling, destroy it completely so it cannot be eaten by people or animals. (See also A Quiet Method of Destroying Botulism Toxin in Chapter 11.)

And about Salt

Salt is merely an optional seasoning in canned meat and poultry; not being an essential ingredient, it can be omitted.

If you use salt, your regular table salt will do.

Salt substitutes for special diets can leave an unwanted aftertaste when used in canning: wait to add these seasonings when the food is heated for serving.

Where to Get Your Meat

More people each year are either raising food animals themselves, or are bespeaking part of a beef or a pig or a lamb grown by a country neighbor, because they hope to save money over what they would pay at supermarket special sales. Whether they do indeed get a bargain is not always predictable, and depends upon how the animal was weighed, the costs for slaughtering and butchering—and if the meat was cut with professional economy and minimal waste. In addition to knowing the grower, therefore, it is equally important to be certain that the meat was handled with adequate refrigeration and scrupulous cleanliness at every stage. *C. botulinum* can be introduced by careless treatment. More often, though, it's the Salmonellae that contaminate meat, with the meat of poultry being the most common source for this nasty form of gastro-intestinal illness.

All in all, your safest buy is from a meat-seller whose goods are purchased under the regulations that now require all meat sold in, as well as between, states to be inspected, and the premises where it was dealt with checked regularly. As for putting it by, weigh the benefits of freezing and Pressure Canning both for time spent and for ultimate convenience.

CONVERSIONS FOR CANNING MEAT

Do look at the conversions for metrics, with workable roundings-off, and for altitude—both in Chapter 3—and apply them.

Warning about Game

Any wild game may be diseased or carry parasites. Bear, for example, often have trichinosis, and a number of cases are reported to the Center for Disease Control each year; and the University of Georgia, Athens, advises handling rabbit with utensils and rubber gloves as safeguards against tularemia.

So if you are a successful hunter, or have been given a present of wild meat *do not eat it cooked Rare.* Roasted or grilled until Well Done, or cooked long in a stew: these are the insurance.

Pressure Canning at 10 pounds (240 F/116 C) at the sea-level zone or at 15 pounds (250 F/121 C) at higher altitudes—and for the length of time required for large pieces, small pieces, or ground meat, respectively—is the only safe way for canning game.

The State Biologist in the headquarters of your Fish and Game Department will tell you of any disease problems with game in your area. A quicker source of such information is likely to be one of your county game wardens.

STEPS IN CANNING MEAT

Cuts of Meat from Large Animals

Know what a whole, half, or quarter of an animal will yield in the way of cuts before you buy one. There is 20 to 25 percent waste to start with, and usually there will be more pounds of stewing and/or ground meat than pounds of steaks and roasts and chops. Various USDA publications are helpful about actual yield.

Cutting the Meat

Have the carcass cut in serving pieces by a professional meat-cutter to get the most out of your investment.

If you wish to tackle the job yourself, ask your County Extension Service office for a copy of *Farmers Bulletin No. 2209, Slaughtering, Cutting and Processing Beef on the Farm,* and for a copy of *Farmers Bulletin No. 2265, Slaughtering, Cutting and Processing Pork on the Farm.*

Audrey Alley Gorton's *The Venison Book, How to Dress, Cut up and Cook Your Deer* is an ongoing helpful guide to field-dressing and butchering large game.

Canning Large Pieces of Meat

Pressure Canning only. *PFB* is leery of Raw pack for any meat, and therefore offers no processing times for it. Nevertheless, if you insist on Raw pack for meat, exhaust each open jar (and cans as a matter of course) to 170 F/ 77 C before continuing with the capping procedure and processing.

Prime cuts of beef, pork, lamb, veal and large game are best canned in large pieces. The less choice parts are good for stews and ground meats.

Major Cuts of Beef (or Elk or Moose)

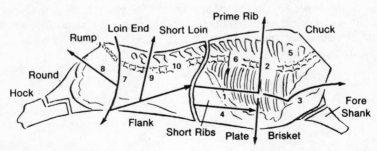

(*Drawing by Norman Rogers*)

Major Cuts of Pork

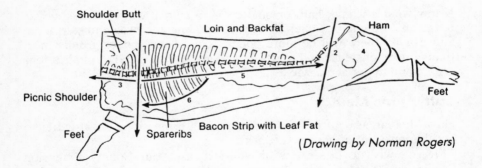

(*Drawing by Norman Rogers*)

Major Cuts of Lamb (or Goat)

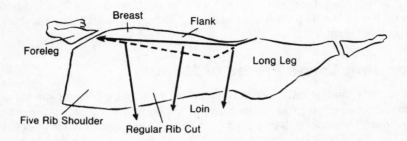

To cut up lamb or goat, the carcass is handled whole, not split lengthwise like the larger beef, pork, elk, etc. (*Drawing by Norman Rogers*)

Packing and Processing Large Cuts

PFB is indebted to the University of Minnesota for underscoring a long-harbored worry of ours about Raw pack for meats. Their *Extension Bulletin 413* (revised 1980) states that research done at the university shows that meat and poultry are *satisfactorily canned ONLY by Hot pack* (italics ours).

HOT PACK (PRECOOKED)

Wipe raw meat with a clean damp cloth. Remove bones *and all surface fat* (fat in canned meat is likely to shorten its storage life, and fat is a No. 1 seal-spoiler).

Put large cut-to-measure pieces of boned, de-fatted meat in a large, shallow pan. Add just enough water to keep meat from sticking; cover, and cook slowly on top of the stove or in a 350 F/177 C oven until the meat is Medium done, turning it now and then so it precooks evenly.

In straight-sided jars. Pack hot meat loosely, leaving 1 inch of headroom. (Optional: add ½ teaspoon salt to pints, 1 teaspoon salt to quarts.) Add boiling meat juice (extended with boiling water if necessary), leaving 1 inch of headroom. Wipe jar rims carefully to remove any fat. Adjust lids. Pressure-process at 10 pounds (240 F/116 C)—pints for 75 minutes, quarts for 90 minutes. Remove jars; complete seals if necessary.

• **Adjustment for my altitude** _____ •

In plain cans. Pack hot meat loosely, leaving ½ inch of headroom. (Optional: add ½ teaspoon salt to No. 303 cans, ¾ teaspoon salt to No. 401 cans.) Fill cans to the top with boiling meat juice (extended with boiling water if necessary), leaving *no* headroom. Wipe can rims carefully to remove any fat; seal (no exhausting is necessary, because long precooking has driven air from the meat). Pressure-process at 10 pounds (240 F/116 C)—No. 303 cans for 75 minutes, No. 401 cans for 90 minutes. Remove cans; cool quickly.

• **Adjustment for my altitude** _____ •

Note: you may also roast large pieces of meat as for the table until it is Medium done; pack as above (extending pan juices with boiling broth or water if necessary), and Pressure-process at 10 pounds (240 F/116 C) for the full time required above.

Canning Small Pieces of Meat

Use the less choice parts of the animal for future use in stews, main-dish pies.

Pressure Canning only, Hot (precooked) pack. Use straight-sided jars or plain cans.

Remove all surface fat from clean meat. Cut the meat off any bones. As you cut in stewing-size pieces, remove any interior bits of fat; cut away any tough muscle-sheath.

Put meat in a large, shallow pan, with just enough water to keep it from sticking; cover. Precook until Medium done. Stewing-size pieces take less tending if you do them in a 350 F/77 C oven; but you can also precook them to Medium on top of the stove, turning or stirring them from time to time.

If you want to brown the meat before canning it, do *not* dredge in flour first—just put it under a hot broiler long enough to brown it on all sides. Slosh a little water around the pan to pick up any juice, and save the water to use in precooking the meat, as above.

In straight-sided jars. Pack hot meat loosely, leaving 1 inch of headroom. (Optional: add ½ teaspoon salt to pints, 1 teaspoon salt to quarts.) Add boiling meat juice (extended with boiling water if necessary), leaving 1 inch of headroom. Wipe jar rims carefully to remove any fat. Adjust lids. Pressure-process at 10 pounds (240 F/116 C)—pints for 75 minutes, quarts for 90 minutes. Remove jars; complete seals if necessary.

• **Adjustment for my altitude** _____ •

In plain cans. Pack hot meat loosely, leaving ½ inch of headroom. (Optional: add ½ teaspoon salt to No. 303 cans, ¾ teaspoon salt to No. 401 cans.) Fill cans to the top with boiling meat juice (extended if necessary), leaving *no* headroom. Wipe can rims carefully to remove any fat. Seal. Pressure-process at 10 pounds (240 F/116 C)—No. 303 cans for 65 minutes, No. 401 cans for 90 minutes. Remove cans; cool quickly.

• **Adjustment for my altitude** _____ •

See Chapter 12, "Canning Convenience Foods," for dealing with ground meat, pork sausage, bologna-style sausage, corned beef, and soup stock (including broth).

Canning Variety Meats

Most of the variety meats—liver, heart, tongue, and sweetbreads and brains—are best cooked and eaten right away. Certainly sweetbreads and brains, the most delicate foods of the lot, should be served when they are fresh. For liver and kidneys, freezing is recommended. Tongue may be canned satisfactorily, as well as frozen.

Canning Beef Tongue

The following procedure of course may be used for smaller tongues.

Soak the tongue in cold water for several hours, scrubbing it thoroughly and changing the water twice. Put it in a deep kettle, cover with fresh water, and bring to boiling. Skim off the foam well, then salt the water lightly; cover, and cook slowly until Medium done—*not tender* in the thickest part. Remove from kettle and plunge into cold water for a moment; peel off skin and trim off remaining gristle, etc., from the root.

Cut in container-size pieces, and pack Hot as for Large Pieces of Meat, or slice evenly and pack Hot as for Small Pieces of Meat. Pressure-process at 10 pounds (240 F/116 C) for the times required above.

Canning Frozen Meat

If you're ever faced with a freezing emergency, you can salvage frozen meat by canning it—provided it is good quality to start with and was correctly frozen and stored.

First, thaw it slowly in the refrigerator below 40 F/4 C; or microwave according to maker's instructions. Then handle it as if it were fresh, using Pressure Canning only, Hot pack only, and the processing times that apply for Canning Large/Small Pieces of Meat, above.

CANNING POULTRY AND SMALL GAME

The following—which use chicken as the example for simplicity's sake—may be applied to canning domestic rabbits, wild birds, and other small game, as well as canning other domestic poultry such as ducks, guinea hens, geese, turkeys, etc. *All these animals may be canned the same way:* general preparation (with specific exceptions as they come along), packing, and processing are the same for all.

If you refrigerate adequately, prevent contamination during handling, work quickly, and don't try short cuts in packing and processing, poultry and small game may be canned satisfactorily.

Freezing of course is easier.

Use Pressure Canning only, and Hot pack. Use straight-sided jars or plain cans.

Where Canning Poultry, etc., Is Different

Unlike the procedures given in the preceding section for packing meat, the methods that follow include canning with the bone left in.

Also, the skin on large pieces of birds—breast, thighs, drumsticks—is left on: processing at 240 F/116 C compacts the surface of meat next to the sides of the container (making a pressure mark), so the skin you leave on acts as a cushion. Breast meat is skinned if packed in the center of jars/cans (surrounded by skin-on pieces that touch the containers' sides); so skin as you pack.

To Dress Poultry

Dressing involves two steps: (1) removing feathers by plucking or removing the fur pelt by skinning and (2) drawing, which is removing the internal organs in one intact mass. Domestic birds are plucked before being drawn because they are handled for food immediately after they're killed.

However, a hunter *field-dresses* his kill by removing the innards on the spot, since they spoil a great deal more quickly than muscle tissue does; and he waits to skin it until after he's home. Immediate drawing therefore reduces the chance of spoiling the rest of the meat en route home, and is especially necessary with mammals. Game birds may be held for several hours before being drawn; but if they cannot be taken home for handling within half a day, they too should be drawn in the field, and plucked later.

Plucking a Chicken

Pluck feathers from the still warm, fresh-killed and bled chicken, being careful to get all the pinfeathers. Hold the bird by its feet and pull the feathers toward the head, in the opposite direction to the way they lie naturally. Scalding the whole bird is not necessary, but if the feathers are resisting enough so you're afraid of tearing the skin, you can spot-scald: lift the chicken by the feet with its head dangling, and pour nearly boiling water into the base of the feathers, where it will be trapped momentarily against the skin.

Dry the bird and singe off the hairs. Wipe it clean.

Drawing a Chicken

Cut off the head of the fresh-killed and plucked bird if it is still on; remove feet at the "ankle" joint just below the drumstick. Cut out the oil sac at the top of the tail (it would flavor the meat unpleasantly, so don't break it).

Lay the chicken on its back, feet toward you. Using a sharp knife, cut a circle around the vent (anus), so it can be removed intact with the internal organs still attached. Cut deeply enough to free it, and be careful not to cut into the intestine that leads to it.

Insert the tip of the knife, with cutting edge upward, at the top of the circle around the vent, and cut through the thin ventral wall toward the bottom of the breast bone, making the slit long enough so you can draw the innards out through it with vent attached.

Reach clear to the front of the body cavity and gently pull out the mass of organs. Separate and save the heart, liver, and gizzard.

Next, turn the chicken over and slit the skin lengthwise at the back of the neck—if you slit down the *front* of the neck you may cut into the crop; push away the skin and remove the crop and windpipe.

Cut the neck bone off close to the body. Wash the whole bird.

From the liver, cut away the green gall sac, roots and all; and be mighty careful not to break it, because gall ruins the flavor of any meat it touches. Split one side of the gizzard, cutting until you see the tough inner lining. Press the gizzard open and peel the lining away and discard it and its contents. Trim the heart.

Promptly refrigerate each dressed chicken, either whole or cut up, until you are ready to can it, within 24 hours.

Cutting Up Poultry

Lay the dressed, clean bird on its back and, using a sharp boning knife, disjoint the legs and wings from the body. Separate thighs from drumsticks at the "knee" joint. If the bird is very large—like a turkey or goose— separate the wing at its two joints, saving the two upper meaty sections for canning with bone in. (Very small birds, such as grouse, etc., may do best merely quartered with poultry shears.)

Turn the chicken crossways to you, hold the bottom of the breast section, and cut under it, through the ribs, until you reach the backbone;

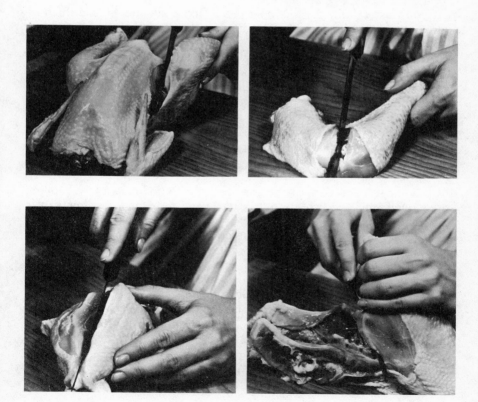

separate it from the backbone by cutting through the ribs. Poultry shears or heavy kitchen shears will be handy for use on the stronger bones at the shoulder joint.

Bone the breast meat by cutting down one side of the breast bone and easing the white meat off in a large piece; repeat on the other side.

Remove lumps of fat and any bits of broken bone from each piece of chicken and wipe it with a clean damp cloth.

Drawing, Skinning, Cutting Up Rabbits

Lay the fresh-killed rabbit on its back, and proceed to draw it as if it were a chicken: cut around the vent carefully; make a slit in the abdominal wall, reach in and pull out the innards with vent attached; save the liver and heart.

Cut off the feet. Working from the hind legs upward, work the rabbit out of its skin, easing the job with your knife where you need to. The head may be skinned, but chances are you'll prefer to remove it with the skin when you reach it. Cut away the gall sac from the liver; trim the heart. Wash the dressed rabbit and pat it dry. *Refrigerate each dressed rabbit until you are ready to can it*.

If you raise rabbits for the table you'll find it simpler to skin them as soon as they are killed and before drawing them.

Because of its anatomy, think of a rabbit as making two forequarters, two hind quarters, and a saddle. Split the saddle down the backbone as you would split the breast of a chicken, boning it if you like. The size of the rabbit has a lot to do with whether you joint the quarters.

Canning Poultry, etc., Bone In

HOT PACK (PRECOOKED), BONE IN

Put raw meaty pieces in a large pan, cover with boiling water or boiling unseasoned (chicken) broth. Cover the pan and cook the meat slowly over moderate heat on top of the stove or in a 350 F / 177 C oven until Medium done. Pack hot meat with breasts preferably in the center (so skin them), surrounded by legs and thighs (unskinned, because they touch the sides of the containers).

In straight-sided jars. Pack hot meat in loosely, leaving 1 inch of headroom. (Optional: add ½ teaspoon salt to pints, 1 teaspoon salt to quarts.) Cover with boiling unseasoned cooking liquid, leaving 1 inch of headroom. Wipe jar rims carefully. Adjust lids. Pressure-process at 10 pounds (240 F / 116 C)—pints for 65 minutes, quarts for 75 minutes. Remove jars; complete seals if necessary.

• **Adjustment for my altitude** _____ •

In plain cans. Pack hot meat in loosely, unskinned if they touch the side of the can; leave ½ inch of headroom. (Optional: add ½ teaspoon salt to No. 303 cans, ¾ teaspoon salt to No. 401 cans.) Add boiling unseasoned cooking water to the top of the cans, leaving *no* headroom. Wipe can rims carefully to remove any fat. Seal. Pressure-process at 10 pounds (240 F/ 116 C)—No. 303 cans for 65 minutes, No. 401 cans for 75 minutes. Remove cans; cool quickly.

   • **Adjustment for my altitude** _____ •

Canning Poultry, etc., without Bones

HOT PACK (PRECOOKED), WITHOUT BONES

Remove bones from good meaty pieces, but leave skin on all pieces of poultry until you're filling the containers: then skin the ones in the center of the pack (usually breasts), leaving skin on the ones that touch the sides of jars/cans (usually legs).

Cover boned meat with boiling water or unseasoned broth and cook slowly on stove or in oven until Medium done as for Hot pack, Bone In, above.

In straight-sided jars. Pack hot boned meat loosely (with outside pieces unskinned). Leave 1 inch of headroom. (Optional: add ½ teaspoon salt to pints, 1 teaspoon to quarts.) Add boiling unseasoned broth, leaving 1 inch of headroom. Wipe jar rims carefully. Adjust lids. Pressure-process at 10 pounds (240 F/116 C)—pints for 75 minutes, quarts for 90 minutes. Remove jars; complete seals if necessary.

   • **Adjustment for my altitude** _____ •

In plain cans. Pack hot boned meat loosely, with outside pieces un-skinned. Leave ½ inch of headroom. (Optional: add ½ teaspoon salt to No. 303 cans, ¾ teaspoon salt to No. 401 cans.) Add boiling unseasoned broth to the top of the cans, leaving *no* headroom. Wipe can rims carefully. Seal. Pressure-process at 10 pounds (240 F/116 C)—No. 303 cans for 75 minutes, No. 401 cans for 90 minutes. Remove cans; cool quickly.

   • **Adjustment for my altitude** _____ •

Canning Poultry Giblets

Giblets are more useful if they are canned together, rather than spread out among the cans of meat (use them chopped in gravies, meat sauces, or spreads, as fillings for main-dish pies, on rice as a supper dish, etc.). Furthermore, the livers are better and handier if they are canned separately; and being so tender, they need much shorter precooking before Hot packing and processing.

Pressure Canning only. Use Hot pack only. Use straight-sided pint jars or No. 303 plain cans.

PREPARE AND PACK HOT (PRECOOKED) ONLY

Cut clean gizzards in half, trimming off the gristle; cut smaller if necessary. Remove tops of hearts where the blood vessels come in; halve hearts if they're very large. Cover gizzards and hearts with hot water or hot unseasoned broth and cook until Medium done.

Remove all fat from the livers; cut away gall sac and connecting tissue between the lobes. Cover livers with hot water or hot unseasoned broth and cook gently until firm and Medium done; stir occasionally to prevent sticking.

Pack gizzards and hearts together, pack livers separately.

In straight-sided pint jars. Fill with hot gizzards and hearts, or hot livers, leaving 1 inch of headroom. (Optional: add ½ teaspoon salt.) Add boiling cooking liquid, leaving 1 inch of headroom. Wipe jar rims carefully. Adjust lids. Pressure-process at 10 pounds (240 F/116 C)—pint jars of either livers, or gizzards and hearts for 75 minutes. Remove jars; complete seals if necessary.

• **Adjustment for my altitude** _____ •

In No. 303 plain cans. Fill with hot gizzards and hearts, or hot livers, leaving ½ inch of headroom. (Optional: add ½ teaspoon salt.) Add boiling cooking liquid to the top of the cans, leaving *no* headroom. Wipe can rims carefully. Seal. Pressure-process at 10 pounds (240 F/116 C)—No. 303 cans of either gizzards and hearts, or livers for 75 minutes. Remove cans; cool quickly.

• **Adjustment for my altitude** _____ •

Canning Frozen Poultry, Etc.

In case your freezer conks out, or you have a windfall of frozen poultry, domestic rabbits, or small game, you may can it—*if:*

(1) It is good quality and was properly frozen (see the table Freezer Storage Life of Various Foods in Chapter 13), and (2) it is thawed slowly in the refrigerator below 40 F/4 C.

Then treat it as if it were fresh, using Pressure Canning only, Hot pack only, straight-sided jars or plain cans. Follow preparation and processing under Bone In/Boned, above.

11
CANNING SEAFOOD

Finned or shell-bearing creatures taken from salt or fresh water are right up among the front-runners in the botulism sweepstakes. Add to this that in general they are the most perishable of all fresh foods and have great density of texture, and you see why fish and shellfish require faultless handling and longer Pressure-processing than do other foods that are canned at home.

So why can them? Why indeed, when proper freezing is an all-round better, and safer, means of putting them by?—or when even salt-curing followed by drying and cold storage is, in the regions that practice this twofold method, less of a hazard?

But maybe you're faced with a surfeit of fresh seafood, and either freezing or curing-and-drying the excess is impossible. If such a bonanza is a repeated occurrence, you could plan ahead and organize a community kitchen, complete with good equipment and a skilled director in charge, as described in Chapter 3. If it's a once-in-a-lifetime event, though, and comes without warning (like the beach full of lobsters cast up by a hurricane that hit the Maine coast)—well, go ahead and can what you and your neighbors aren't able to eat, or swap for staples, or give to a public-service group near by.

The following procedures are for canning, without frills, some representative varieties of fresh fish and shellfish.

CONVERSIONS FOR CANNING SEAFOOD

Do look at the conversions for metrics, with workable roundings-off, and for altitude—both in Chapter 3—and apply them.

Equipment for Canning Seafood

You cannot can fish or shellfish at home without an honest-to-goodness Pressure Canner. Not a pressure *saucepan:* the leading manufacturers do not recommend use of these small utensils for the extra long processing required for such low-acid food as fish.

And, just as for canning meats and poultry, you'll need a pencil-shaped glass thermometer because you'll be exhausting your jars here too.

In addition to the standard kitchen furnishings and the sharp good knives and cutting-boards you used for preparing meat, and the where-withal to keep everything properly sanitary, you need:

Modern straight-sided ½-pint and pint canning jars in perfect condition, their two-piece screwband lids ditto.
Inexpensive styrofoam chest(s) in which to hold fish on ice.
Hose or sprayer connected to your sink's drinking-water tap, for washing fish or shellfish.
Fish-scaler.
Small wire brush for scrubbing shells, etc.
Big crocks or enameled vessels for soaking fish in salted water to remove blood from the tissues (dishpans will do).
Large enameled kettle for boiling shellfish, treating crabmeat, etc., to a mild salt/acid "blanch," or steaming open clams (your B-W Bath kettle is fine).
Special blunt knife for opening clams (if you shuck them raw).
Wire basket or rack for steaming.
Shallow pans with perforated bottoms that will fit inside your Pressure Canner (for the so-called "tuna pack").

General Handling—Plus Reasons Why

From the sources mentioned earlier, and others, we have compiled the following stipulations, which must be followed by everyone who undertakes to can fish or shellfish at home.

All seafoods must be processed in a regular Pressure Canner for the *full long time* required, and *at the pressure given* (which of course is corrected for altitudes higher than 1000 feet). If the pressure drops below the recommended level at any time during the processing period, for safety's sake you must raise the pressure to the correct number of pounds, and start retiming as if you were starting the entire processing period from scratch.

Reason: the average natural acidity of seafood is so low that it flirts with Neutral on the *pH* scale (see Chapter 2). Therefore, constant pressure for the full time is needed if enough heat is to penetrate the dense pack, thus destroying dangerous bacteria.

Use only modern jars, manufactured for home-canning under pressure, that have two-piece screwband lids. And use only ½-pint or pint jars (preferably straight-sided ones so the contents can slip out easily).

Reasons: for seafoods—or all home-canned products—it doesn't make sense to use makeshifts, or old-style jars and closures, or any other containers that have not been tested under the conditions required for safety by independent food scientists, and O.K.'d. As for the two-piece screwband lids, the flat metal disks indicate, by having snapped down to be concave, that you have obtained a proper seal. And finally, adequate processing cannot be assured for jars larger than 1 pint—or larger than ½-pint for certain fish or shellfish.

(We recommend *against cans* for home-processing of seafoods: (1) the correct sizes—like the commercial ones—are different from the cans used for other foods in this book; and (2) especially with the meat of lobsters and crabs, parchment-paper liners are usually needed to make an attractive product.)

All home-canned fish must be exhausted to a minimum of 170 F/77 C at the center of the packed jar before it is Pressure-processed.

Reason: before actual processing begins, we must drive air from the tissues of raw fish as well as from the pack to help ensure the seal and to prevent unwanted shrinkage of the food during processing. Fish in the so-called "tuna pack"—i.e., fully precooked—are cooled completely before packing; these packs also must be exhausted. (The completely precooked, and picked, meat of lobsters, crab, shrimp, and clams is not exhausted when packed in small jars.)

Exhausting jars of fish is done best in the Pressure Canner at Zero pounds. Place filled jars on the rack in the bottom of the canner and pour hot water around them until it comes halfway up their sides. *Lay* the cover on and *leave the vent open.* Turn the heat up high, and when you hear the water boiling hard inside the canner and steam flows strongly in a steady stream from the vent—indicating that the temperature has reached 212 F/100 C inside (see Correcting for Altitude, in Chapter 3 if you're canning above 1000 feet)—when the steam flows strongly, start counting the exhaust time. It will take 10 to 20 minutes for the center of the filled jars to reach the desired minimum of 170 F/77 C, depending on the size of the jar and the size of the solidness of the fish pieces; always insert your pencil thermometer deep in a test jar to make sure.

Water used in cleaning seafoods and preparing them for packing *must be of drinking quality*—whether it's the running water for washing them (which is always done under a tap, or with a spray or hose), or the water in a brine or antidiscoloration solution, or the canning liquid that goes in the jar.

Reason: it's easy to introduce dangerous bacteria, including *C. Botulinum* itself, into the flesh by using polluted or contaminated water at any stage. Do not rinse fish in stream or lake water. Do not precook shellfish in seawater. If your household drinking water contains a lot of minerals, use bottled water at least for the canning liquid (iron, especially, reacts with the sulfur in the meat of shellfish and causes the product to darken).

All seafoods to be canned must be as fresh as is possible. Fish must be gutted as they're caught, and refrigerated or packed in ice immediately, to be kept cold until they are precooked or packed. Head shrimp immediately as they come from the water, refrigerate, or hold on ice. Keep lobsters, crabs, and clams alive and cool until you prepare them for packing.

Reason: it takes only a couple of hours at room temperature to make dead seafood unfit to can; and spoilage is hastened if intestines and body wastes are not removed.

The flesh of all dressed and cleaned fish and shelled shrimp is given a preliminary brining; lobsters and crabs are precooked in brine and well rinsed before shelling. (The picked meat of lobsters, crabs, and clams is given a further anti-darkening treatment in a mild acid "blanch" before packing.)

Reason: brining draws diffused blood from the tissues, and reduces the chance that white curds of coagulated protein will occur in the processed jars. (Brines must be made up just before use, and should be *used only once*.)

The day after the seafood has been canned, store the jars in a cool, dry, dark place.

Reason: storage that lets the jars freeze can also break the seals; storage over 50 F/10 C courts spoilage; damp storage rusts metal closures and endangers the seals. During the 24 hours between processing and storing, check all seals, clean, and label the jars.

Canned Seafood Troubles

Do NOT reprocess jars of seafood found to have poor seals during the 24 hours of grace between canning and storage. And even if the contents are decanted into fresh containers and done over from scratch, the result is likely to be unsatisfactory (all the more reason for taking care in the first place).

After jars are stored you must be super-critical in examining them for *external* signs of spoilage: broken seal (flat lid no longer concave)—seepage around the closure—gassiness in the contents—cloudy, yeasty liquid or sediment at the bottom of the jar—contents an unnatural color or texture. *If any of these signs are present, destroy the food so it cannot be eaten by people or animals, sterilize the container and closure by boiling, and discard the sterilized closure.*

Even when the seal seems good and none of the trouble symptoms just listed is apparent, these are the signs of spoilage when you open a *jar of seafood:* pressure inside the container (instead of a vacuum), or spurting contents—fermentation—sour, cheesy odor—soft, mushy contents. If any of these signs is present, *destroy the food and sterilize the container and closure, as above.*

Before tasting home-canned seafood: you must be unshakably certain that your Pressure Canner was operated correctly—pressure gauge accurate and deadweight/weighted gauge signaling properly—and that requirements for times and corrections for altitude were followed.

Boil the food hard for 20 minutes to destroy possible hidden toxins, stirring to distribute the heat and adding water if necessary. If the food foams unduly or smells bad during boiling, destroy it completely so it cannot be eaten by people or animals.

A Quiet Method of Destroying Botulism Toxin

The instructions in the special warning above are standard—standard because they have been proved to be (1) capable of disclosing the presence of spoilers other than/in addition to the toxin of *C. botulinum*—which by itself is not certain to create undue foaming or an unpleasant smell; and (2) destroying the botulinum toxin if it is present. This vigorous boiling of from 10 to 20 minutes (depending on the density of the pieces of food), stirring all the while, does not do great harm to the texture of, say, mixed vegetables; but it does make a jar of canned fish into a fairly sad ingredient for salmon mayonnaise on a buffet supper table.

Therefore, Dr. Margy Woodburn, microbiologist of the Department of Foods and Nutrition, Oregon State University, Corvallis, and Edward J. Schantz and Jennifer Rodriguez, both of the Food Research Institute, University of Wisconsin, Madison, performed a series of tests whereby toxins from several types of *C. botulinum* were introduced into glass jars of home-canned salmon; the jars were then incubated. Here is what the research team says to do to inactivate by heat the toxin(s) in home-canned fish:

Open the jar. If there are no signs of spoilage as noted in the warning above, insert the pencil (or meat) thermometer you use for exhausting jars so the tip is as near the center of the contents as you can judge; cover the jar with foil. Put the jar in an oven at 350 F/177 C for 30 to 40 minutes for either ½-pint or 1-pint jars—*or until the thermometer registers 185 F/ 85 C* (the time taken to reach the required temperature is not the important factor: the 185 F/85 C *is*).

Do not speed the cooling. Let the hot jar stand at room temperature for 30 minutes to complete the heat treatment. At the end of the 30 minutes, serve hot (or reheated if necessary), or refrigerate for use later.

Specific Seafood Products

Salmon and Shad

Pressure Canning only. Use Raw pack and exhaust. Use pint jars (preferably straight-sided) with two-piece screwband lids only.

Twenty-five pounds of round fish (i.e., whole and not dressed) will fill about twelve 1-pint jars.

Dress, scale, scrub perfectly fresh fish; cut away the thin belly flap. Using a jar laid on its side as a measure, cut the fish across the grain in jar-length pieces—and *not one whit longer* lest they interfere with the seal (the fish will shrink in the jar to leave headroom). Prepare a cold brine of $\frac{3}{4}$ cup pure pickling salt dissolved in 1 gallon of water, an amount of brine that will do 25 pounds of prepared fish. Use enameled or non-metal tubs; use brine only once. Weight the fish pieces down in the brine for 60 minutes to draw out diffused blood and firm the flesh. Drain the pieces for 10 minutes; do not rinse.

Fill the jars solidly and in effect just to the top—this means no more than $\frac{1}{16}$ inch below the sealing rim—packing the pieces upright, skin side next to the glass, and carefully inserting slimmer pieces to fill vertical gaps. Don't pack the pieces so tight that they spring back and hurt the seal.

Next, *half*-close the filled jars. To do this, place the flat lid on the sealing rim of the jar, and screw the band down *just until the band cannot be pulled up off the threads*. (Practice on an empty jar to get the feel of this half-closure.) Exhaust the jars in the Pressure Canner at Zero pounds (as described in General Handling above); after 15 minutes, open the canner, then open one jar and test the center of its contents with your pencil thermometer. By this time it should read at least 170 F/77 C; if not, give it 5 minutes more.

When jars are exhausted, lift the canner off heat and finish screwing the bands firmly tight as for any processing. Return the canner to heat, put on the lid and let steam vent in a strong, steady flow for 10 minutes before closing the petcock/vent and starting to time the processing period. The amount of very hot water remaining in the canner after exhausting the jars should be ample for Pressure-processing.

Pressure-process at 10 pounds (240 F/116 C) for 1 hour and 50 minutes. Remove jars; air-cool naturally.

• **Adjustment for my altitude** _____ •

Tunas, Large and King Mackerel

Glassy crystals of magnesium ammonium phosphate may form in tuna canned at home. No worries: they are safe to eat, and actually may dissolve when contents are heated.

Pressure Canning only. Precook completely, cool, then exhaust. Use only $\frac{1}{2}$-pint jars with two-piece screwband lids.

Estimate 12 $\frac{1}{2}$-pint jars for every 25 pounds of round fish. Dress, gut, scrub the fish; cut away the thin belly flap. Cut fish crossways in good-sized chunks (it will be cut for the jars after it's precooked).

For the precooking stage you will need several large round pans with perforated bottoms that can be stacked inside your Pressure Canner. Put $2\frac{1}{2}$ to 3 inches of hot water in the canner; put a perforated support on the bottom of the canner—a wire cake rack, laid on some retired screwbands to help it take the weight of the fish; or an inverted metal pie pan with holes punched in its bottom; stack the pan-loads of fish in the canner. Put the lid on the canner; vent it (a strong, steady flow for 10 minutes); close the vent and Pressure-cook the fish at 10 pounds (240 F/116 C) for 2 hours.

Remove fish, cool on large beds of cracked ice for several hours to ensure firm texture and good flavor. Then scrape away the skin, lift out the bones, remove dark streaks of flesh along the sides. Cut the cold chunks of fish ¾ inch shorter than the height of the containers (½-pints are about 4 inches tall, so the steak rounds of fish would be about 3¼ inches thick). Put ½ teaspoon *optional* salt and 3 tablespoons fresh water (or salad oil) in the bottom of each jar. Pack solidly with fish, and leave ½ inch of headroom.

To exhaust, *half*-close the lids as for Salmon, and boil hard in the Pressure Canner at Zero pounds (see the section on Salmon again) for 10 minutes: check center of contents of a test jar to make sure it has reached 170 F/77 C. When jars are exhausted, finish screwing the bands down firmly tight.

Pressure-process at 15 pounds (250 F/121 C) for 100 minutes. Remove jars; air-cool naturally.

• **Adjustment for my altitude** ——————— •

Lake Trout, Whitefish, Small Mackerel, Florida Mullet

Pressure Canning only. Raw pack and exhaust. Use pint jars (preferably straight-sided) with two-piece screwband lids only.

About 35 pounds of fish, round weight, will fill 12 pint jars.

Dress, clean, scale, scrub perfectly fresh fish. Split the fish, leaving in the backbone; cut away the thin belly strip. Cut in jar-length pieces (as for Salmon, above), and brine the pieces for 60 minutes in a cold solution of ¾ cup pure pickling salt dissolved in 1 gallon of water. Remove the fish pieces, drain, and pack solidly just to the top of the jars—not more than 1/16 inch below the sealing rim—alternating head and tail sections upright in the jars for a firm pack, with skin sides next to the glass.

Use the kettle of your big enameled Boiling–Water Bath for exhausting the jars, because they will be boiled in a weak brine that would mar the metal surface of your Pressure Canner. Do not cap the filled jars; put them, open, on the rack in the B–W Bath kettle; pour in a fresh hot brine of ⅓ cup pure pickling salt dissolved in 1 gallon of water, until the brine comes 1 inch above the top of the jars. Bring to boiling, and boil the jars briskly for 15 minutes, which should be enough to raise the temperature deep inside the jars to a minimum of 170 F/77 C (check with your thermometer to make sure).

Remove the jars and invert them to drain on a wire cake-cooling rack for about 3 minutes. (Clap the slotted blade of a metal spatula over the mouth of the jar before you up-end it, and you won't have to worry about bits of fish sliding out.) Right the jars, wipe their rims carefully to remove any speck of material that would interfere with the seal; put on the lids and screw the bands down firmly tight.

Pressure-process at 10 pounds (240 F/116 C) for 1 hour and 40 minutes. Remove jars; air-cool naturally.

• **Adjustment for my altitude** _____ •

Crab—Dungeness (Pacific) and Blue (Atlantic)

Pressure Canning only. Precook and exhaust. Use only ½-pint jars with two-piece screwband lids.

To fill twelve ½-pint jars it will take about 25 pounds live weight of average-size Atlantic crabs, or 13 to 15 average Pacific crabs.

Use enameled or stainless-steel ware for boiling or acid-blanching shellfish—*never use* copper or iron. And *never use* seawater: use fresh drinking water to which you've added salt, etc.

For canning use only fresh-caught, frisky crabs in prime condition (not recently molted, not feeble or sickly). To avoid needless contamination of the meat by visceral matter you should butcher and clean them before precooking. Stun the live crab by holding it in the freezer for half an hour, or submerging in ice water, then quickly twist off the legs, take off the back, remove gills and "butter" and the rest of the innards; clean out the body cavity under a strong flow of fresh, cool drinking water. Save claws and bodies of Atlantic crabs, discarding their legs as too small to bother with; save legs as well as claws and bodies of Pacific crabs, which are larger.

Meanwhile prepare and have heating in your biggest enameled B–W Bath kettle enough brine to cover the broken crabs you're dealing with, made in the proportion of 1 cup pure pickling salt and ¼ cup lemon juice to each 1 gallon of fresh water. Dump crab pieces in the brine, bring back to boiling, and boil hard for 15 minutes by your timer. Quickly dip out the crab pieces and cool them quickly under cold running water *just until* they're cool enough to handle (the meat comes more easily from the shells if it's still warm). Pick out the meat, keeping body meat separate from leg and claw meat. Wash the meat piecemeal under a gentle spray to get rid of any curds of coagulated protein, etc., and press excess moisture out with your hands.

The following acid-blanch is designed to prevent natural—and harmless but unsightly—sulfur compounds present in shellfish from darkening the meat during processing. Therefore prepare beforehand and have ready a cold mixture in the proportions of 1 cup lemon juice (or of the citric-acid solution described in Chapter 5, or of distilled white vinegar in a pinch) to 1 cup salt dissolved in 1 gallon of water. This amount will treat 15 pounds of picked meat; make up fresh brine for each batch. Dealing with a colander-ful at a time, immerse leg meat for 2 minutes, body meat for 1 minute. Drain well, pressing out excess moisture with your hand.

Fill ½-pint jars firmly with meat, making a solid pack with attractive pieces next to the glass; leave ½ inch of headroom. Add boiling water to cover the meat—it won't take much—leaving ½ inch of headroom. Half-close jars and exhaust at Zero pounds (see Salmon, above) until the inside of the pack reaches a minimum of 170 F/77 C on your thermometer—about 10 minutes. Finish screwing down bands firmly tight.

Pressure-process ½-pint jars at 10 pounds (240 F/116 C) for 70 minutes. Remove jars; air-cool naturally.

> • **Adjustment for my altitude** _____ •

Lobsters

Pressure Canning only. Precook and exhaust. Use only ½-pint jars with two-piece screwband lids.

To fill twelve ½-pint jars, figure on 7 to 10 lobsters—depending on size and whether they're the huge-clawed Atlantic lobster of cold North American waters or the bigger-tailed spiny lobster without claws.

Can only fresh-caught, healthy, lively lobsters. Cook and then cool them in separate containers of brine made of 2 tablespoons pure pickling salt to each 1 gallon of fresh drinking water: *never cook or cool lobsters in seawater;* make up fresh cooking/cooling brines for each batch.

In your biggest B–W Bath kettle, bring to boiling 3 to 4 gallons of the brine just described. Plunge live lobsters head first into the boiling salted water and, when it returns to boiling, boil them until their entire shells are bright red—about 20 minutes on the average. Lift them from the kettle and immerse them immediately in a tub of very cold brine (also made as above) to cool as fast as possible. When well cooled, each lobster is split, cleaned under running water, and the meat picked from the shell. Quickly and gently spray the picked meat as necessary to remove curds of co-agulated protein. Press out excess liquid. Dip the picked meat, a small amount at a time, in a fresh acid-blanch as for Crab (in the proportions of 1 cup lemon juice and 1 cup salt dissolved in 1 gallon of water). Press out extra liquid and pack attractively in ½-pint jars, fitting claw and tail meat carefully to get a firm, solid pack; leave ½ inch of headroom. Just cover the meat with boiling fresh brine made of 1¼ teaspoons salt to each 1 quart of water; leave ½ inch of headroom. Exhaust as for Crab, above. Finish screwing down bands firmly tight.

Pressure-process ½-pint jars at 10 pounds (240 F/116 C) for 70 minutes. Remove jars; air-cool naturally.

> • **Adjustment for my altitude** _____ •

Shrimp

Pressure Canning only. Precook and exhaust. Use only ½-pint jars with two-piece screwband lids.

About 10 pounds of fresh-caught headless shrimp will fill twelve ½-pint jars.

If shrimp are headed within 30 minutes after catching, the "sand vein" will come out with the head section. At any rate, pack raw shrimp immediately in crushed ice and hold them on ice to retard spoilage—and also to make peeling them easier. Remove heads, peel off shells, take out the sand vein; wash the meat quickly in fresh water and drain thoroughly.

In a large enameled kettle make enough cold brine to cover the meat, in the proportion of 2 cups of pure pickling salt dissolved in each 1 gallon of water. Hold the meat in the brine 20 to 30 minutes depending on the size of the shrimp, stirring from time to time to make this first brining uniform. Remove the meats and drain thoroughly.

Meanwhile prepare an acid-blanch of 1 cup lemon juice and 1 cup pure pickling salt for each 1 gallon of fresh water (the same as for Crab and Lobster, above). Make enough to deal with all the shrimp you're working with, because you must use a fresh lot of the solution for each blanching-basket's worth of shrimp meat; otherwise the liquid will become ropy from diffused blood, etc. Bring to boiling enough of the acid-blanch to cover a household deep-frying basket *half*-filled with shrimp. Boil the meat 6 to 8 minutes (again depending on size of the shrimp) after the liquid returns to the boil. Lift out the shrimp, spread them on wire racks to air-dry and cool. An electric fan blowing across the shrimp will hurry the process: shrimp must be cool and surface-dry when packed.

Fill ½-pint jars with shrimp, fitting them in carefully to get a solid pack—but *don't crush them down;* leave ½ inch of headroom. Add boiling water just to cover the shrimp, leaving ½ inch of headroom. Half-close and exhaust as for Salmon, Crab, and Lobster. Finish screwing down bands firmly tight.

Pressure-process ½-pint jars at 10 pounds (240 F/116 C) for 35 minutes. Remove jars; air-cool naturally.

• **Adjustment for my altitude** _____ •

Hard-Shell Clams Whole or Minced (Littleneck, Butter, Razor, Quahogs)

Pressure Canning only. Blanch, pack, and exhaust. Use only ½-pint jars with two-piece screwband lids.

To fill twelve ½-pint jars with *whole* clam meats (including their juice), you'll need about 6 quarts of raw shucked meats; about 12 quarts of raw shucked meats will fill twelve ½-pint jars with *minced* clams.

The early steps for preparing clams for canning are the same for whole or minced meats. We'll describe the complete procedure for canning whole clams, and give the variations for minced clams separately later.

Have ready some large vessels filled with clean salt water in which to hold your clams from 12 to 24 hours, so they'll have time to get rid of any sand in their stomachs. *Do not use seawater:* instead, approximate the necessary salinity by making a mild brine in the proportion of ¼ cup pure pickling salt for every 1 gallon of drinking water; and make enough to cover them by several inches.

Pick over the clams, choosing only those with tightly closed shells or ones that quickly retract their siphons (necks) when touched; discard any with broken or open shells. Scrub them with a stiff brush, rinse quickly, and put them in your mild holding brine. Sprinkle a few handfuls of corn-meal in the brine and swish it around; during the night the critters will eat the cornmeal and spit out sand.

Take the clams from their holding brine, throwing away any that don't have shells closed fiercely tight. Open the clams by steaming, or by shuck-ing the live clams with a blunt knife as described in Freezing Oysters, Clams, etc. in Chapter 16. If you open them live, work over a bowl to catch the juice (which you save, strain through cheesecloth, boil down to ⅔ its original volume, and use for canning liquid in the jars). *To steam open:* take clams from their holding brine, spray-rinse, and pile them wet on a rack in the bottom of a big steel or enameled kettle with a tight lid; work with about ½ peck at a time, because their volume increases as they open. Cover the kettle, put it on High heat; reduce the heat to Medium when the liquid from the clams starts to boil, and let them steam until their shells are part way open.

From each opened clam, remove the dark gasket-like membrane that runs around the inside edge of the shell and encloses the siphon/neck; snip off the dark tip of the neck. Keep the dark stomach mass if you like: it's nutritious, but it also could give the canned product an unappetizing color or odor. Wash the dressed meats thoroughly in a fresh brine made in the proportion of 3 tablespoons pickling salt to each 1 gallon of water; make enough so you can change the washing-brine often.

Acid-blanch the meats in a boiling solution of 2 teaspoons pure citric acid powder dissolved in 1 gallon of water, making enough so you can change the blanch often. Half-fill a deep-frying basket with clam meats and hold them submerged in the acid-blanch for 2 minutes after the liquid returns to boiling. Lift out the basket of meats; drain.

Pack in ½-pint jars, leaving ¾ inch of headroom; do not add salt. Add boiling-hot, reduced clam juice to cover the meats, leaving ½ inch of head-room. Half-close the lids and exhaust as for Crab, Lobster, Shrimp, above. Finish screwing bands down firmly tight.

Pressure-process ½-pint jars at 10 pounds (240 F/116 C) for 1 hour. Remove jars; air-cool naturally.

• **Adjustment for my altitude** _____ •

MINCED CLAMS

Remove the dark stomach mass as you dress the meats after the clams are opened. Proceed with washing the meats in brine and acid-blanching, as above. Drain. Put the meats through a food grinder, using a plate with ⅜-inch holes. Strain the clam broth through cheesecloth, bring to boiling. Pack minced clams in ½-pint jars, leaving ¾ inch of headroom; do not add salt. Add hot clam broth, leaving ½ inch of headroom.

Exhaust and Pressure-process as for whole meats.

12
CANNING CONVENIENCE FOODS

Because we at *PFB* live in quite small households, we see the value of having a variety of just-open-it foods at hand. Some of them: soup bases, broths, hearty small stews; vegetables and fruit canned in ½-pints—and even quart jars of pie fillings. In one farmhouse, a "company dinner" for four, two—or oneself—can be put together from jars brought up from the cellar. And large families can benefit from canned main- or side-dishes—in conjunction with what's in the freezers.

Many of these foods were canned from small immediate supplies of ingredients: with the exception of Corn Relish benders, or a several days' rhubarb binge, no large supplies were brought in to make a killer bout of canning. Enough. Here's a sampling.

CONVERSIONS FOR CANNING CONVENIENCE FOODS

Do look at the conversions for metrics, with workable roundings-off, and for altitude—both in Chapter 3—and apply them. **Note: this chapter involves Canning. Make your altitude adjustments accordingly.**

There are no single vegetables given in the following recipes, because there are canned vegetables aplenty in Chapter 9. And there are canned meats ready to serve in Chapter 10, and seafood in Chapter 11. However, we have included several recipes-plus-procedures for using tomatoes in combination with a variety of vegetables; these are intended as bases for pasta sauces, or for whatever your own recipe file indicates, and are included under Meats and the Makings of Main Dishes.

Where flavorings usually included in such recipes are omitted, it is because some herbs and spices produce an unwanted taste after Pressure-processing. This is especially true of artificial sweeteners and salt substitutes: add these at the time the food is to be served. And note that salt may be left out, since it is optional.

Finally, the amounts canned are limited to pint-size jars, with only a couple of exceptions. Cans are omitted from most of the specific instructions for individual foods even though they are included throughout Chapters 7 to 11 on canning. However, it is unlikely that a householder cooking for just one or two people would crank up for using cans. So: jars are preferred. And pint jars or smaller (except for some of the Meat Broth/ Stocks in quarts).

For ½-pint jars leave the same headroom as for pints. Process ½-pints the same length of time as is given for pints.

Remember that you will need more headroom at altitudes above 3000 ft/914 m. See Chapter 3, under the section Correcting for Altitude to save trouble and loss of liquid (or eventual loss of seal).

Before tasting any canned food with low-acid ingredients: you must be unshakably certain that your Pressure Canner was operated correctly—pressure gauge accurate and deadweight gauge signaling properly—and that requirements for times and corrections for altitude were followed.

Unless you are certain that all safeguards were observed, boil the canned low-acid food hard for 15 minutes to destroy any hidden toxins (corn, greens, meat, and seafood require 20 minutes), stirring to distribute the heat and adding water as necessary. If the food foams unduly or smells bad during boiling, destroy it completely so it cannot be eaten by people or animals.

Canning Soup Stock and Broth

Good broth or soup stock, whether all-vegetable or all-meat or a savory combination, is one of the most valuable staples to have on hand—partly because of its versatility, partly because of its cost at the supermarket, and a great deal because you are the monitor for the amount of salt that goes into it.

The following are the simplest of broths to make. We wish there were room to include delicious variations and some of the consommés, etc., but there's always your own inventiveness or the fun of hunting for specialties in gourmet cookbooks. Another reason for keeping the seasonings simple here: their flavors change too much in Pressure-processing (and you probably like to add your own touches, anyway, when you use the broth).

Broth/Stock from Meat or Poultry

Meaty bones from beef, veal, or lamb and from turkey, chicken, or duck all make good soup bases; beef, veal, and chicken are especially good all-purpose broths for such pleasures as adding your own homemade noodles. Stocks with more pronounced flavors of their own are excellent for sauces or combination main-dish soups.

Freezing also is satisfactory for putting by these broths or stocks, but they really should be boiled down to more concentrated form to save valuable freezer space.

Pressure Canning only. Use Hot pack only. Use pint or quart jars.

PREPARE

Cover the meaty bones with water, and salt only very lightly *if you use salt at all:* there's no certain way to remove too much salt after the broth has reduced to the strength you want. You could add a whole peeled onion, a few ribs of celery; some dill if for chicken—all optional. Bring to a slow boil over medium heat, and skim carefully to remove the grey protein froth that will collect: this will take only a couple of minutes. Then reduce the heat to low, and *simmer* until meat falls from the bones, because simmering is the main way to obtain clear broth.

Lift out the bones and onion, celery, etc.; and when the bones are cool enough, remove any good little bits of meat. Meanwhile let the pot simmer a while longer until the broth is nearly the strength you want it. Pour the broth through a sieve into another large kettle or large bowl; save any meat residues in the colander, discarding gristle, skin, bits of bone, etc.

If there is an accumulation of sediment and little edible pieces, do not throw it away: instead, strain the broth again, and reserve the sedimentary things for a special container that will be used for gravies or casseroles or the like. The simplest way to clarify small batches of stock is to heat it again, and pour it by limited amounts through a good coffee filter into a large ceramic or glass coffee pot (which can be emptied as needed into another bowl for de-fatting).

If there is not much fat, you can get it by laying absorbent paper towels gently and briefly on the surface of the stock; change the towels each time they absorb the fat. If there is a good deal of fat, chill the bowls of broth until the fat can be lifted off in a hard sheet. Chicken fat should be saved and used for the pastry or biscuit toppings for meat pies or for meat tarts; but the other fats would be welcome treats for birds in cold weather.

PACK HOT IN JARS

Pour boiling hot stock into jars, leaving 1 inch of headroom. Wipe jar rims carefully. Adjust lids.

Pressure-process at 10 pounds (240 F/116 C)—pints for 20 minutes, quarts for 25 minutes. Remove jars; complete seals if necessary.

• **Adjustment for my altitude** _____ •

All-Vegetable Broth / Stock

Useful, light, with a bouquet that should never be smothered; what is often designated as vegetarian broth should be a staple for meat-eaters too. The main rule to follow—of course in addition to Pressure-processing—is that the vegetables and their trimmings that are used *must be impeccable—no blemishes, scars, cuts, spoiled spots, blossom or stem ends.*

A favorite broth is based on quite thick potato peelings and leeks. Scrub about 4 pounds of fresh baking potatoes (around 8), cut out all eyes and blemishes, then cut off quite thick peels. Put these in a good-sized enameled or stainless steel kettle with 1 medium leek, split and cleaned, then cut in 4-inch chunks (or 2 medium onions, peeled and quartered); 2 ribs of celery and 1 large scraped carrot, both cut in chunks; about 1 cup of carefully cleaned mushroom trimmings (stems and peels); $\frac{1}{2}$ teaspoon dried rosemary leaves. Cover the vegetables with 2 quarts of water ($\frac{1}{2}$ *teaspoon salt optional*), bring just to the boil, skim for a couple of minutes; reduce heat and simmer, *covered,* for 1 hour. Check the liquid level, add 1 cup water (or more) if needed. Put lid on again, and continue cooking gently for 1 hour more.

Strain off most of the broth, correct the salt as desired; put remaining broth and soft vegetables through a food mill to make a purée. Keep clear stock and puréed base separate in canning. Together they will yield about 4 pints.

Pressure Canning only. Use Hot pack only. Use pint jars.

Follow processing procedures and times for Broth / Stock from Meat or Poultry, above.

Shrimp Stock

Fresh raw shrimp, either peeled or in their shells, are just covered with cold water, brought to boiling; then the heat is reduced and they are cooked more gently only until they curl, turn pink, and the flesh is opaque. Lengthy, brisk boiling toughens them. *Salt*—$\frac{1}{4}$ teaspoon for each 1 pound of shrimp—helps their flavor, but *is optional.*

Remove shrimp from water, cool, and handle as desired. Meanwhile turn up heat, reduce the stock to $\frac{3}{4}$ or $\frac{1}{2}$ its original volume. Strain through a sieve to remove coagulated material. Reheat and pack in clean, hot jars. This stock makes a superior base for seafood soufflés and sauces.

Pressure Canning only. Use Hot pack only. Use $\frac{1}{2}$-pint jars with two-piece screwband lids.

In $\frac{1}{2}$-pint jars. Pour hot stock into clean, hot jars, leaving $\frac{1}{2}$ inch of headroom. Wipe jar rims carefully; adjust lids. *Pressure-process* at 10 pounds, sea level (240 F/116 C) for 20 minutes. Remove jars to cool naturally.

• **Adjustment for my altitude** _____ •

Clam Broth

This is a tasty by-product of steamer clams, whether they are the soft-shell type of the New England and Maritime shores, or the hard-shell variety generally steamed on the Pacific Coast.

Scrub clams thoroughly, put them in a heavy enameled or stainless steel kettle, or regular clam-steamer designed for large batches. If they're loose in a kettle, they'll need only about ¼ cup water per 1 pound in addition to the water clinging to their shells; you may add ¼ cup per 1 pound more water if you want extra broth to can, after cups of broth are served with the clams.

Over moderate heat, bring the covered pot to boiling; turn the clams over once with a large spoon to ensure that all are exposed to steam. Cook just until they are opened—5 to 10 minutes, depending on size and the shell. Remove clams and eat.

Reduce remaining broth to ¾ or ½ its volume. Pour through a good coffee filter into a ceramic or glass coffee pot to clarify and remove sediment. Return broth to heat, and pack in clean, hot ½-pint jars.

The broth can be diluted and heated to serve before a meal, or to combine with other juices for an appetizer. It is a great help in boosting the flavor of oyster stew or scalloped oysters—which can be a little shy on liquid. All these aside from its virtue in clam dishes—bisques, chowders, and bouillabaisses.

Pressure Canning only. Use Hot pack only. Use ½-pint jars with two-piece screwband lids.

In ½-pint jars. Pour hot broth into clean, hot jars, leaving ½ inch of headroom. Wipe jar rims carefully; adjust lids. *Pressure-process* at 10 pounds, sea level (240 F/116 C) for 20 minutes. Remove jars and cool naturally.

• **Adjustment for my altitude** _____ •

Tennessee Vegetable Soup Base

In this and in any other vegetable combination, Pressure-process the length of time stipulated for the vegetable(s) *requiring the longest processing time.* In this case the critical vegetable is corn, which is lower-acid than fresh lima beans or okra. If you substitute lima beans for *all* the corn, the processing time would be 40 minutes for pints. But if some corn is used, the whole recipe must be processed for corn's 55 minutes for pints. Sounds more complicated than it is—just look up the individual vegetables. And *don't use short-cut processing.*

5 quarts peeled, cored, and quartered unblemished ripe tomatoes
2 quarts corn kernels OR 2 quarts fresh lima beans
2 quarts prepared okra
(Optional 2 tablespoons salt)

Save all the tomato juice possible while preparing the tomatoes. Blanch ears of corn to set the milk, then cut from the cob at ⅔ the depth of the kernel (this, to avoid the points of the hulls); shell fresh lima beans. Wash and trim tender pods of okra, boil for 1 minute; cut pods in 1-inch pieces. Combine the vegetables (and salt, if wanted) and cook to the consistency of thick soup.

Pressure Canning only. Use Hot pack only. Use pint jars. Ladle bubbling-hot soup into clean, hot jars, leaving 1 inch of headroom; remove any trapped air and adjust lids. Process at 10 pounds, sea level (240 F/116 C) for 55 minutes. Remove jars; complete seals if necessary.

• **Adjustment for my altitude** _____ •

Chili Salsa
About twelve ½-pints

Flautas, tacos, burritos, enchiladas—just the beginnings for this versatile Mexican accent. Be careful with the chiles until the sauce is developed to your liking. The friendly Anaheim peppers are the 6-inch, pointed ones that turn red; jalapeños are about 1½ inches long and jolting, turning mahogany at the tips—use sparingly; most savage of the lot are the serranos, smaller, slender versions of the jalapeños.

   1 pound onions
   2 pounds Anaheim (rather hot) peppers
   5 pounds tomatoes
   3 teaspoons salt
   ½ teaspoon pepper
¾ to 1 cup vinegar

Finely chop or coarsely grind onions and peppers.

Peel, if desired, and chop fresh tomatoes or canned, whole, peeled tomatoes into small pieces.

Add onions, peppers, and other ingredients to chopped tomatoes. Heat to simmering; simmer 10 minutes.

Pack into clean, hot pint or half-pint jars with ¼ inch headroom. Seal.

Process 15 minutes in simmering Hot-Water Bath—180–185 F/82–85 C.

• **Adjustment for my altitude** _____ •

Country Tomato Soup
About 4½ quarts

This plain and good soup is not diluted for serving.

Pressure Canning only (onions and peppers are low-acid). Use Hot pack only. Use jars only; for a very small household pints are more useful.

Wash 1 peck (8 quarts) of ripe red tomatoes; remove blossom and stem ends and cores; cut in pieces. In a large kettle, cook and stir the tomatoes until soft—about 15 minutes. Push the pulp and juice through a wire strainer or food mill to remove skins and seeds; return the purée to the kettle.

Cook together until soft (in enough water just to cover) 3 large onions and 2 green peppers—all finely chopped. Sieve, and add to the puréed tomatoes in the kettle. Mix together ¾ cup of sugar, and 8 tablespoons cornstarch; blend in 3 tablespoons white vinegar and just enough more water or cool tomato juice to make a smooth paste. (Optional: add 2 tablespoons salt to the paste.) Pour slowly into the sieved tomatoes, stirring all the while. Heat to boiling and stir until the liquid clears.

Pack Hot only, in jars. Pour boiling-hot soup into clean, hot jars, leaving ¾ inch of headroom for pints, 1¼ inches for quarts. Adjust the lids. Pressure-process at 10 pounds (240 F/116 C)—pints for 20 minutes, quarts for 30 minutes. Remove jars; complete seals if necessary.

• **Adjustment for my altitude** _____ •

Stewed Tomatoes with Added Vegetables

The addition of lower-acid vegetables to tomatoes means that you must Pressure-process the mixture *according to the specific rule for the lowest-acid vegetable in the combination.*

How much of which of the usual vegetables is added for interest of course depends on the family's taste. However, *density* of the pack is an important factor in any timetable for processing. Therefore we say that the total amount of several added vegetables *should not exceed one-fourth the volume of tomatoes in the mixture.* For example, to 8 cups of prepared cut tomatoes we would add no more than 1 cup chopped celery, ½ cup chopped onion, and ½ cup chopped green pepper. Incidentally, this balance of added vegetables also makes for good flavor.

(Although tomatoes with zucchini squash are a popular side dish, the amount of squash added is generally so large that this mixture is better if you combine canned Squash with as much canned Stewed Tomatoes as you like, and the two are heated together just before serving.)

GENERAL HANDLING

Pressure Canning only. Use Hot pack only.

To avoid diluting acidity or flavor, it's a good idea to prepare 3 or 4 cups of Tomato Juice (see Chapter 8) to have ready in case you need extra hot liquid when filling the containers; or use canned juice, heated.

Wash, peel, core, and cut the tomatoes in quarters or smaller, saving the juice; measure. Add the desired proportion of well-washed coarsely chopped celery, finely chopped onions, or chopped seeded green peppers. Combine the vegetables in a large enameled kettle and boil them gently in their own juice *without added water* for 10 minutes, stirring to prevent sticking.

HOT PACK ONLY, IN PINT JARS

Ladle boiling hot into clean hot pint jars, leaving ½ inch of headroom. Add ¼ teaspoon citric acid (or 1 tablespoon white vinegar). (Optional: add ½ teaspoon salt to pints.) If there is too little free liquid, make up the difference with boiling tomato juice, *not water*. Adjust lids; process. After processing, remove jars; complete seals if necessary.

With only onion added, Pressure-process at 10 pounds (240 F/116 C)—25 minutes for pints.

With celery added, Pressure-process at 10 pounds (240 F/116 C)—30 minutes for pints.

With green peppers added, Pressure-process at 10 pounds (240 F/116 C)—35 minutes for pints.

• **Adjustment for my altitude** _____ •

Annette Pestle's Tomato Cocktail
About 7 quarts

This is delicious as an appetizer or for aspic. The herb seasonings may be varied, and *of course the salt is optional,* but NEVER decrease the proportion of tomato-juice-plus-added-acid to the total amount of vegetables. Annette says she has never had this cocktail separate after it sits in properly cool storage; but in case yours does, just give it a good shake.

Quart jars are worthwhile to have on hand with this, even for a small family. Remember: *never* process different sizes of containers in the same batch.

Use a Boiling–Water Bath. Use Hot pack only.

To make about 7 quarts of cocktail, you'll need 8 quarts of cut-up tomatoes. Wash thoroughly the firm-ripe unblemished tomatoes; remove stems, blossom ends, and cores; cut in small pieces. In a large enameled kettle simmer the tomatoes over low heat until soft; put through a fine sieve or food mill to remove skins and seeds, and set the strained juice aside. Rinse the kettle, and into it measure 2 cups of the strained juice, add 2 diced medium onions, 1¼ cups diced celery (including a few leaves), 1 large seeded and chopped green pepper, 3 bay leaves, 8 or 10 fresh basil leaves (or 2 teaspoons dried basil), ½ teaspoon ground pepper, 3 tablespoons sugar, and 2 teaspoons Worcestershire sauce. (Optional: 4 teaspoons salt.) Boil over medium heat—stirring, and adding extra juice as needed to keep the mixture from sticking—until soft, about 30 minutes. Pick out the bay leaves, then press the vegetables through a fine sieve or food mill. Add 3½ teaspoons crystalline citric acid (or ⅔ cup bottled lemon juice) and the rest of the tomato juice. Bring to simmering. Pack Hot.

Hot pack only, in quart jars. Fill jars with hot juice, leaving ½ inch of headroom; adjust lids. Process in a Boiling–Water Bath (212 F/100 C)—35 minutes. Remove jars; complete seals if necessary.

• **Adjustment for my altitude** _____ •

Tomato Ketchup
6 to 7 pints

---

  24 pounds ripe tomatoes
   3 cups onions, chopped
   ¾ teaspoons cayenne pepper
   3 cups cider vinegar
   4 teaspoons whole cloves
   3 sticks cinnamon, crushed
  1½ teaspoons whole allspice
   3 tablespoons celery seeds
  1½ cups sugar
   ¼ cup salt

Wash and peel tomatoes; core; quarter. Combine tomatoes, onions, and cayenne pepper in a large stainless steel kettle. Bring to boiling; and simmer 20 minutes. Meanwhile in a small pot bring vinegar and spices tied in a bag to boiling, remove from heat. After 20 minutes, add to tomato mixture, and boil all together about 30 minutes. Put through a food mill, return to pot. Add sugar and salt, boil gently, stirring until volume is reduced by half and mixture rounds on spoon without separating. Fill pint jars, leaving ⅛-inch headroom; adjust lids. Process in a B–W Bath for 15 minutes. Remove jars, cool upright.

• **Adjustment for my altitude** ——————— •

Meatless Spaghetti Sauce
About 9 pints

---

  30 pounds tomatoes
   1 cup onions, chopped
   5 cloves garlic, minced
   1 cup celery or green pepper, chopped
   1 pound fresh mushrooms, sliced (optional)
  4½ tablespoons salt
   2 tablespoons oregano
   4 tablespoons parsley, minced
   2 teaspoons black pepper
   ¼ cup brown sugar
   ¼ cup vegetable oil

Wash tomatoes and scald, remove skins; core, cut in quarters. Bring to boiling in a large kettle and cook uncovered for 20 minutes. Sauté onions, garlic, celery or peppers, and mushrooms in vegetable oil until tender; Add spices, salt, sugar, and tomatoes. Bring to a boil, then simmer uncovered until reduced by nearly one-half. Stir frequently to avoid burning. Fill jars, leaving 1-inch headroom. Adjust lids, process at 10 psig (240 F/116 C) for 20 minutes.

• **Adjustment for my altitude** ——————— •

Minnesota Tomato Mixture
7 pints

This combination has been developed by the Agricultural Extension Service of the University of Minnesota to be in such accurate *pH* balance that, although the celery, onions and green pepper are all low-acid foods, the extremely large proportion of tomatoes guarantees that the over-all acidity remains intense enough to allow for processing in a Boiling–Water Bath.

But it cannot be stressed too strongly that the proportion of green pepper, onion, and celery to tomatoes CANNOT BE INCREASED—unless the mixture is to be Pressure-processed for the full time required to deal with the vegetable with the lowest amount of acid.

12 cups peeled, cored, and quartered tomatoes
 1 cup chopped celery
 ½ cup chopped onion
 ½ cup chopped pepper
 (Optional 3 teaspoons salt)
 (Optional ⅓ cup bottled lemon juice)

Simmer the vegetables for 10 minutes, then ladle into clean, hot pint jars: the vegetables will have released their juices to provide canning liquid, so *do not add any water*. Leave ¾ inch of headroom; adjust lids.

Process in a B–W Bath for 40 minutes for the pints. Remove jars; complete seals if necessary.

• **Adjustment for my altitude** _____ •

Meats and the Makings for Main Dishes

With the exceptions of the simplified Beef Stew with Vegetables—which invites titivating *after it is opened and is being prepared for serving*—there are no complete main dishes given here.

The reason is simple and has been stressed before: varying degrees of acidity, as well as varying densities and textures, dictate Pressure-processing times that would subdue completely some ingredients in order to deal safely with another ingredient in the same combination. Therefore, the components are canned in separate groups, to be put together for heating and serving.

Ground Meat

Freezing ground meat gives a much better result than canning does. But if circumstances oblige you to can it, *do not can it in bulk in a solid mass*. Make it up into meatballs by your favorite recipe (omitting herbs and spices that change character in canning); brown lightly under a medium broiler, because frying is a very poor way to precook meat for canning: the meat is case-hardened, and tastes overcooked. Or shape it in thin patties; these you can always break up and add to the Minnesota Tomato Mixture, say, or Tomatoes Stewed with Vegetables to use with pasta.

Pressure Canning only. Use Raw or Hot pack. Use straight-sided pint jars.

PREPARE AND PACK HOT (PRECOOKED)

Trim and grind *lean* meat *adding no fat*. Make meatballs, browning their surface and draining well; or shape thin patties, slightly smaller in diameter than the containers. In a slow oven (325 F/163 C) precook meatballs or patties until Medium done. Skim off all fat from the drippings in the pan, saving the pan juices.

In straight-sided pint jars. Pack hot patties (in layers) or hot precooked meatballs, leaving 1 inch of headroom. (Optional: add ½ teaspoon salt to pints.) Cover with boiling *fat-free* pan juices (extended with boiling meat broth if necessary), leaving 1 inch of headroom. Wipe jar rims carefully to remove any fat. Adjust lids. *Pressure-process* at 10 pounds (240 F/116 C)—pints for 75 minutes. Remove jars; complete seals if necessary.

• **Adjustment for my altitude** _____ •

Corned Beef

Corn the beef as recommended in Chapter 20.

Pressure Canning only. Use Hot pack only (but in this case the meat is not precooked as for fresh meat: it is still raw, but some of the salt cure has been removed in freshening). Use straight-sided jars.

PREPARE AND PACK HOT (FRESHENED)

Wash the corned beef and cut it in chunks or thick strips to fit your containers, removing all fat. Put the pieces of meat in cold water and bring to boiling. Taste the broth in the kettle: if it's unpleasantly salty, drain the meat, cover it with fresh cold water, and bring again to boiling. This boiling merely freshens (removes salt), *it does not cook the corned beef appreciably*.

In straight-sided pint jars. Fit hot freshened meat in jars, leaving 1 inch of headroom. Add boiling broth in which the meat was freshened, leaving 1 inch of headroom. Wipe jar rims carefully to remove any fat. Adjust lids. *Pressure-process* at 10 pounds (240 F/116 C) 75 minutes. Remove jars; complete seals if necessary.

• **Adjustment for my altitude** _____ •

Pork Sausage

Freezing is better for pork sausage, especially in view of the large amount of fat (but remember that fatty food has short freezer-storage life).

Make your sausage by any tested recipe, *but use your seasonings lightly* because such flavorings change during canning and storage; and *omit sage*—it makes canned pork sausage bitter. And note that precooking is done in the oven.

Pressure Canning only. Use Hot pack only. Use straight-sided pint jars.

PREPARE AND PACK HOT (PRECOOKED)

Shape raw sausage in thin patties, slightly smaller in diameter than the containers. In a slow oven (325 F/163 C) precook patties until Medium done. Skim off all fat from the drippings in the pan, saving pan juices.

In straight-sided pint jars. Pack hot sausage patties in layers, leaving 1 inch of headroom. Cover with boiling *fat-free* pan juices (extended with boiling meat broth if necessary), leaving 1 inch of headroom. Wipe jar rims carefully to remove any fat. Adjust lids. Process at 10 pounds psig (240 F/116 C) for 75 minutes. Cool jars upright and naturally.

- **Adjustment for my altitude** _____ •

Vegetable-Beef Stew
Makes 7 or 8 pints

Because of the long Pressure-processing and its effect on some of the seasonings that might be used if the stew were served from its pot, this is not a dish with gourmet touches. But this is a virtue, under the circumstances; and titivating touches can be added when the jars are opened and their contents are being heated for a meal. Being packed Raw, the ingredients provide a good deal of the canning liquid needed for convection as well as conduction heat through the contents.

    4 cups lean beef (chuck is good) cut in 1-inch cubes
    4 cups new potatoes cut in ½-inch cubes
    4 cups carrots cut in ½-inch pieces
    4 cups small whole onions, peeled
    1½ cups coarsely chopped celery

Put prepared meat in a very large bowl; add vegetables and mix with the meat. Pack firmly into clean, hot pint jars, leaving 1 inch of headroom, and apportion juices collected in the bowl among the jars, adding ¼ cup boiling water to each jar if needed. (Add an optional ½ teaspoon salt to each pint jar.) Wipe sealing rims of jars. Adjust two-piece screwband lids firmly tight.

Process in a Pressure Canner at 10 psig (240 F/116 C) for 60 minutes.

- **Adjustment for my altitude** _____ •

Pie Fillings

Following are well-tested pie fillings: the first two are good friends and fine performers from CES-Nebraska. Tapioca as the firming agent is more stable than cornstarch, which formerly was used here—⅓ cup—and still could be, in a pinch. (So could all-purpose flour, in twice the amount of cornstarch; the result would not be clear, and would be stiff. Mochiko, the sweet rice flour *PFB* sometimes recommended for sauces, cannot withstand processing; and indeed is not so good a thickener as Clearjel A, used in Apple Pie II, Cherry Pie II, and Blueberry Pie.)

Fillings may be canned in pints for pie-for-two, or in ½-pints (processing times are the same as for pints). **Make adjustments for your altitude throughout.**

Nebraska's Apple Pie Filling
6 or 7 quarts

          4½ cups sugar
           1 cup quick-cooking tapioca
           1 teaspoon salt (optional)
           2 teaspoons ground cinnamon
           ¼ teaspoon ground nutmeg
          10 cups water
           3 tablespoons lemon juice
      2 or 3 drops yellow or red food coloring (optional)
      8 to 9 pounds of tart apples, peeled, cored, and sliced (about 16 cups of
                 sliced apples)

In a large saucepan blend the first five *dry* ingredients. Stir in 10 cups water, cook on medium-high heat; stir until thickened and bubbly; cook an additional 2 minutes. Add lemon juice, and if desired, food coloring. Add the apples to the sauce. Stir constantly and bring mixture up to a rolling boil; cook 1 minute. Promptly ladle mixture into jars, leaving ½ inch headroom. Adjust lids, process in Boiling–Water Bath: pints 25 minutes, quarts 25 minutes.

Prepare pastry for a two-crust, 9-inch pie. Line pie plate with pastry; add 1 quart of apple pie filling. Adjust top crust, cutting slits for escape of steam; seal. Bake at 400 F/205 C for 50 minutes, or until crust is nicely browned.

Nebraska's Cherry Pie Filling
6 or 7 quarts

The relative amount of sugar needs to be increased because cherries are more sour than apples. Almond flavoring or kirsch are the flavorings commonly used in cherry pie. The following recipe should work reasonably well.

       7 cups sugar
       1 cup quick-cooking tapioca
       1 teaspoon salt (optional)
      14 drops almond flavoring or ¾ cup kirsch
       7 cups water
       3 tablespoons lemon juice
       9 pounds pitted cherries (sour)

Blend the dry ingredients and add water and flavoring. Cook on medium-high heat and stir until thickened and bubbly. Cook an additional 2 minutes and add lemon juice. Add the cherries to the sauce. Gently stir and bring mixture up to a rolling boil for 1 minute, ladle mixture into hot jars, leaving ½ inch headroom. Adjust lids; process in a Boiling–Water Bath: pints 25 minutes, quarts 25 minutes.

Cherry Pie Filling II: USDA-CES
1 or 7 quarts

	1 Quart	7 Quarts
Fresh sour cherries	3⅓ cups	6 quarts
Granulated sugar	1 cup	7 cups
Clearjel A	¼ cup + 1 tablespoon	1¾ cups
Cold water	1⅓ cups	9⅓ cups
Bottled lemon juice	1 tbsp. + 1 tsp.	½ cup
Ground cinnamon (optional)	⅛ tsp.	1 teaspoon
Almond extract (optional)	¼ teaspoon	2 teaspoons
Red food coloring (optional)	6 drops	¼ teaspoon

Rinse and pit cherries, and hold in cold water. To prevent stem end browning, use ascorbic acid solution. Combine sugar, Clearjel A, and cinnamon in a large saucepan; add water, and optional food coloring or almond extract. Stir mixture and cook over medium-high heat until it thickens and begins to bubble. Add lemon juice and boil 1 minute, stirring constantly. Fold in the cherries immediately and fill jars without delay, leaving ½ inch headroom. Adjust lids, process in a Boiling–Water Bath, 30 minutes for either pints or quarts (and a maverick ½-pint). Remove jars; cool upright and naturally.

Blueberry Pie Filling: USDA-CES
1 or 7 quarts

	1 Quart	7 Quarts
Fresh blueberries	3½ cups	6 quarts
Granulated sugar	¾ cup + 2 tablespoons	6 cups
Clearjel A	¼ cup + 1 tablespoon	2 cups
Cold water	1 cup	7 cups
Bottled lemon juice	3½ teaspoons	½ cup
Blue food coloring (optional)	3 drops	20 drops
Red food coloring (optional)	1 drop	7 drops

(Recipe continued next page)

Pick over, wash, drain blueberries, set aside. Combine sugar and Clearjel A in a large kettle, stir; add water and optional food coloring. Place over medium-high heat and cook until the mixture thickens and begins to bubble. Add lemon juice and boil 1 minute, stirring hard. Immediately fold in the berries and fill jars without delay, leaving ½ inch of headroom (¼ inch for ½ pints). Adjust lids, process in a Boiling–Water Bath 30 minutes for pints (and ½-pints) and quarts. Remove jars; cool upright and naturally.

Apple Pie Filling II: USDA-CES
1 or 7 quarts

This rule, giving one-pie and whole-batch quantities, calls for the apple slices to be blanched, drained, and kept warm until they are added to the sugar syrup. Simplest ascorbic acid anti-oxidant holding treatment: 1 teaspoon ascorbic acid crystals dissolved in 2 cups warm water—and sprinkle, turn freshly cut slices in it to coat.

Have some ½-pint jars ready, to hold any extra filling: handy for tarts and toppings.

	1 Quart	7 Quarts
Blanched sliced apples	3½ cups	6 quarts
Granulated sugar	¾ cup + 2 tablespoons	6 cups
Clearjel A	¼ cup	1¾ cups
Ground cinnamon	½ teaspoon	1 tablespoon
Cold water	½ cup	4 cups
Bottled apple juice (*not* cider)	¾ cup	5 cups
Bottled lemon juice	2 tablespoons	¾ cup
Nutmeg (optional)	⅛ teaspoon	1 teaspoon
Yellow food coloring (optional)	1 drop	7 drops

Use firm, crisp apples, and, if they lack tartness, add an extra ¼ cup of lemon juice for each 6 quarts of slices. Wash, peel, core apples; slice ½-inch wide and treat with ascorbic acid while they wait to be blanched. In a blanching basket, dip 2 quarts at a time in boiling water for 1 minute. Hold blanched apples in a large covered pot to stay warm until all are treated. Meanwhile combine sugar, Clearjel A, cinnamon, and nutmeg in a large kettle with the water, apple juice, and optional food color; stir and cook on medium-high heat until the mixture thickens and begins to bubble. Drain apple slices. Add lemon juice and boil 1 minute, stirring constantly. Drain warm, blanched slices and immediately fold them into the sweet mixture, and fill jars, leaving ½ inch headroom. Adjust lids and process immediately: 25 minutes either pints or quarts (and for ½-pints). Remove jars, cool upright and naturally.

13
GETTING AND USING A FREEZER

Decent freezing requires a mechanical freezer. It used to be that one could rent space, if just a drawer, in a freezer locker plant, and this was a good start for young families on a strict budget. Nowadays, though, such lockers are few and far between. Instead, one can learn a good deal about freezer management through using well the separate freezer compartment in the usual modern refrigerator/freezer. From this can be developed a workable food-storage program and an idea of what size and kind of freezer would be feasible to buy later on.

What Freezing Does

Freezing does NOT destroy the organisms that cause spoilage, as canning does—it merely stops their growth temporarily. When they become suitably warm again, they multiply as quickly as ever. This matter will come up repeatedly.

Freezing correctly means subjecting each sealed-from-the-air parcel/ container of food to the sharpest cold we can manage to give it—ideally −20 F/−29 C—for 24 hours, and then storing it at a sustained Zero F/ −18 C for as long as its quality holds well; thereafter it is wasting expensive storage space, even though it has not become dangerous to eat.

Food does carry a startling bacterial load, a load that increases geometrically if it is held too long at thawing (usually when it is allowed to thaw at room temperature, especially if it's low-acid, and occasionally if an unrecognized power failure has ruined whatever was stored in the freezer).

The freezers that *PFB* knows best are held at Sharp Freeze, with the result that the temperature in the chest model in the cellar registers − 10 F / − 23 C near the top, and the upright freezer in the pantry is Zero to 5 F / − 18 C to − 15 C depending on how much more often the door is opened.

CONVERSIONS FOR FREEZING

Do look at the conversions for metrics, with their workable roundings-off, and for altitude—in Chapter 3—and apply them.

How Big a Freezer?

Some advisers recommend 6 cubic feet of freezer space for each person in the family. A cubic foot holds about 35 pounds of food ideally, but actually a good deal less because some foods are more dense than others (pork loin roast takes only a little more space than a loaf of Italian bread, which of course weighs less). Further, manufacturers' sales brochures describing cubic-foot sizes in freezers are sometimes approximate, so check the specifications and measurements to determine the actual usable interior space. See the entries in the Appendix.

Of all the ways to put food by, freezing limits storage room most severely, so what you freeze should be given careful thinking-through beforehand. If your freezer is to be more than a place to stash random things you're not going to use right away, figure out a system of priorities. A good rule-of-thumb is to assign freezer space first to the more expensive and heavier foods, and to ones that can't be preserved so well any other way. Therefore, plan to freeze meats and seafood; and plan to freeze certain vegetables (prime example, broccoli) and certain fruits (prime example, strawberries—assuming you don't want all of them in jam), and mentally allot room for some favorite main-dish combinations or desserts.

What Type?

To do its job, the freezer must have adequate controls, no warm spots ("warm" being a constant temperature higher than the rest of the recommended Zero F / − 18 C storage area). *The small enclosed space for ice cubes or below 32 F / Zero C storage in some refrigerators is not adequate unless it has its own controls for Sharp Freeze and proper storage, and has its own outside door.*

CHEST

The chest type with a top opening offers the best use of the space within it, and it holds its temperature better than do the uprights. Cold air sinks downward, so you don't spill cold air when you lift the lid of a chest freezer, but cold air tumbles out from an upright whenever the door is opened.

A chest should be set up on 2-by-4's or some other supports at the corners to make it easier to draw off water when defrosting. It must be in a dry and cool place.

Chest freezers are not built to have automatic defrosting features. Chests are more of a chore to keep reasonably clear of frost than upright models; however, frost builds up in them less readily than in uprights.

Also, more planning-ahead is needed in filling the chest freezer, because you must bend over the freezer to paw through contents that were piled in any which way. Separate dessert materials from vegetables, meats, convenience foods, and oddments, and keep track of the more perishable foods in each category.

Chest freezers are less expensive to buy, and less expensive to run, than uprights are. Of two freezers of about 15-cubic-foot capacity, the upright is likely to cost as much as 30 percent more with the self-defrost feature. As for operation, a manually defrosting upright costs about 14 percent more to run than its chest counterpart, and sometimes even up to 60 percent more to operate if it has automatic defrosting.

UPRIGHT

The upright ones with a refrigerator-type door of course take up less space in the room, but plenty of room must be allowed for the side-opening door: the angle between front of shelves and fully opened door must be *at least* 90 degrees. Uprights are easier to load and unload than the chests, and the little shelves on the door are handy for temporary storage of dabs and snippets.

More cold spills out when the door is opened than is lost with opening a chest; and irregular-shaped packages may tumble out at the same time.

TWO-DOOR COMBINATIONS

Side-by-side or stacked (one above the other) freezer-refrigerator combinations save floor space, and the freezer section is an adequate, though usually smaller, variation of the upright type with its own controls.

Frozen Food/Freezer Plans

It's a good idea to hold off on joining any of the time-payment plans that involve buying a freezer and its contents (delivered seasonally). Wait until you know what, actually, you'll need. And make a routine check with your Better Business Bureau.

Where to Put It?

Locate the freezer near the kitchen: you'll use it more (which means using it better) than if it's in the cellar or other remote spot.

If you have a choice of convenient locations, choose the cooler one—so long as the place isn't actually freezing.

Put it in a relatively dry room, because moisture rusts the mechanism and can build up frost inside non-defrosting models, particularly upright ones.

And *please* place its back a few inches away from the wall, so there's adequate air circulation and you can get under and around it to get out the fluff. Lack of air and a build-up of dust can make the motor overheat and even cause fire. Mounting it on small rollers is a help in cleaning.

Using a Freezer

Operating costs per pound of food are less if the freezer is kept at least ¾ full at all times.

For the initial sharp freeze, set the control at the lowest possible point: at or below − 20 F/ − 29 C, the temperature that makes smaller ice crystals in the food and gives a better finished product.

Place packages in single layers in contact with the parts of the box that cover the freezing coils—these would be certain shelves or walls or parts of the floor—and leave the food spread there for 24 hours before stacking the packages compactly for storage. For best results, don't try to sharp-freeze, at one time, more than 2 to 3 pounds of food for each 1 cubic foot of available freezer space. Later, when you're not sharp-freezing but simply storing, turn the controls back to no higher than Zero F/ − 18 C.

For storing food after it's sharp-frozen, stack packages close together, and keep the storage section temperature at Zero F/ − 18 C or lower.

How Long in the Freezer?

Frozen foods lose quality when subjected to freezer temperatures above Zero F/ − 18 C. While the storage life of different products varies, it can be stated generally that each rise of 10 degrees Fahrenheit lowers the storage quality by half. (Thus, if a food has good quality for 8 months maintained at Zero F/ − 18 C, its top quality will be only 4 months maintained at 10 F/ −23 C, and only 2 months if maintained at 20 F/ −7 C.) Foods maintained at − 10 F/ −23 C *in general* will keep their quality longer—although keeping an item more than 12 months is uneconomical use of freezer space. The table at the end of this chapter is a workable guide to quality.

Keeping an Inventory

First, label and date each package of frozen food so you know how much of what is in each parcel; what ingredient it may be for whichever final product; and when it was put by.

Store similar foods together, and you won't end up with a hodgepodge you have to paw through to find what you're looking for.

Devise some sort of inventory sheet or board that lets you keep track of food going in and coming out of your freezer.

Check the contents of your freezer every so often, and put maverick or to-be-used-soon items in places where you can't overlook them.

Caring for a Freezer

Freezers need little care—just respect.

Treat the outside the same way you do your refrigerator. Keep the surface and the door gaskets wiped clean—taking care to clean the condenser coils—and keep the protective grid over the motor free of dust.

Many are self-defrosting, others you defrost yourself. Do it once or twice a year at times when the food supply is low. Disconnect the freezer, remove the food, and wrap it in newspapers and blankets to keep it frozen. Use a scraper that won't puncture the freezing element or scratch the casing. Remove condensation (frost) from the walls of the cabinet onto papers or towels. Wipe the box with a clean cloth wrung out in water and baking soda. Dry the box before restarting the motor. About once a year, really wash the inside of the freezer.

By the way, have you read the manufacturer's instruction book lately?

When the Freezer Stops Freezing

Now and again everybody's electricity fails for a time. Or, Heaven forbid, someone accidentally disconnects the freezer. Or the motor isn't working properly. Resist the impulse to open the door to check everything: make a plan of action first.

Find out, if you can, how long your freezer has been, or is likely to be, stopped. If it can be running in a few hours, don't worry. Food in a fully loaded, closed freezer will keep for two days; if it's less than half loaded, the food won't keep longer than one day.

A freezer full of meat does not warm up so fast as a freezer full of baked foods. Reason: the meat is denser, and so is more like a block of solid ice.

The colder the food, the longer it will keep. The larger a well-stacked freezer, the longer the food will stay frozen. *But you must not open the door*.

Emergency Measures for a Long Stoppage

If you foresee that your freezer will be out of running order for more than 48 hours, try to locate a supply of dry ice—which is carbon dioxide in its solid state (it's solid at -107 F). It has no liquid state, but becomes a gas when it is in the presence of oxygen, and hence evaporates into nothing after several days in the freezer. Look under "Dry Ice" or "Ice and Fuel" in the Yellow Pages; and try places that sell welders' supplies too: in some machine shops, the welders use carbon dioxide to freeze— and thereby shrink—metal that they are working on. A 50-pound cake of dry ice is $10 \times 10 \times 10$ inches. A 10-pound piece of dry ice will hold 20 pounds of food frozen for around 24 hours, so do your arithmetic and order accordingly.

When you get your dry ice, wrap it in many layers of newspaper (that great insulator) *and use lineman's gloves to handle it, because just touching the stuff can cause severe frostbite.*

Consolidate the food packages into a compact pile. Put heavy cardboard directly on the food packages and lay the dry ice on the cardboard. Then cover the entire freezer with blankets, but leave the air-vent openings free so the motor won't overheat in case the current comes on unexpectedly.

But If All That Food Thaws . . .

If, despite your emergency measures, the food in your freezer thaws, there are several things you can do. You can refreeze some of it, you can cook up some of it and freeze the cooked dishes from scratch; some of it you may can. And some you may have to destroy.

WHEN TO REFREEZE

When food has *thawed,* it still contains many ice crystals; individual pieces may be able to be separated, but they still contain ice in their tissues; and dense foods, or ones that pack solidly, might have a firm-to-hard core of ice in the middle of the package, in addition to the crystals in the tissues. Many thawed foods may be refrozen. Be sure to re-label them for limited storage.

When food has *defrosted,* all the ice crystals in its tissues have warmed to liquid. *No foods that have warmed above refrigeration temperature— except very strong-acid fruits—should be refrozen.* Reason: defrosted low-acid foods, if refrozen, are possible sources of food poisoning. (Technically, if the foods have just reached refrigeration temperature, they are still safe to refreeze. But the problem is, how can we know for sure what their temperature is, and how long it has held at an acceptable level? We can't.)

If the foods are still icy—see below—they may still be safe.

Remember, however, that defrosted low-acid foods, vegetables, shellfish, and precooked dishes all may be spoiled although they have no telltale odor, and could be downright dangerous if cooked up and served.

WHAT CAN BE SAFELY REFROZEN

The first check of thawed food in a package or a non-rigid container is to squeeze it. *Don't open it.* Squeeze it: if you can feel good, firm crystals inside, the package is O.K. to refreeze—provided that the food is *not highly perishable in the first place,* and of course that its quality is appealing.

Of course food in rigid containers must be opened to be inspected for adequate ice crystals.

Even though they're defrosted, strong-acid fruits may be refrozen if they're still cold; there will be definite loss in quality, however.

Refreeze thawed vegetables only if they contain plenty of ice crystals.

Give wrapped meat packages the squeeze test. Beef, pork, veal, lamb and poultry that are firm with ice crystals may be refrozen, but you can always cook them up in convenience dishes and freeze them from scratch. The salt in merely thawed short-storage cured pork helps the ice crystals, and it can be refrozen, but with a noticeable loss in quality. Thawed seafoods, being extremely perishable, should be cooked and served instead of being refrozen, because they lose quality so quickly.

Never refreeze melted ice cream. Never refreeze cream pies, eclairs, or similar foods. But you can refreeze unfrosted cakes, uncooked fruit pies, bread, rolls, etc.

Freezer-Packaging Materials

Although a manufacturer might not say on the label that his product is *not* for the freezer, he surely will announce, loud and clear, that it *is* "ideal for freezing" because it has been passed by the FDA as food grade.

The prime purpose of freezer packaging is to keep frozen food from drying out ("freezer-burning"), and to preserve nutritive value, flavor, texture, and color. To do this, packaging should be moisture/vapor-*proof*—or at least *vapor-resistant*—and be easy to seal. And the seal should do its own job too.

And Don't Be Sad If . . .

Don't expect perfect results from all your work if you package the food for your freezer in household waxed paper or regular aluminum foil or wrappings that are intended for short-term storage in the refrigerator.

And don't expect perfect results if you make-do with those coated-paper cartons that cottage cheese or milk or ice cream came from the store in.

And don't expect perfect results if you seal your good food with the sticky tape you use on Christmas parcels. The adhesive used on made-for-the-freezer tape remains effective at temperatures well below Zero F/ − 18 C, and the stuff on regular household tapes does not.

Rigid Containers

As the name implies, these hold their shape and may be stood upright, and are suitable for all foods except those with irregular shape (a whole chicken, say); and they are the best packaging for liquids.

Made of aluminum, glass, plastic, or plastic-coated cardboard, these boxes, tubs, jars, and pans come fitted with tight-sealing covers. If the rims and lids remain smooth they often may be re-used; however, the aluminum ones have a tendency to bend as the packages are opened.

Some modern glass canning jars may also be used for freezing most fruits and vegetables. The wide-top jars with tapered sides are advised for liquid packs: the contents will slide out easily without having to be fully defrosted.

Non-Rigid Containers

Non-rigid containers are the moisture/vapor-resistant bags and sheet materials used for Dry pack fruits and vegetables, meats and poultry, fish, and sometimes liquids. They are made of cellophane, heavy aluminum foil, plio-film, polyethylene or laminates of paper, metal foil, glassine, cellophane, and rubber latex.

The best ways of using sheet wrapping are the *butcher wrap* and the *drugstore fold*—both shown in the Freezing Roasts subsection of Chapter 16.

Usually food in bags, and sometimes sheet-wrapped foods, are stored in a cardboard carton (re-usable) for protection and easier stacking in the freezer.

And then there are the so-called "cook-in" or "boil-in" pouches or bags, which are not as readily available as the regular non-boilable ones. Made of a tougher plastic to withstand hard boiling for up to 30 minutes, they are a good deal more expensive than conventional freezer bags; in addition, they come only in relatively small sizes. Also, because they're too stiff to twist and tie tight like the regular bags, they must be heat-sealed with a special appliance. Cook-in-pouch foods are described at the end of Chapter 17, "Freezing Convenience Foods."

Well established are the pressure-"locking" plastic freezer bags of extra-thick plastic. The "lock" occurs when the ridge-in-groove closure is pressed together. The maker of the original self-locking bag has come out with a notably sturdy sealing arrangement. Look for it.

Sealing and Labeling

The packaging is no better than the sealing that closes it.

SEALING RIGID CONTAINERS

Some rigid freezing containers are automatically sealed by their lids, or by screw-type bands, or by flanged snap-on plastic covers.

Air is drawn from a moisture/vapor-proof freezer bag. And the bag is promptly twisted, then looped down, and tied. (*Photographs by Jeffery V. Baird*)

Then there's the waxed-cardboard freezer box with a tuck-in top that is sealed tight shut with freezer tape. If the contents are already sealed in an inner bag, though, you don't have to seal the box top.

There is also the re-usable container like a coffee can with an extra plastic lid: for your seal, tape the lid to the can all around.

The lids for glass jars must have an attached rubber-composition ring or a separate rubber ring to make a seal.

The lids for cans are put on with their own special sealing machine, as effectively for the freezer as for the Pressure Canner.

SEALING NON-RIGID CONTAINERS

Freezing bags are best sealed by twisting and folding the top, and fastening them with string, a rubber band, or a strip of coated wire.

Heat-sealing is possible—but it's tricky unless you have the special equipment for doing it. (Again, see Chapter 17.)

Non-rigid sheet wrappings sometimes can be heat-sealed, but people more often seal all the edges with freezer tape.

LABELING

Use a wide, indelible marking pen to label each package with the name of the contents, the amount, and the date packaged. Also tuck a slip bearing this information inside a clear bag of food.

MAXIMUM FREEZER STORAGE FOR BEST QUALITY MAINTAINED AT ZERO F/−18 C

FOOD	MONTHS AT ZERO F	FOOD	MONTHS AT ZERO F
Fruits		*Fish and Shellfish*	
Apricots	12	Fatty Fish (Mackerel,	3
Peaches	12	Salmon, Swordfish,	
Raspberries	12	etc.)	
Strawberries	12	Lean fish (Haddock, Cod,	6
		Flounder, Trout, etc.)	
		Lobster, Crabs	2
		Shrimp	6
Vegetables		Oysters	3 to 4
Asparagus	8 to 12	Scallops	3 to 4
Beans, green	8 to 12	Clams	3 to 4
Beans, Lima	12		
Broccoli	12		
Cauliflower	12	*Bakery Goods*	
Corn on the cob	8 to 10	*(Precooked)*	
Corn, cut	12	Bread:	
Carrots	12	Quick	2 to 4
Mushrooms	8 to 10	Yeast	6 to 12
Peas	12	Rolls	2 to 4
Spinach	12	Cakes:	
Squash	12	Angel	4 to 6
		Gingerbread	4 to 6
		Sponge	4 to 6
		Chiffon	4 to 6
Meats		Cheese	4 to 6
Beef:		Fruit	12
Roasts, Steaks	12	Cookies	4 to 6
Ground	8	Pies:	
Cubed, Pies	10 to 12	Fruit	12
Veal:		Mince	4 to 8
Roasts, Chops	10 to 12	Chiffon	1
Cutlets, Cubes	8 to 10	Pumpkin	1
Lamb:			
Roasts, Chops	12	*Other Precooked*	
Pork:		*Foods*	
Roasts, Chops	6 to 8	Combination dishes	4 to 8
Ground, Sausage	4	(Stews, Casseroles, etc.)	
Pork or Ham, smoked	5 to 7	Potatoes:	
Bacon	3	French-fried	4 to 8
Variety meats (Liver,	Up to 4	Scalloped	1
Kidneys, Brains, etc.)		Soups	4 to 6
Poultry	6 to 12	Sandwiches	2

14
FREEZING FRUITS

Use only perfect fruit, and treat it with respect. You've invested a good deal of money in a freezer, your packaging materials cost a bit if they're good quality food grade re-usable containers; add your time, your freezer's portion of the monthly bill for electricity, and all these items require that your raw materials be top quality to start with.

Handle only a small quantity at a time—2 or 3 quarts. A good way to wash fruits: put them in a wire basket and dunk it up and down several times in deep, cold water. After peeling, trimming, pitting, and such, fix fruits much as you would for serving. Cut large fruits to convenient size, or crush. Small ones, such as berries, usually are left whole, or just crushed.

Crush soft fruits with a wire potato masher or pastry blender, firm ones in a food chopper. To make purée, press fruits through a colander, food mill, or strainer. (Blenders can liquify too much; food processors liquify even more unless you're careful—the directions for your processor are reliable here.)

Headroom for Fruits

If you use rigid containers—as against bags alone or inside protective paper boxes—you must leave ample headroom so that the expansion of the food during freezing doesn't force off the closures.

The wide-top containers referred to in the following chart are tall, with sides either straight or slightly flared. The narrow-top containers mentioned include canning jars, which may be used for freezing most fruits that are *not packed in liquid*.

TYPE OF PACK	WIDE-TOP CONTAINER		NARROW-TOP CONTAINER	
	Pint	Quart	Pint	Quart
Liquid pack:				
Fruit in juice, sugar, syrup or water; or crushed or puréed	½ inch	1 inch	¾ inch	1½ inches
Fruit juice	½ inch	1 inch	1½ inches	1½ inches
Dry pack:				
Fruit packed without added sugar or liquid	½ inch	½ inch	½ inch	½ inch

CONVERSIONS FOR FREEZING FRUITS

Do look at the conversions for metrics, with their workable roundings-off, and for altitude—both in Chapter 3—and apply them.

To Prevent Darkening

Apples, apricots, peaches, nectarines, pears, and other oxidizing foods are kept from darkening by the addition of ascorbic acid (Vitamin C) either crystalline or tablets; commercial mixtures with an ascorbic acid base; crystalline citric acid; or plain lemon juice. *Or steam-blanching for 3 to 5 minutes, depending on size of the pieces, may be used.*

Fruit has a tendency to float to the top, where it changes color when exposed to the trapped air; so crumple some moisture-resistant food-grade sheet wrapping and put it on the top of the packaged fruit to hold it below the syrup. Seal and freeze.

Ascorbic Acid, Crystalline

Usually available from natural food stores and specialty shops or drugstores. If bought in ounces, figure 1 ounce will give roughly forty ¼-teaspoons or twenty ½-teaspoons (these being the most common amounts called for in freezing). Less expensive are Vitamin C tablets of 500 mg to crush; see Chapter 5 on "Common Ingredients." *There is no record*

YIELDS IN FROZEN FRUIT

Since the legal weight of a bushel of fruits differs among states, the weights given below are average; the yields are approximate.

FRUITS	FRESH	PINTS FROZEN
Apples	1 bu (48 lbs)	32—40
	2½—3 lbs	2
Applesauce	1 bu (48 lbs)	30—36
	2½—3½ lbs	2
Apricots	1 bu (50 lbs)	60—72
	2—2½ lbs	2—3
Berries (excluding strawberries)	24-qt crate	32—36
	5—8 cups	2
Cherries, as picked	1 bu (56 lbs)	36—44
	2—2½ lbs	2
Cranberries	1 peck (8 lbs)	16
Figs	2—2½ lbs	2
Grapes	28-lb lug	14—16
	4 lbs	2
Grapefruit	4—6 fruit	2
Nectarines	18-lb flat	12—18
	2—3 lbs	2
Peaches	1 bu (48 lbs)	32—48
	2—2½ lbs	2
Pears	1 bu (50 lbs)	40—50
	2—2½ lbs	2
Pineapple	2 average	2
	5 lbs	4
Plums and Prunes	1 bu (56 lbs)	38—56
	2—2½ lbs	2
Rhubarb	15 lbs	15—22
	2 lbs	2
Strawberries	24-qt crate	38
	6—8 cups	2

of known undesirable side effects from using ascorbic acid to hold the color of processed foods. It is Vitamin C.

It is the most effective of the agents employed in freezing to prevent darkening, because it will not change the flavor of the food, as the larger amounts needed of citric acid or lemon juice will do.

It dissolves easily in cold water or juices. Figure how much you'll need for one session at a time (see individual instructions), and prepare enough.

For Wet pack with syrup. Add dissolved ascorbic acid to cold syrup and stir.

For Wet pack with sugar. Just before packing, sprinkle the needed amount of ascorbic acid—dissolved in 2 or 3 tablespoons of cold water—over the fruit before you add the sugar.

For Wet pack in crushed fruits and purées. Add dissolved ascorbic acid to the prepared fruit; stir well.

In fruit juices. Add dry ascorbic acid to the juice and stir to dissolve.

In Dry pack (no sugar). Just before packing, sprinkle dissolved ascorbic acid over the fruit; mix gently but thoroughly to coat each piece.

Ascorbic Acid Tablets

It takes 3000 milligrams (mg) worth of ascorbic acid to equal 1 teaspoon of the crystalline form. Crush the tablets between the nested bowls of two large spoons, and dissolve in a little water. The fillers present in all such tablets are no problem.

Citric Acid

Drugstores carry citric acid in pure crystalline form—but again, it is expensive bought this way. National and ethnic grocery stores (and what treasures they offer!) sell it in bulk, chunked like chopped nuts, or cut fine as "sour salt" for kosher cooking, or as "lemon salt" for Greek cooking.

You need three times more citric acid than ascorbic acid to help prevent discoloration. Dissolve the required amount in 2 or 3 tablespoons of cold water. For the individual fruits, add it as for dissolved ascorbic acid, above.

Lemon Juice

A long-time favorite, it contains both citric and ascorbic acids. An equal amount of crystalline ascorbic acid is six times more effective than lemon juice—which also imparts its own flavor to the food.

Steaming

Steaming in a single layer over boiling water is enough to retard darkening in some fruits (for example, apples). The treatments described above are easier, though: they take 3 to 5 minutes, depending on altitude.

Important Note About Sugar: it is too cumbersome to indicate in every individual instance later on that sweetening with sugar—or with any other natural or non-nutritive sweetener—is OPTIONAL. But it is. The amounts and types of sweet syrups and sugar are intended as maximum indications only, and are given for the use of people who are used to added sweetening for a special purpose, and want to include it in the pack.

THE VARIOUS PACKS

A few fruits freeze well without sweetening, but most have a better texture and flavor when packed in sugar or a sugar syrup.

The size and texture of the fruit influences the form in which you pack it for freezing. The intended future use is your deciding factor.

Fruit to freeze whole, in pieces, juiced, crushed or puréed, is packed Dry or Wet.

Dry Pack (Always Sugarless)

The simplest way is just to put whole or cut-up firm fruits in containers (do not add a thing), seal, and freeze. This is especially good for blueberries, cranberries, currants, figs, gooseberries, and rhubarb.

If you have the space, spread raspberries, blueberries, currants, or other similar berries one layer deep on a tray or cookie sheet and set in the freezer. When berries are frozen hard, pour them into polyethylene bags and seal. They won't stick together. Later the bag may be opened; the needed amount taken out, the bag reclosed and returned to the freezer.

Versatile Dry pack lets you use the fruits as if they were fresh.

Wet Packs

This means adding some liquid—such as its own natural juice, sugar syrup, crushed fruit, or water.

Wet pack with sugar. Plain sugar is sprinkled over and gently mixed with the prepared fruit until juice is drawn out and sugar is dissolved. Then you pack and freeze. Fruit fixed this way is especially good for cooked dishes and fruit cocktails. This has less liquid than the Wet pack with syrup.

Wet pack with syrup. Fruit, whole or in pieces, is packed in containers and covered with cold sugar syrup to improve their flavor and make a delightful sauce around the fruit. Generally best for dessert dishes. Plan on using ⅓ to ½ cup of syrup for each pint package of fruit. A 40 percent Syrup (see Sugar Syrups for Freezing Fruit) is used for most fruits, but to keep the delicate flavor of the milder ones, use a thinner syrup. A 50 to 60 percent Syrup is best for sour fruits such as pie cherries.

Wet pack with fruit juice. The fruit—whole, crushed, or in pieces—is packed in the container and covered with juice extracted from good parts of less perfect fruit, and treated with ascorbic acid to prevent darkening. Pack with a piece of crumpled moisture-resistant sheet wrapping on top to hold fruit below the liquid. Seal and freeze. People on sugar-restricted diets can enjoy this unsweetened fruit. Or artificial sweeteners may be added at serving time: see Sweeteners in Chapter 5.

Sugar Syrups for Freezing Fruits

DESIGNATION	SUGAR	WATER	YIELD
30 percent (Thin)	2 cups	4 cups	5 cups
35 percent	2½ cups	4 cups	5⅓ cups
40 percent (Medium)	3 cups	4 cups	5½ cups
50 percent	4¾ cups	4 cups	6½ cups
60 percent (Very Heavy)	7 cups	4 cups	7¾ cups
65 percent	8¾ cups	4 cups	8⅔ cups

Dissolve the sugar thoroughly in cold or hot water (if hot, chill it thoroughly before packing). Syrup can be made the day before and stored in the refrigerator: it must be kept quite cold.

Roughly, estimate ½ to ⅔ cup of syrup for each pint container of fruit.

Substitutions: Generally, ¼ of the sugar may be replaced by light corn syrup without affecting the flavor of the fruit; indeed, the additional blandness is often desirable for delicately flavored fruits, and some cooks prefer substituting even more corn syrup.

Honey or maple syrup may also replace ¼ of the sugar—if the family likes the different flavor either imparts. Brown sugar of course affects the color and, to some degree, the flavor.

For general use, sugar may be dissolved in the juice in the same proportion used in making a sugar syrup suitable for the particular fruit.

A greater degree of natural flavor is kept in the Juice pack, either sweetened or unsweetened, than in the Syrup pack.

Wet pack, purée. Fruit is puréed by forcing it through a food mill, strainer, or colander. Dissolved ascorbic acid or lemon juice is mixed with the purée before packing to prevent darkening. Sweetening may or may not be added.

Wet pack with water. This is similar to the Juice and Syrup packs, except that the added liquid is cold water in which ascorbic acid has been dissolved. The flavor is not as satisfactory as it is in Juice or Syrup packs.

Apples

Apples, more so than most produce, store well by several methods: fresh in a root cellar, or dried, or as canned applesauce or dessert slices. But you may want to freeze a few for late-season cooked dishes—especially in a package shaped for a pie.

SLICES

Peel, core, and slice. As you go, treat against darkening by coating the slices with ½ teaspoon pure ascorbic acid dissolved in each 3 tablespoons of cold water. Or steam-blanch. Less satisfactory but easier: drop slices in a solution of 2 tablespoons salt to each 1 gallon of water (no vinegar) for no longer than 20 minutes; rinse well and drain before packing.

Dry pack, no sugar (for pies). Arrange in a pie plate as for a pie, slip the filled plate into a plastic bag and freeze. Remove the solid chunk of slices from the plate as soon as frozen and overwrap it tightly in moisture/vapor-proof material—as if it were a piece of meat—and return to the freezer. (Handy at pie-making time because you lay the pie-shaped chunk of slices right in your pastry, put on the sugar and seasonings, top with a crust, and bake.)

Wet pack, sugar. Sprinkle ¼ cup sugar over each 1 quart of slices for pie-making. Leave appropriate headroom. Seal; freeze.

Wet pack, syrup. Cover with 40 percent Syrup for use in fruit cocktail or serving uncooked. Leave appropriate headroom. Seal; freeze.

SAUCE

Make applesauce as you like it—strained, chunky, sweetened, or unsweetened.

Wet pack, puréed. Fill containers. Leave ½ inch of headroom. Seal; freeze.

Apricots

HALVES AND SLICES

If apricots do not need peeling, heat them for 30 seconds in boiling water to keep their skins from toughening. Cool immediately in cold water. Cut up as you like.

Wet pack, syrup. Pack in container, cover with 40 percent Syrup to which has been added ¾ teaspoon ascorbic acid to each 1 quart of syrup. Leave appropriate headroom. Seal; freeze.

Wet pack, sugar. Sprinkle ¼ teaspoon ascorbic acid dissolved in ¼ cup water over each 1 quart of apricots. Mix ½ cup of sugar with each 1 quart of fruit, stir until sugar dissolves. Pack in containers with appropriate headroom. Seal; freeze.

CRUSHED OR PURÉED

Wash and treat skins as for whole fruit. Crush or put through a sieve or food mill.

Wet pack, sugar. Mix 1 cup sugar with each 1 quart of crushed or sieved fruit. Add anti-darkening agent. Pack, leaving ½ inch of headroom. Seal; freeze.

Avocados

Most versatile if frozen unflavored but with some anti-discoloration protection. Sweetening for milk shakes or ice cream, etc., may be added when you make them up. And for Guacamole, the further seasonings—minced onion, tomatoes, peppers, etc.—are also better added shortly before serving this delicious dip/spread.

Peel and mash. If intended for future sweet dishes, add ⅛ teaspoon crystalline ascorbic acid to each 1 quart of purée to prevent darkening. If for Guacamole, add 1 tablespoon lemon juice (anti-darkening plus flavoring) and a dash of salt for each 2 avocados as you mash them.

Wet pack, puréed. Leave ½ inch of headroom. Seal; freeze.

Most Soft Berries

WHOLE

Sort, wash gently, and drain: blackberries, boysenberries, dewberries, loganberries, youngberries.

Dry pack, no sugar. Pack in containers, leaving ½ inch of headroom. Or see the alternate Dry method in The Various Packs. Seal; freeze.

Wet pack, syrup. (For berries to be served uncooked.) Pack and cover with 40 to 50 percent Syrup. Leave ½ inch of headroom. Seal; freeze.

Wet pack, sugar. (For berries to be used in cooked dishes.) In a bowl mix ¾ cup sugar with each 1 quart of berries. Mix until sugar dissolves. Pack; leave ½ inch of headroom. Seal; freeze.

CRUSHED OR PURÉED

Wet pack, puréed. Add 1 cup sugar to each 1 quart crushed or puréed berries. Mix well. Pack; leave ½ inch of headroom. Seal; freeze.

Most Firm Berries

Sort and wash blueberries, elderberries, huckleberries. Optional: steam berries for 1 minute to tenderize skins.

WHOLE

Dry pack, no sugar. (For berries to be used in cooked dishes.) Pack; leave ½ inch of headroom. Seal; freeze.

Wet pack, syrup. (For berries to be served uncooked.) Pack; cover with 40 percent Syrup, leaving ½ inch of headroom. Seal; freeze.

CRUSHED OR PURÉED

Wet pack, puréed. Add 1 to 1½ cups sugar to each 1 quart of crushed or puréed berries; stir to dissolve. Leave appropriate headroom. Seal; freeze.

Cranberries

WHOLE

Wash and drain.

Dry pack, no sugar. Fill containers with clean berries. Leave ½ inch of headroom. Seal; freeze.

Wet pack, syrup. Cover with 50 percent Syrup. Leave appropriate headroom. Seal; freeze.

PURÉED

Wash and drain berries. Add 2 cups water to each 1 quart (1 pound) berries and boil until skins burst. Press through a sieve and add 2 cups sugar to each 1 quart purée. Mix.

Wet pack, puréed. Pack; leave appropriate headroom. Seal; freeze.

Currants

Wash; remove stems.

WHOLE

Dry pack, no sugar. Treat like Cranberries. Seal; freeze.

Wet pack, sugar. Add ¾ cup sugar to each 1 quart of fruit; stir gently to dissolve. Pack; leave appropriate headroom. Seal; freeze.

Wet pack, syrup. Treat like Cranberries. Seal; freeze.

CRUSHED

Wet pack, crushed. Add 1⅛ cups sugar to each 1 quart crushed currants; stir to dissolve sugar. Pack; leave appropriate headroom. Seal; freeze.

JUICE

For beverages, use ripe currants. For future jellies, mix in some slightly underripe currants for added pectin.

Crush currants and warm to 165 F/74 C over low heat. Drain through a jelly bag. Cool.

Wet pack, juice. Sweeten with ¾ to 1 cup sugar to each 1 quart of juice, or pack unsweetened. Leave appropriate headroom. Seal; freeze.

Gooseberries

Wash; remove stems and tails.

Dry pack, no sugar. (Best for future pies and preserves.) Pack whole berries; leave ½ inch of headroom. Seal; freeze.

Wet pack, syrup. Cover whole berries with 50 percent Syrup. Leave appropriate headroom. Seal; freeze.

Raspberries

The versatile and very tender raspberries freeze even better than strawberries do. The wild ones, though small, have fine flavor. Real seedy berries are best used in purée or as juice.

WHOLE

Sort; wash very carefully in cold water and drain thoroughly.

Dry pack, no sugar. Fill containers gently, leaving ½ inch of headroom. Seal; freeze.

Wet pack, sugar. In a shallow pan, carefully mix ¾ cup sugar with each 1 quart of berries so as to avoid crushing. Pack; leave ½ inch of headroom. Seal; freeze.

Wet pack, syrup. Cover with 40 percent Syrup. Leave appropriate headroom. Seal; freeze.

CRUSHED OR PURÉED

Crush or sieve washed berries.

Wet pack, juice. Add ¾ to 1 cup sugar to each 1 quart of berry pulp; mix to dissolve sugar. Pack; leave appropriate headroom. Seal; freeze.

JUICE

Select fully ripe raspberries. Crush and slightly heat berries to start juice flowing. Strain through a jelly bag.

Wet pack, juice. For beverage, sweeten with ½ to 1 cup sugar to each 1 quart of juice. (For future jelly, do not sweeten.) Pour into containers; leave appropriate headroom. Seal; freeze.

Strawberries

Choose slightly tart firm berries with solid red centers. Plan to slice or crush the very large ones. Sweetened strawberries hold better than unsweetened.

WHOLE

Sort; wash in cold water; drain. Remove hulls.

Wet pack, sugar. In a shallow pan, add ¾ cup sugar to each 1 quart of berries and mix thoroughly. Pack; leave ½ inch of headroom. Seal; freeze.

Wet pack, syrup. Cover berries with cold 50 percent Syrup. Leave appropriate headroom. Seal; freeze.

Wet pack, water (unsweetened). To protect the color of the berries, cover them with water in which 1 teaspoon crystalline ascorbic acid to each 1 quart of water has been dissolved. Leave appropriate headroom. Seal; freeze.

SLICED OR CRUSHED

Wash and hull as for whole berries, then slice or crush partially or completely.

Wet pack, sugar. Add ¾ cup sugar to each 1 quart of berries in a shallow pan. Mix thoroughly. Pack; leave appropriate headroom. Seal; freeze.

JUICE

Crush berries; drain juice through a jelly bag.

Wet pack, juice. Add ⅔ to 1 cup sugar to each 1 quart of juice—or omit sugar if you wish. Pour into containers; leave appropriate headroom. Seal; freeze.

Cherries, Sour (for Pie)

As pie timber these are better canned ready to bake (see two recipes in Chapter 12); but if you want to freeze them, here's how.

WHOLE

Use only tree-ripened cherries. Stem, wash, drain, and pit. (The rounded end of a clean paper clip makes a good cherry-pitter.)

Wet pack, sugar. Add ¾ cup sugar to 1 quart of pitted cherries; stir until dissolved. Pack, leaving appropriate headroom. Seal; freeze.

Wet pack, syrup. Cover pitted cherries with cold 60 to 65 percent Syrup. Leave appropriate headroom. Seal; freeze.

CRUSHED

Wet pack, juice. Add 1 to 1½ cups sugar to each 1 quart of crushed cherries. Mix well. Pack; leave appropriate headroom. Seal; freeze.

PURÉED

Wet pack, juice. Crush cherries; heat just to boiling and press through a sieve or food mill. Add ¾ cup sugar to each 1 quart of purée. Pack; leave appropriate headroom. Seal; freeze.

JUICE

Home-made cherry juice in a party punch makes it exceptional!

Wet pack, juice. Crush cherries, heat slightly (*do not boil*) to start juice flowing. Strain through a jelly bag. Add 1½ to 2 cups sugar to each 1 quart of juice; or pack unsweetened. Pour into containers; leave appropriate headroom. Seal; freeze.

Cherries, Sweet

The dark and "black" varieties are best for freezing—but do handle them quickly to prevent color and flavor changes. Use only tree-ripened fruit and remove the pits: they give an almond flavor to cherries when frozen.

WHOLE

Wet pack, syrup. Cover pitted cherries with 40 percent Syrup in which you've dissolved ½ teaspoon crystalline ascorbic acid to each 1 quart of syrup. Leave appropriate headroom. Seal; freeze.

CRUSHED

Wet pack, juice. To each 1 quart of crushed cherries add 1½ cups sugar and ¼ teaspoon crystalline ascorbic acid; mix well. Pack; leave appropriate headroom. Seal; freeze.

JUICE

Sweet red cherries and sweet white cherries are handled differently for juice.

Heat sweet *red* cherries slightly (to 165 F/74 C) to start the juice. Strain through a jelly bag.

Crush sweet *white* cherries *without heating*. Strain through a jelly bag. Then warm this juice in a double boiler or over low heat to 165 F/74 C. Cool the red or white juice and let it stand covered overnight.

Wet pack, juice. Pour off the clear juice into containers, being careful not to include any sediment from the bottom of the kettle. Add 1 cup sugar to each 1 quart of juice; or leave unsweetened if you prefer. Leave appropriate headroom. Seal; freeze. (Sweet cherry juice by itself is pretty blah. So mix some *sour* cherry juice with the sweet to make a better beverage.)

Coconut

If you have a windfall, freeze some simply for fun—you may want to have a Mainland luau!

Puncture the "eye" of the coconut; drain out and save the milk. Remove the meat from the broken-open shell. Shred it, or put it through a food chopper.

Wet pack, juice. Cover shredded meat with coconut milk. Leave appropriate headroom. Seal; freeze.

Dates

Wash, if necessary, and dry on paper toweling; remove pits.

Dry pack, no sugar. Pack in containers with no headroom. Seal; freeze.

Figs

WHOLE OR SLICED

Only tree-ripened, soft-ripe fruit, please; and check a sample for good flavor clear through the flesh. Sort, wash, and cut off stems. Peeling is optional.

Dry pack, no sugar. Fill containers with the prepared figs; leave appropriate headroom. Seal; freeze.

Wet pack, syrup. Cover with 35 percent Syrup to which you have added ¾ teaspoon crystalline ascorbic acid—or ½ cup lemon juice—to each 1 quart of syrup. Seal; freeze.

Wet pack, water. Pack figs; cover with water to which you have added ¾ teaspoon crystalline ascorbic acid to each 1 quart of water. Leave appropriate headroom. Seal; freeze.

CRUSHED

Wet pack, juice. Crush prepared figs. Mix ⅔ cup sugar and ¼ teaspoon crystalline ascorbic acid with each 1 quart of crushed fruit. Leave appropriate headroom. Seal; freeze.

Fruit Cocktail (or Compôte)

Freezing is excellent for your favorite combinations of fruit to serve either as an appetizer or dessert. A few added blueberries or dark sweet cherries make a nice color contrast.

Use any combination of fruits peeled, cored, etc., and cut to suitable size.

Wet pack, syrup. Pack. Cover with cold 30 to 40 percent Syrup in which ¾ teaspoon crystalline ascorbic acid to each 1 quart of syrup has been dissolved. If cut-up oranges are in the mixture, the ascorbic acid may be omitted. Leave appropriate headroom. Seal; freeze.

Grapefruit (and Oranges)

Commercial processors do a fine job with citrus fruits. It's hardly worthwhile to compete unless you've a surplus of grapefruit and/or oranges.

Use heavy, blemish-free, tree-ripened fruits.

SECTIONS OR SLICES

Wash; peel, cutting off the outside membranes. Cut a thin slice from each end. With a sharp, thin-bladed knife, cut down each side of the membranes and lift out the whole sections. Work over a large bowl to catch the juice. Remove seeds. Oranges may be sliced.

Wet pack, syrup. Cover fruit with 40 percent Syrup made with excess fruit juice, and water if needed. (For better quality, add ½ teaspoon crystalline ascorbic acid to each 1 quart of syrup before packing.) Leave appropriate headroom. Seal; freeze.

JUICE

Use good tree-ripened fruits. Squeeze, using a squeezer that does not press oil from the rind.

Wet pack, juice. Either sweetened with 2 tablespoons sugar to each 1 quart of juice, or pack unsweetened. (For best quality, add ¾ teaspoon crystalline ascorbic acid to each 1 gallon of juice before packing.) Pour into glass freezing jars. Leave appropriate headroom. Seal; freeze.

Grapes

WHOLE OR HALVES

Use firm-ripe grapes with tender skins and nice color and flavor. Wash and stem. Leave seedless grapes whole; cut other varieties in half and remove their seeds.

Dry pack, no sugar. Leave appropriate headroom. Seal; freeze.

Wet pack, syrup. Cover grapes with cold 40 percent Syrup. Leave appropriate headroom. Seal; freeze.

JUICE

For a beverage or future jelly-making, use firm-ripe grapes.

Wash, stem, crush. Do *not* heat. Strain through a jelly bag. Allow juice to stand overnight in the refrigerator while sediment settles to the bottom. Carefully pour off the clear juice, leaving tartaric crystals behind.

Wet pack, juice. Pour into containers; leave appropriate headroom. Seal; freeze. (If tartrate crystals—the basis for cream of tartar—form in frozen juice, strain them out after the juice thaws.)

Melons

SLICES, CUBES, OR BALLS

Cut firm-ripe melons in half; remove seeds and soft tissues holding them. If for slices or cubes, cut off all rind; cut to shape. If for balls, do not cut off rind, but scoop out with a baller, taking care not to include any rind.

Wet pack, syrup. Cover with 30 percent Syrup. Leave appropriate headroom. Seal; freeze.

CRUSHED (NOT FOR WATERMELON)

Halve, cut off rind; remove seeds and their soft tissue. Crush or put through the food chopper, using a coarse knife.

Wet pack, juice. Add 1 tablespoon sugar to each 1 quart of crushed melon, if you wish (and an added 1 teaspoon lemon juice points up the flavor). Stir to dissolve. Pack; leave appropriate headroom. Seal; freeze.

Nectarines

These are not as satisfactory frozen as most other fruits are.

HALVES, QUARTERS, OR SLICES

Choose only firm, fully ripe nectarines—avoiding overripe ones, which often develop a disagreeable flavor in the freezer.

Wash and pit. Peeling is optional.

Wet pack, syrup. Put ½ cup of 40 percent Syrup in each container and cut fruit directly into it. (For a better product add ½ teaspoon crystalline ascorbic acid to each 1 quart of syrup before packing.) Gently press fruit down and add extra syrup to cover. Top with crumpled moisture-resistant wrap to hold fruit in place. Leave appropriate headroom. Seal; freeze.

PURÉED

Treat like Peach Purée.

Peaches

Peaches are excellent canned; they freeze well.

HALVES AND SLICES

Use firm, ripe peaches without any green color on their skins. Wash, pit, and peel. (They are less ragged if peeled without the boiling-water dip.)

Wet pack, sugar. Coat cut peaches with a solution of $\frac{1}{4}$ teaspoon crystalline ascorbic acid dissolved in each $\frac{1}{4}$ cup of water to prevent darkening. Add $\frac{2}{3}$ cup of sugar to each 1 quart of fruit, and mix gently. Pack, leaving appropriate headroom.

Wet pack, syrup. Put $\frac{1}{2}$ cup 40 percent Syrup in the bottom of each container. Cut peaches directly into it. (For better product add $\frac{1}{2}$ teaspoon crystalline ascorbic acid to each 1 quart of the syrup before packing.) Gently press fruit down and add extra syrup to cover. Top with crumpled moisture-resistant wrap to hold fruit in place. Leave appropriate headroom. Seal; freeze.

Wet pack, water. Cover cut peaches with water in which 1 teaspoon crystalline ascorbic acid has been dissolved in each 1 quart of water. Leave appropriate headroom. Seal; freeze.

CRUSHED OR PURÉED

Loosen skins by dipping peaches in boiling water for 30 to 60 seconds. Cool immediately in cold water; peel and pit.

Crush coarsely. For purée, press through a sieve or food mill; it's easier to make the purée if you heat the peaches in a very little water for 4 minutes before you sieve them.

Wet pack, juice. Mix 1 cup sugar and $\frac{1}{8}$ teaspoon crystalline ascorbic acid with each 1 quart of peaches. Pack; leave appropriate headroom. Seal; freeze.

Pears

Use Bartlett or a similar variety—not any of the so-called winter pears, which keep in cold storage (see "Root-Cellaring," Chapter 22).

HALVES OR QUARTERS

Choose well-ripened pears, firm but not hard. Wash, cut in halves and quarters; core. Cover them with cold water to prevent their oxidizing during preparation (leaching is negligible because immersion time is so short).

Wet pack, syrup. Handling no more than 3 pints at a time in a deep-fry basket, lower cut-up pears into boiling 40 percent Syrup for 1 to 2 minutes. Drain; cool. (Save the hot syrup for another load of fruit.) To pack, cover cooled pears with cold 40 percent Syrup to which has been added $\frac{3}{4}$ teaspoon crystalline ascorbic acid to each 1 quart of syrup. Leave appropriate headroom. Seal; freeze.

PURÉED

Wash well-ripened pears that are not hard or gritty. Peeling is optional. Proceed as for Peach Purée.

Persimmons

PURÉED

Purée made from late-ripening native ones needs no sweetening, but nursery varieties may be packed with or without sugar.

Choose orange-colored, soft-ripe persimmons. Sort, wash, peel, and cut in sections. Press through a sieve or food mill. Mix ⅛ teaspoon crystalline ascorbic acid—or 1½ teaspoons crystalline citric acid—with each 1 quart of purée.

Wet pack, juice (unsweetened). Pack unsweetened purée. Leave appropriate headroom. Seal; freeze.

Wet pack, juice (sweetened). Mix 1 cup sugar with each 1 quart of purée. Pack; leave appropriate headroom. Seal; freeze.

Pineapple

Use firm, ripe pineapple with full flavor and aroma. Pare, removing eyes, and core. Slice, dice, crush, or cut in wedges or sticks.

Wet pack, syrup. Pack fruit tightly. Cover with 30 percent Syrup made with pineapple juice, if available, or water. Leave appropriate headroom. Seal; freeze.

Wet pack, juice (unsweetened). Pack fruit tightly without sugar: enough juice will squeeze out to fill the crevices. Leave appropriate headroom. Seal; freeze.

Plums (and Prunes)

Frozen plums and prunes are good in pies and jams, salads, and desserts. Use the unsweetened pack for future jams. To serve unsweetened whole plums raw, see below.

WHOLE, HALVES, OR QUARTERS

Choose tree-ripened fruit with deep color. Wash. Cut as desired. Leave pits in fruits you freeze whole.

Wet pack, syrup. Cover with cold 40 to 50 percent Syrup in which is dissolved ½ teaspoon crystalline ascorbic acid to each 1 quart of syrup. Leave appropriate headroom. Seal; freeze.

Wet pack, juice (unsweetened). Pack plums tightly. Leave appropriate headroom. Seal; freeze. (To serve whole plums uncooked, dip them in cold water for 5 to 10 seconds; remove skins, and cover with 40 percent Syrup to thaw. Serve in the syrup.)

PURÉED

Purée may be made from heated or unheated fruit, depending on its softness.

Wash plums, cut in half and pit. *Unheated fruit:* press raw through a sieve or food mill. Add ¼ teaspoon crystalline ascorbic acid—or ½ teaspoon crystalline citric acid—to each 1 quart of purée. *Heated fruit* (the firm ones): add 1 cup water to each 4 quarts of plums; boil for 2 minutes; cool, and press through a sieve or food mill.

Wet pack, juice. Mix ½ to 1 cup sugar with each 1 quart of purée. Pack; leave appropriate headroom. Seal; freeze.

JUICE

Wash plums, simmer until soft in enough water barely to cover. Strain through a jelly bag and cool the juice.

Wet pack, juice. Add 1 to 2 cups sugar to each 1 quart of juice. Pour into containers; leave appropriate headroom. Seal; freeze.

Rhubarb

Freeze only firm, young, well-colored stalks with good flavor and few fibers. (See also "Canning Fruits," Chapter 7.)

PIECES

Wash, trim, and cut in 1- to 2-inch pieces, or longer to fit the package. Heating rhubarb in boiling water for 1 minute and cooling immediately in cold water helps to set the color and flavor.

Dry pack, no sugar. Pack either raw or preheated (and now cold) rhubarb tightly in containers. Leave appropriate headroom. Seal; freeze.

Wet pack, syrup. Pack either raw or preheated (and now cold) rhubarb tightly. Cover with cold 40 percent Syrup. Leave appropriate headroom. Seal; freeze.

PURÉED

Prepare as pieces. Add 1 cup water to each 6 cups of rhubarb and boil 2 minutes. Cool immediately; press through a sieve or food mill.

Wet pack, juice. Add ⅔ cup sugar to each 1 quart of purée. Pack; leave appropriate headroom. Seal; freeze.

JUICE

Select as for pieces. Wash, trim, and cut in 4- or 5-inch lengths. Add 4 cups water to each 4 quarts of rhubarb, and bring just to a boil. Strain through a jelly bag.

Wet pack, juice. Pour into containers. Leave appropriate headroom. Seal; freeze.

Tomatoes

See Chapter 15, "Freezing Vegetables"; Chapter 17, "Freezing Convenience Foods"; and of course, Chapter 8, "Canning Tomatoes".

15
FREEZING VEGETABLES

Freeze only prime vegetables that are garden-fresh and tender-young—younger, usually, than for canning. Freeze them in small batches, refrigerating overnight if you can't freeze them promptly the day they're picked. Any vegetable that cans well freezes equally well at home, with only several exceptions. These, in their raw state, are whole tomatoes, greens for salads, white ("Irish") potatoes, and cabbage. Because they have a high water content, home-freezing allows large ice crystals to form and rupture their flesh; the result, defrosted, is flabby or shapeless.

Certain vegetable varieties are better for freezing than others, so read your seed catalogs to see which ones you'll have the most luck with. Or ask your County Agent for good performers in your area. Or, a truck gardener can tell you (but sometimes the person tending his roadside stand cannot).

CONVERSIONS FOR FREEZING VEGETABLES

Do look at the conversions for metrics, with workable roundings-off, and for altitude—both in Chapter 3—and apply them.

General Preparation

The first step, after you've gathered your packaging, etc., is to wash the vegetables. Use cold water and lift the vegetables out of it to leave any grit in the bottom of the pan.

You may need to take a further step to draw out possible insects in broccoli, Brussels sprouts, and cauliflower: simply soak them for ½ hour in a solution of 1 tablespoon salt to each 1 quart of cold water; insects will float to the surface, to be skimmed off. Wash vegetables again in fresh cold water to get rid of the salt. Sort the vegetables according to size; peel, trim, and cut to size as needed.

YIELDS IN FROZEN VEGETABLES

Since the legal weight of a bushel of vegetables differs among states, the weights given below are average; the yields are approximate.

VEGETABLES	FRESH	PINTS FROZEN
Asparagus	1 bu (45 lbs)	8–11
	3–4 lbs	3–4
Beans, Lima, in pods	1 bu (32 lbs)	12–16
	4–5 lbs	2
Beans, snap/green/wax	1 bu (30 lbs)	30–45
	1½–2 lbs	2
Beets, without tops	1 bu (52 lbs)	35–42
	2½–3 lbs	2
Broccoli	25-lb crate	24
	2–3 lbs	2
Brussels sprouts	4 qts	6
	1 lb	1
Carrots, without tops	1 bu (50 lbs)	32–40
	2½–3 lbs	2
Cauliflower	2 medium heads	3
	1 bu (12 lbs)	8–12
Corn, in husks	1 bu (35 lbs)	14–17
	4–5 lbs	2
Eggplant	2 average	2
Kale	1 bu (18 lbs)	12–18
	2–3 lbs	2
Okra	1 bu (26 lbs)	34–40
	1½ lbs	2
Peas, green, in pods	1 bu (30 lbs)	12–15
	2–2½ lbs	1
Peppers, sweet	⅔ lb (3 peppers)	1
Pumpkin	50 lbs	30
	3 lbs	2
Spinach (most greens)	1 bu (18 lbs)	12–18
	2–3 lbs	2
Squash, summer	1 bu (40 lbs)	32–40
	2–2½ lbs	2
Squash, winter (strained)	3 lbs	2

Blanching

Even after vegetables are picked, the enzymes in them make them lose flavor and color and sometimes make them tough—*even at freezer temperatures*. Therefore the enzymes must be stopped in their tracks by being heated for a few minutes (how many minutes depends on the size and texture of the vegetable) before the vegetables are cooled quickly and packed. This preheating is necessary for virtually all vegetables: green (sweet) peppers are the notable exception.

In Boiling Water

Most vegetables are easily blanched in boiling water. Do no more than *1 pound* of prepared vegetables at a time, submerged in *4 quarts* of briskly boiling water: these proportions let water keep the boil and agitate food for uniform treatment. The kettle should be large; ideally it has a wire basket that holds the vegetables and fits down into it. Otherwise, gather the food loosely in a large square of cotton cheesecloth, knot the corners together, and plop the bundle into boiling water. Shake the basket twice, or slosh the wrapped food up and down. When the specified time is up, lift the food out and dunk it promptly in icy water to cool it fast. Spread it on clean paper toweling, pat off the water, and pack.

Altitude Note: Dr. Pat Kendall of Colorado State University Extension offers a good rule-of-thumb for the "Centennial State," where average altitude is 4000 ft / 1219 m above sea level. She says that dwellers on higher ground should preheat vegetables for *one minute longer* than the sea-level-zone requirement—and should *not increase blanching time further* at greater altitude. Workable advice for any such hill-country anywhere. (See also Correcting for Altitude in Chapter 3.)

Blanching in Steam

A few vegetables are better if heated in steam, and some may be done in either steam or boiling water.

For steaming, use a large kettle with a tight lid and a rack that holds a steaming basket at least 3 inches above the bottom of the kettle. Put in 1 or 2 inches of water and bring it to a boil.

Put your prepared vegetables in the basket in only a single layer, so the steam can reach all parts quickly. Cover the kettle and keep heat high. Start counting the time as soon as the cover is on.

As in the altitude note added to the boiling-water blanch, add 1 minute to steaming time if you live 4000 ft / 1219 m above sea level.

Blanching in a Microwave Oven

Follow the instructions that come with your oven. This is good sense, not laziness on our part, because, aside from their varied sophistication, ovens differ in capacity and in the wattage that runs them: both factors affect the way each oven treats food.

Microwave blanching can be a fussy business, but you can't go badly wrong with the process. Vegetables' color is superior to that retained in other blanching methods.

Other Ways to Preheat

Pumpkins, squash and sweet potatoes are best fully cooked in a pressure *cooker* (if you use one) or baked in an oven; when done, they are scooped out, mashed/strained, cooled, and frozen. Cleaned, trimmed, whole, or sliced mushrooms may be pan-broiled in a nonstick skillet, or in a little butter/margarine (a little, because fat acts as an insulator and generous amounts of it can reduce storage life of a food); sautéed this way, they produce some juice, which is frozen with them.

Cool after Blanching

Cool all vegetables as quickly as possible after they've been preheated. Use plenty of ice water, and change it often to keep it cold. It takes time to chill vegetables properly.

When they are completely cooled, drain them well on clean, absorbent toweling: you want as little dampness as possible in the pack.

The Packs

Vegetables for freezing may be packed either dry or in brine. The Dry pack is easier and lets you use the vegetables as if they were fresh, so Dry pack is the method we'll use the most. To make packaging dry-packed vegetables (and some fruits) easier, just place a single layer of any freezer-ready small vegetable on a tray and sharp-freeze it fast (near −20 F/ −29 C). Then pour the frozen vegetable into a freezer-type container and seal. Because the pieces are not stuck to each other, you can pour out the amount needed, reclose and seal the container, and return it and its partial contents to the freezer.

Cooking Frozen Vegetables

The secret of cooking frozen vegetables well, if it is indeed a secret, is to cook them in a minimum amount of water—*no water* if they're to be microwaved—and only until the texture suits you. A bit of the cooking already happened in blanching.

Most are best cooked *without* thawing, but defrost greens enough to separate the leaves.

Generally you bring to the boil ½ cup of water for each 2 cups of frozen vegetables. Add the food, cover, and begin to count cooking time when water returns to the boil. Exceptions: *1 cup* of water for each 2 cups of lima beans; water to cover for corn-on-the-cob.

Remember that at high altitudes, water boils at lower temperatures the more you rise above sea level.

Cooking times. Spinach—3 minutes; turnip greens—15 to 20 minutes; all other greens—8 to 12 minutes.

Depending on size of pieces: large lima beans, cut green/snap/wax beans, broccoli, carrots, cauliflower, corn in all forms, green peas—all from 3 to 10 minutes.

Kohlrabi (and similar-textured vegetables)—8 to 10 minutes.

Summer squash—up to 12 minutes.

Microwave Cooking

Microwaving is ideal for cooking any vegetable, fresh or frozen. Follow the maker's directions for your particular microwave oven.

GETTING DOWN TO FREEZING

Asparagus

Sort for size, wash well. Peel ends back about 2 inches from the bottom, cut off tough ends. Leave spears in uniform length to fit package, or cut in 2-inch pieces.

Blanch. By boiling thin stalks 2 minutes, medium 3 minutes, thick 4 minutes.

Pack. With headroom; for spears, alternate tips and ends down (wide containers, *all* tips down). Seal; freeze.

Beans, Lima

Handier canned, but freeze tenderest ones if you can afford the space. Shell, wash, sort for size.

Blanch. In boiling water, small beans 2 minutes, medium 3 minutes, large 4 minutes; cool promptly and drain well.

Pack. With $\frac{1}{2}$ inch of headroom. Seal; freeze.

Beans—Snap/String/Green/Italian

These also can well. Fancy young tender ones are better frozen.

Cut in 1- or 2-inch pieces, or in lengthwise strips (frenching), or leave whole if they're very young and tender.

Blanch. In *soft* boiling water—for 3 minutes. Cool immediately, drain.

Pack. Leave $\frac{1}{2}$ inch of headroom. Seal; freeze.

Beets

Baby ones are worth freezer space. (Why not can larger ones plain or pickled?)

Wash and sort for size—maximum 3 inches, small are best. Leave on tails and ½ inch of stem so their juice won't bleed out while boiling.

Boil. Until tender—25 to 30 minutes for small beets, 45 to 50 for medium. Cool quickly. Slip off skins; trim and cut in slices or cubes.

Pack. Leave ½ inch of headroom for cubes; no headroom for whole or sliced. Seal; freeze.

Broccoli

Peel coarse stalks, trimming off leaves and blemishes; split if necessary. Salt-soak for ½ hour (1 tablespoon salt for each 1 quart cold water) to drive out bugs; wash well. Sort for uniform spears, or cut up.

Blanch. In steam—5 minutes for stalks; in boiling water—3 minutes for stalks. (Reduce blanching time for cut-up or chopped.) Cool immediately; drain.

Pack. Leave no headroom for spears or large chunks; arrange stalks so blossom ends are divided between either end of the container. Leave ½ inch of headroom for cut-up or chopped (they have less air space). Seal; freeze.

Brussels Sprouts

Give freezer space only to the best heads.

Salt-soak as for Broccoli. Wash well. Trim off outer leaves. Sort for size.

Blanch. In boiling water—small heads for 3 minutes, medium heads for 4 minutes, large heads for 5 minutes. Cool immediately, drain well.

Pack. Leave no headroom. Seal; freeze.

Cabbage (and Chinese Cabbage)

Plan to use these only in cooked dishes: after being frozen they aren't crisp enough for salads.

Trim off coarse outer leaves; cut heads in medium or coarse shreds or thin wedges, or separate the leaves.

Blanch. In boiling water—1½ minutes. Cool immediately and drain.

Pack. Leave ½ inch of headroom. Seal; freeze.

Carrots

These cold-store and can well, so freeze only the fancy young ones (preferably whole).

Remove tops, wash, and peel. Leave baby ones whole; cut others into ¼-inch cubes, thin slices or lengthwise strips.

Blanch. In boiling water—tiny whole ones for 5 minutes; dice, slices, or lengthwise strips for 2 minutes. Cool immediately; drain.

Pack. Leave ½ inch of headroom. Seal; freeze.

Cauliflower

Infinitely better frozen than canned.

Break or cut flowerets apart in pieces about 1 inch across. Salt-soak as for Broccoli for ½ hour to get rid of bugs, etc. Wash thoroughly; drain.

Blanch. In boiling salted water (1 teaspoon salt to each 1 quart of water)—3 minutes. Cool immediately; drain.

Pack. Leave no headroom. Seal; freeze.

Celery

Usable only in cooked dishes, so assign it freezer space accordingly. Tender leaves may be cut small and frozen in small packets for flavoring soups and stews; finely minced, it is a basic for ragú sauces or braising mixtures. A number of condiments use celery: see Corn Relish in Chapter 19, and freeze it accordingly in recipe-size amounts.

Strip any coarse strings from any young stalks; wash well, trim, and cut in 1-inch pieces.

Blanch. In boiling water—3 minutes. Cool immediately; drain.

Pack. Leave ½ inch of headroom. Seal; freeze.

Corn

Feasibility for freezing sweet corn: whole-kernel, Yes (it's better than canning); cream-style, Maybe (it's certainly handier canned, and there's not much difference in the product); on-the-cob, No—unless you've got loads of freezer space and don't mind thawing it before cooking it for the table (it shouldn't be popped frozen into the pot because the kernels will be cooked to death by the time the core of the cob is hot through).

WHOLE-KERNEL

Choose ears with thin, sweet milk; husk, de-silk, and wash. (Cut from cob *after* blanching.)

Blanch. In boiling water—4 minutes. Cool ears immediately; drain.

Pack. Cut from cob about ⅔ the depth of the kernels, and don't scrape in any milk. Leave ½ inch of headroom. Seal; freeze.

CREAM-STYLE

Choose ears with thick and starchy milk. Husk, de-silk, and wash. (Cut from cob *after* blanching.)

Blanch. In boiling water—4 minutes. Cool immediately; drain.

Pack. Cut from the cob at about the center of the kernels, then scrape the cobs with the back of the knife to force out the hearts of the kernels and the juice (milk); mix with cut corn. Pack, leaving ½ inch of headroom. Seal; freeze.

ON-THE-COB

Choose ears with thin, sweet milk (as for whole-kernel). Husk, de-silk, wash; sort for size.

Blanch. In boiling water—small ears (1¼ inches or less in diameter) for 7 minutes, medium ears (to 1½ inches) for 9 minutes, large ears (over 1½ inches) for 11 minutes. Drain on terrycloth, and refrigerate immediately on dry toweling, in a single layer.

Pack. In containers, or wrap in moisture/vapor-resistant material. Seal; freeze.

MAVERICK FREEZING IN THE HUSK

People who know the *Why*s and the *How*s of freezing say: "Never freeze corn without blanching it first to stop enzymatic action." But one hears of corn frozen successfully in its husk (though de-silked), without blanching.

Without husking, pull out the silk; and, to save freezer space, remove a little of the outer husk. Do not blanch. Pack in freezer bags; freeze.

Eggplant

Choose glossy, rather small fruits whose seeds are tender. Because most people are watching their sodium intake, we opt for steam-blanching without any salting to draw out juice (blanching also reduces oxidation).

Very young eggplant need not be peeled. If to be fried, cut in ¾-inch slices; for casseroles or in mixed vegetables, dice or cut in strips. Steam-blanch 2 minutes for small dice/thin slices, up to 5 minutes for thick slices. Chill in cold water to which 4 teaspoons of lemon juice have been added to each 1 gallon of water. Drain, pat dry.

Pack. Leave ½ inch of headroom. Seal; freeze.

Greens, Garden

Remove imperfect leaves, trim away tough midribs and tough stems; cut large leaves (like chard) in pieces. Wash carefully, lifting the leaves from the water to let silt settle.

Blanch. In boiling water, and shake the pot to keep the leaves separated—spinach, New Zealand spinach, kale, chard, mustard and beet and turnip greens: all for 2 minutes; collards for 3 minutes. (Steam-blanching causes leaf vegetables to mat, and thus prevents correct blanching.) Cool immediately; drain.

Pack. Leave ½ inch of headroom. Seal; freeze.

Greens, Wild

Collect and clean fiddleheads (ostrich fern).

Blanch. For 2 minutes. Cool and drain.

Dandelions: if you like slightly bitter taste, merely blanch the very tenderest leaves for 1½ minutes; otherwise boil in two or more waters. Cool and drain. Milkweed: boil in several waters. Cool and drain.

Pack. Leave ½ inch of headroom. Seal; freeze.

Jerusalem Artichokes

Treat like Kohlrabi or small Turnips.

Kohlrabi

Cut off the tops and roots of small to medium kohlrabi. Wash, peel; leave whole or dice in ½-inch cubes.

Blanch. In boiling water—whole for 3 minutes, cubes 1 minute. Cool immediately and drain.

Pack. Whole in containers or wrap in moisture/vapor-resistant material. Cubes in containers, leaving ½ inch of headroom. Seal; freeze.

Mushrooms

Wash carefully in cold water. Cut off ends of stems. Leave stems on fancy small buttons if you like; if mushrooms are larger than 1 inch across the caps, slice or quarter them. If serving cold (in salads, etc.), blanch in steam; if serving hot (as garnish for meats, or in combination dishes), pre-cook.

Blanch. In one layer, over steam—whole for 5 minutes, quarters or small caps for 3½ minutes, slices for 3 minutes. (This also prevents darkening; see To Prevent Darkening early in Chapter 14.) Cool immediately; drain.

Precooking. In table fat—sauté in a skillet until nearly done. Air-cool, or set the skillet in cold water (you'll freeze them in the good buttery juice from the pan).

Pack. Leave ½ inch of headroom. Seal; freeze.

Okra (Gumbo)

Use in soups and stews.

Wash. Cut off stems, being careful not to open the seed cells.

Blanch. In boiling water—small pods 3 minutes, large pods 4 minutes. Cool immediately; drain. Leave whole, or cut in crosswise slices.

Pack. Leave ½ inch of headroom. Seal; freeze.

Parsnips

Really best left in the ground over winter for the first fresh treat of spring—freezing is only a second choice. Treat like Carrots.

Peas, Black-eyed (Cowpeas, Black-eyed Beans)

Shell; save only the tender peas (see Sorting Trick, coming soon).
   Blanch. In boiling water—for 2 minutes. Cool immediately, drain well.
   Pack. Leave ½ inch of headroom. Seal; freeze.

Peas, Green

Shell; use only sweet, tender peas (see Sorting, coming soon). For Edible-pod/Snow types, see Peas, Green, in Chapter 9; continue as below for freezing.
   Blanch. In boiling water—for 1½ minutes. Cool immediately; drain.
   Pack. Leave ½ inch of headroom. Seal; freeze.
Sorting trick for peas: To size peas as a guide to their tenderness or maturity, make a solution in the proportions of 1½ cups regular canning-pickling salt to 1 gallon of cool water, and put the peas in it: floating peas are likely to be the very tender ones, while peas that sink are usually older or more mature. Lift out the floaters in a strainer, rinse well in cold water to get rid of the salt; collect the sinkers by pouring off the salt solution, and rinse in cold water.
   Ohio State's Co-operative Extension Service recommends that shelled peas be washed in shallow pans before sorting, by the way, because un-formed peas and bits of skin will float, and may be skimmed off and discarded.

Peppers, Green (Bell, Sweet)

Here is a vegetable that *does not require* blanching: the brief precooking described below is designed to make them more limp, so you can pack more peppers in the container—and it's for large-ish pieces you plan to use in cooked dishes, at that.
   If you plan to serve them raw (for instance in thin rings as a garnish, or diced in a salad), don't bother to blanch.
   Wash; cut out stems, remove seeds and white "partitioning" material. Cut in halves, or cut in slices, strips, rings, or dice (depending on future use).
   If blanched. In boiling water—halves for 3 minutes, slices for 2 minutes. Cool immediately; drain.
   Pack. Blanched, leave ½ inch of headroom. Raw, leave no headroom. Seal; freeze.

Peppers, Hot

Wash and stem.
  Blanch. No.
  Pack. Leave no headroom. Seal; freeze.

Pimientos

Wash and dry crisp, thick-walled pimientos.
  Roast. In a 400 F/205 C oven—for 3 to 4 minutes. Rinse and rub off charred skins in cold water. Drain.
  Pack. Leave ½ inch of headroom. Seal; freeze.

Pumpkin

Pumpkin makes fine pies and breads, but is seldom used as a table vegetable. Why not can it *cubed* instead?
  Wash whole pumpkin; cut or break in pieces. Remove seeds. Do not peel.
  Precook. Until soft—in boiling water, steam, a pressure cooker, or in the oven. Scrape pulp from rind; mash through a sieve. Cool immediately.
  Pack. Leave ½ inch of headroom. Seal; freeze.

Rutabagas

Cut off tops of young, medium-sized rutabagas; wash and peel. Cut in cubes to freeze merely blanched, or in large chunks to cook and mash before freezing.
  Blanch (for cubes). In boiling water—for 2 minutes. Cool immediately; drain.
  Cook (chunks to mash). In boiling water until tender. Drain; mash or sieve. Cool immediately.
  Pack. Leave ½ inch of headroom for either cubed or mashed. Seal; freeze.

Soybeans

To serve as a vegetable, wash firm, well-filled, bright-green pods (shell *after* blanching).
  Blanch. In boiling water—5 minutes. Cool quickly. Squeeze beans out of pods.
  Pack. Leave ½ inch of headroom. Seal; freeze.

Squash, Summer (and Zucchini)

Only young squash with small seeds and tender rinds are suitable for freezing.

Cut off blossom and stem ends; wash and cut in slices.

Blanch. In boiling water—for 3 minutes. Cool immediately in ice water; drain well.

Pack. Leave ½ inch of headroom. Seal; freeze.

Squash, Winter

Root-cellar mature squash with hard rinds. Treat it like Pumpkin if you do freeze it, though.

Sweet Potatoes (and Yams)

Use medium to large sweet potatoes that have air-dried (to cure) after being dug. Pack whole, sliced, or mashed.

Sort for size; wash. Leave skins on.

Precook. Cook, until almost tender, in water, steam, a pressure cooker or an oven. Cool at room temperature. Peel; cut in halves, slices, or mash.

Prevent darkening. Dip whole peeled sweet potatoes or slices for 5 seconds in a solution of 1 tablespoon citric acid or ½ cup lemon juice to 1 quart of water. For mashed sweet potatoes mix 2 tablespoons orange or lemon juice with each quart.

Pack. Leave ½ inch of headroom. Seal; freeze.

Pack variations. Roll slices in sugar. Or cover whole or sliced with a cold 50 percent Syrup. In either case, leave appropriate headroom. Seal; freeze.

Tomatoes

Aside from taking up a good deal of freezer space, a frozen whole tomato has limited appeal: its tender flesh is ruptured by ice crystals, and you have a deflated mush when you defrost it.

Ruth Hertzberg has a fine sauce to freeze—see Chapter 17.

STEWED TOMATOES

Remove stem ends and cores of ripe tomatoes; peel and quarter.

Cook. In a covered enameled or stainless steel kettle, cook gently in their own juice until tender—10 to 20 minutes. Set the kettle bodily in cold water to cool the contents.

Pack. Leave appropriate headroom. Seal; freeze.

TOMATO JUICE

Cut vine-ripened tomatoes in quarters or smaller. In an enameled or stainless steel kettle start to simmer them piecemeal as you go, in their own juice, for 5 to 10 minutes—or until tender with a good deal of liquid. Put through a sieve or food mill. Season with ½ teaspoon salt to each pint of juice, or 1 teaspoon to each quart if liked.

Pack. Leave appropriate headroom. Seal; freeze.

Turnips, White

Turnips are similar to rutabagas, but they mature more quickly. Freeze them in cubes or fully cooked and mashed. They also keep well in the root cellar.

Cubes: treat like Rutabagas. Mashed: treat like Winter Squash or Pumpkin.

16
FREEZING MEATS AND SEAFOOD

The same kitchen equipment, methods, and safeguards apply to meats and seafood that are to be frozen as applied to meats and seafood that were canned. The big difference between canning and freezing is that freezing does not kill off any dangerous bacteria. It merely holds them inactivated until they warm up and are ready to make trouble.

CONVERSIONS FOR FREEZING MEATS AND SEAFOOD

Do look at the conversions for metrics, with their workable roundings-off, and for altitude—both in Chapter 3—and apply them.

Cooking Frozen Meat

Generally, any cut of meat may be cooked either frozen or thawed—which leaves the decision up to you. How do you plan to serve it?

APPROXIMATE TIMES FOR COOKING FROZEN MEAT

Before we start, remember *never* defrost or thaw meat outside the refrigerator unless you're popping it right into the oven to cook.

Juices rich in B vitamins seep out of all frozen meat and poultry as they defrost. Therefore, if possible thaw meat completely before cooking it, and save the dripped-out juice for the pan gravy. However, juice can be kept in chops or ground-meat patties if they are cooked as soon as ice crystals have disappeared from their *surfaces*. Also, large pieces (roasts) may be put in a preheated oven when the surface yields to the pressure of your hand.

Roasting. If you're caught short of time and must roast a big piece of frozen meat, do it in a preheated oven *about 25 degrees lower* than generally used for roasting unfrozen meat (that is, do it in an oven not more than 300 F / 149 C), and *increase the roasting time by one-half*.

Broiling. Broil frozen meat of any thickness *at least 5 to 6 inches below* the heat source, and *increase broiling time by one-half*.

Pan-broiling. Cook frozen *thin* hamburgers, chops and steaks in a *very hot skillet* with a small amount of fat swished around to keep meat from sticking.

Start to cook frozen *thicker* patties, chops and steaks in a *warm skillet* with 1 tablespoon of fat. Heat the meat slowly and turn it until thawed. Then *increase the heat* and pan-broil the meat as for unfrozen thin cuts.

Freezing Roasts

Trim away excess fat. Wipe with a clean damp cloth. Pad protruding sharp bones with rescued pieces of clean though crinkled aluminum foil or with extra wrapping, so they can't pierce the package.

Pack and seal. Package individual roasts tightly in sheet wrapping, using either the butcher wrap or drugstore fold. Label; freeze.

Freezing Chops and Steaks

Trim away excess fat. Wipe with a clean damp cloth.

Pack and seal. Package, in sheet wrapping, the number needed for one meal. Put a double layer of wrapping between individual chops/steaks or layers of chops/steaks. Press outer sheet wrapping closely to the bundle of meat to exclude air. Use either the butcher wrap or drugstore fold. Label; freeze.

Freezing Ground Meat

Use only freshly ground meat to freeze as patties, loaves, or in bulk. Freshly made Pork Sausage also may be frozen; but its freezer life is short because of its high fat content.

PATTIES

Make up ready to cook.

Pack and seal. Put double layers of lightweight freezer wrap between patties for easy separation when you are ready to cook them. In each bundle, tightly wrap enough patties for one meal, using either the butcher wrap or drugstore fold. Label; freeze.

THE DRUGSTORE FOLD

Roll folded edge down, turn over

Fold ends of roll down

Sides over end

Fold tip of point over

Fold up and tape

THE BUTCHER WRAP

Ends up and over, tuck tight

Tuck sides in

Roll to end of paper—
seal open edges with tape

MEAT LOAVES, COOKED

Cool cooked loaves, remove from baking pan.

Pack and seal. Wrap tightly, using either the butcher wrap or drugstore fold. Label; freeze.

LOAVES, UNCOOKED

Because the onions lose strength and some herbs—especially sage—get bitter when held in the freezer, raw loaves should be stored only for several weeks, to avoid disappointment. Mix loaves as for baking. Line loaf pans with foil; fill with meat-loaf mixture and fold ends of foil over meat. *Freeze.*

Pack and seal. Remove loaves from pans when frozen, and overwrap tightly, using either the butcher wrap or drugstore fold. Label; store in freezer for only a short time.

BULK

Pack and seal. Put meal-size quantities in freezer boxes or bags, excluding air. Seal tightly. Label; freeze.

Freezing Stew Meat

Cut in cubes. They may be packed without browning, but for easier use later, sear them under a hot broiler; when the surfaces of the meat are browned nicely, rinse the pan juices out with a small amount of boiling water (which you'll reduce, cool, and add to the pack). *Frying* meat is usually *not* a good browning treatment for meat to be frozen or canned.

Pack and seal. Fill rigid containers with meal-size portions of browned cubes. Cover with pan liquid or broth, leaving ½ inch of headroom. Seal. Pack unbrowned cubes in rigid containers, freezer bags or sheet wrapping, excluding air. Seal tightly. Label; freeze.

Freezing Cooked Meat

It's better to freeze cooked meat or poultry in large pieces (so less surface may be exposed to air). Slices of meat or poultry keep best if covered with broth or gravy. Do read about the best thickeners for frozen gravy—Clearjel–A or mochiko—in Chapter 5, "Common Ingredients and How To Use Them".

Pack and seal. Large pieces are wrapped tightly, using either the butcher wrap or drugstore fold. Slices are stored in rigid containers of suitable size and covered with broth or gravy, then closely covered and sealed. Label; freeze.

Freezing Store-bought Cuts

(Meaning those prepackaged fresh meats from the market's display case.)

Pack and seal. Remove the store wrapping—even though it is well sealed; discard the tray, and rewrap and seal the meat closely in your own freezing materials. This will close out air and give the meat a more durable cover. (There's too much air held in store packages—and this causes freezer burn; also, the clear film that's O.K. to sell it in is not strong enough for freezer storage.) Label; freeze.

FREEZING POULTRY AND SMALL GAME

Once more it is necessary to start off with a warning against allowing the food to become contaminated by micro-organisms that cause illness, but we shall keep it mercifully brief and point out that certain game is likely to carry tularemia, and that some of the commonest "food poisoning" bacteria are the Salmonellae—which dearly love poultry. They are able to grow and multiply at a stunning rate once they start getting warm again. Handle food with scrupulous care.

And here again, "poultry" applies to domestic and wild birds, domestic rabbits, and small game.

Cooking Frozen Poultry

All freshly killed and dressed birds are better if stored in the refrigerator for 12 hours to develop their greatest tenderness before freezing.

For best results, thaw before cooking (unless you're boiling it to use in a fricassee or such): roasting or broiling is more uniform if the poultry is thawed first, and the meat is less likely to be dry or rubbery. Pieces to be coated before frying, or browned before stewing, should always be thawed beforehand. It's easier to stuff a thawed bird than one that's still frozen.

Cook all poultry soon after thawing, for best quality.

APPROXIMATE THAWING TIMES FOR FROZEN POULTRY

Thaw it in its freezer wrappings.

In the refrigerator: 2 hours per pound.

At room temperature NOT recommended.

Under cold running water: small individually frozen birds, or large joints, etc., may be bagged and sealed in waterproof plastic so they do not get waterlogged.

Freezing Birds Whole

Any bird may be frozen whole for future stuffing and roasting. But stuff it *later*. Even if the cook is careful at every step, dangerous bacteria causing food spoilage can develop in poultry stuffed at home and then frozen: *the slow cooling as it freezes in the center of a densely packed cavity will produce spoilers in the stuffing, and normal roasting will not destroy such products.* Prestuffed frozen birds sold by big commercial processors are prepared under controlled conditions of temperature and humidity, etc., that cannot be duplicated in the home.

Tie legs of dressed, washed birds together with thighs close to body; press wings snugly against breast.

Pack and seal. Put bird in a heavy-duty food-grade plastic freezer bag, press out air and tightly close the bag top. Or wrap the bird in moisture/ vapor-resistant material (see wrapping illustrations); seal tightly. Pack and freeze giblets separately. Label; freeze.

Freezing Birds in Halves

Split dressed, washed birds lengthwise and cut off the backbone (use it in soup stock).
Pack and seal. Put a double layer of freezer paper between the halves. Pack and seal in a freezer bag or wrap as for Whole. Label; freeze.

Freezing Birds in Smaller Pieces

Cut in pieces suitable for intended use (see Canning Poultry in Chapter 10).
Pack and seal. Put a double layer of lightweight wrap between meaty pieces and pack them snugly together in freezer bag or carton; wrap tightly in sheet material. Seal. Label; freeze.

FREEZING SEAFOOD

Fish and shellfish—seafood—are the most perishable of all fresh foods, and therefore are the most vulnerable to careless treatment. Fish must be cleaned immediately and washed in fresh, running water; ocean fish may be kept alive in sea water, but neither fish nor shellfish should be cleaned or cooked in sea water. Ice-pack refrigeration or an accepted substitute method of chilling is a must, especially if you catch your own. You will be meticulous about sanitation and sterilizing surfaces. The packaging materials will be adequate for preventing ice crystals or freezer burn. The seafood will be sharply frozen, stored at minimum temperature, and used relatively soon (compared with a frozen beefsteak).

But it's all worth the trouble. And compared with canning, drying, and curing, the actual freezing procedure is simplicity itself.

Detailed instructions for preparing seafood for processing are given in Chapter 11, "Canning Seafood": do read them.

All fish and shellfish must be stored at ZERO FAHRENHEIT (−18 C) after initial sharp freezing at approx. −20 F/−29 C.

Preliminaries to Freezing Fish

For handling, fish may be divided into two categories: Lean and Fat.

The Fat—mackerel, pink and chum salmon, ocean perch, smelt, herring, lake trout, flounder, shad, and tuna—are more perishable than the leaner varieties; plan to freezer-store these not more than 3 months.

The Lean fish—cod, haddock, halibut, yellow pike, yellow perch, freshwater herring, Coho and King and red salmon—all keep well in frozen storage up to 6 months.

Dressing (Cleaning)

Scale the fish (or skin it, depending on the variety); remove fins and tail. Slit the belly with a thin-bladed sharp knife and remove the entrails, saving any roe; remove head (optional). Wash fish in cold, drinkable *running water*.

Flavor-Protecting Dips

The Fat fish (and roe) are given a 20-second dip in an *ascorbic-acid* solution—2 teaspoons crystalline ascorbic acid dissolved in 1 quart of cold water—to lessen the chance of rancidity and flavor change during storage.

The Lean fish are dipped for 20 seconds in a *brine* of 1 cup salt to 1 gallon of cold water; this firms the flesh and reduces leakage when the fish thaws.

Glazing with Ice

Sometimes whole fish or pieces of fish are ice-glazed before wrapping. This helps keep the air away, thus saving the flavor. The fish is frozen until solid, then dipped quickly in and out of ice-cold water, whereupon a thin coat of ice will form on the fish. Repeat several times to thicken the ice, then wrap the fish for storage.

Cutting to Size

Fish are frozen whole if they are small enough (under 2 pounds); or are cut in steaks—crosswise slices about 1 inch thick, or are filleted. Exception: largish fish you expect to bake whole, you freeze whole.

Fillets are made usually from fish weighing 2 to 4 pounds. Lay the cleaned fish on its side on a clean cutting-board. Run a thin-bladed sharp knife the length of the backbone and slightly above it, and continue cutting to separate the side of the fish from the backbone and ribs; repeat on the opposite side. (This works on most fish; but not on shad—whose build is so complicated that it takes special skill to fillet them.)

Cooking Frozen Fish and Shellfish

With two exceptions, frozen seafood may be cooked when still frozen—the exceptions being a large whole fish you're baking and pieces that are to be crumbed or coated with batter before cooking.

Small whole fish—under ½ pound—may be defrosted just enough to separate them before they're fried (without crumbs or batter coating) or broiled on a greased broiler.

Fish fillets and steaks are baked or poached from the frozen state; they're partially defrosted before boiling or frying (without crumbs or batter coating).

Shellfish and fish for stews, chowders, and Newburgs are cooked still frozen.

Freezing Large Whole Fish

Dress (clean) as above, removing the head if you wish.

Pack and seal. Freeze-glaze with ice. Wrap snugly with moisture/vapor-proof covering, using the butcher wrap or drugstore fold—then overwrap for security. Seal. Label; store in freezer.

Freezing Small Whole Fish

Dress as for large whole fish, leaving on heads if you like.

Pack and seal. Small whole fish are most often packed in rigid containers with added cold water to fill crevices between the fish. Hold the lid tightly on the container with freezer tape wrapped around the rim, and overwrap with moisture/vapor-proof freezer paper, using the butcher wrap or drugstore fold. Seal.

For easy separation in thawing, individual small fish may be enclosed in a household plastic bag or other clear wrapping before going into the rigid freezer containers. Proceed with overwrap, and seal.

Sport fishermen often freeze their catch covered with water in large bread pans or the like, the whole thing sealed in a freezer bag, and frozen. When solid, the block of fish-in-ice is removed from the pan and tightly wrapped in moisture/vapor-proof material, closed with the butcher wrap or drugstore fold, then sealed. Label; store in freezer.

Freezing Fish Fillets and Steaks

Dress and cut up strictly fresh fish. Treat Fat fish pieces with the ascorbic-acid dip, or Lean fish pieces with the brine dip. Fillets may also be glazed with ice before wrapping.

Pack and seal. Fill rigid containers with layers of fillets or steaks, dividing layers with double sheets of freezer wrap for easy separation when frozen. Cover and seal.

For even greater odor prevention, overwrap the container with sheet freezer material, using the butcher wrap or drugstore fold.

Layers of fillets and steaks may also be wrapped in bundles and sealed instead of going into rigid containers; the bundles are then overwrapped and sealed. Label; freeze.

Freezing Fish Roe

Roe is more perishable than the rest of the fish, so it should be frozen and stored separately from the fish. Carefully wash each set of roe from strictly fresh fish and prick the covering membrane in several places with a sterilized fine needle. Treat the roe with an ascorbic-acid dip, even though it may come from a Lean fish.

Pack and seal. Wrap each set of roe closely in lightweight plastic for easy separation when frozen, smoothing out all air. Pack in flat layers in rigid containers and seal; then overwrap the containers in moisture/vapor-proof material, using the butcher wrap or drugstore fold. Seal. Label; freeze. (Sharp-freeze at −20 F/−29 C; store at Zero F/−18 C or below for not more than 3 months before using.)

Freezing Eels

Skin the eel. Tie a stout cord tightly around the fish below the head and secure the end of the cord to a strong, fixed support (a post or whatever). About 3 inches behind the head, cut completely through the skin around the body of the eel, necklace fashion. Grip the cut edge of the skin and pull it downward, removing the entire skin inside out.

Remove the entrails; wash the eel. Cut in fillets or in the more usual steak-type rounds. Because eel is a Fat fish, treat the pieces with an ascorbic-acid dip. Pack as for fillets or steaks of other fish, above. Seal. Label; freeze.

Freezing Crab and Lobster Meat

Scrub frisky live crabs and lobsters, butcher; cook the meat as described fully in "Canning Seafood," Chapter 11. Rinse and cool under drinkable running water; pick the meat carefully, removing all bits of shell and tendon.

Pack and seal. Fill rigid containers solidly with meal-size amounts; add no liquid, but leave ½ inch of headroom. Seal. Label; freeze.

Freezing Shrimp

As with other shellfish, shrimp you freeze must be absolutely fresh. They are best frozen raw, though they may be precooked as for the table before you freeze them.

RAW SHRIMP

Wash, cut off the heads and take out the sand vein. Shelling is optional. Wash again in a mild salt solution of 1 teaspoon salt to each 1 quart of water. Drain well.

Pack and seal. Pack snugly in rigid freezer containers without any headroom. Seal tightly. Label; freeze.

COOKED SHRIMP

Wash in a mild salt solution of 1 teaspoon salt to each 1 quart of water; remove heads. Boil *gently* in lightly salted water until pink and curled tight—average size for 5 minutes, up to 10 minutes for large to jumbo sizes. Cool. Slit the shell and remove the sand vein (for table-ready use remove shells and vein). Rinse quickly. Drain.

Pack and seal. Pack snugly in rigid freezer containers, without any headroom. Seal tightly. Label; freeze.

Freezing Oysters, Clams, Mussels, and Scallops

Probably the most perishable of the shellfish, these should be frozen within hours of the time they leave the sea or *held at refrigerator temperature* (about 36 F/2 C) during any waiting period. Cooked oysters, clams, and mussels toughen in the freezer: freeze them raw.

Wash in cold water while still in their shells to rid them of sand. Shuck them over a bowl to catch the natural liquid. Wash them quickly again in a brine of 4 tablespoons salt to 1 gallon of water.

Shucking is removing the shells. Since shucking bivalves (oysters, clams, etc.) involves severing the two strong muscles that close the two halves of the shell, you can cut yourself badly if you go about it wrong. *DO NOT USE a sharp or pointed knife.* Instead, use a dull blade with a rounded tip; insert it between the lips of the shell just beyond one end of the hinge, twist to cut the muscle at that point, and repeat at the other end of the hinge. A good shucker does it in one continuous *safe* motion: get someone who knows how, to show you.

Easy alternative: put live, scrubbed, tightly closed bivalves in the freezer for 10 minutes or so, until hinges relax and allow shells to open slightly; then just insert the dull blade, etc.

Pack and seal. Put in rigid containers and cover with their own juice, extended with a weak brine of 1 teaspoon salt to 1 cup of water. Scallops are packed tightly, then covered with the brine (they have little juice). Leave appropriate headroom. Seal tightly. Label; freeze.

17
FREEZING CONVENIENCE FOODS

A well-managed freezer need not be the symbol of only a big self-reliant family that figures on a side of beef and half a pig and tiers of neat boxes filled with all manner of produce from the nearby garden. Nor, if it is the special freezer section of a modern refrigerator in a city apartment, need it be mainly the cache for ice cream, several TV dinners, anonymous leftovers, and an emergency supply of ice cubes.

If you already use freezing, generally the larger your family the more likely you are to have a 15- or 20-cubic-foot freezer, and to have in it a greater proportion of raw materials than if your household is small. On the other hand, an older couple, or a single-member household, can get along fine with a 2½-cubic-foot freezing compartment. But the interesting difference is how these freezers are handled, not their sizes. If it also is managed well, the little one will concentrate more on short-storage items like precooked heat-and-serve foods, or the prepared components of main dishes. And in both freezers will be a special place for those small packets of extras that can make a stolid combination sparkle.

CONVERSIONS FOR FREEZING CONVENIENCE FOODS

Do look at the conversions for metrics, with workable roundings-off, and for altitude—both in Chapter 3—and apply them. **Note: this chapter involves some Canning. Make your altitude adjustments accordingly.**

With microwave ovens so popular, and convection ovens increasing in use, householders are often likely to freeze in the container they will be reheating and cooking in. It depends, really, on how formal you want to be. You can always freeze the prepared makings of a casserole or stew or hash in the well-lined and straight-sided dish in which you plan to serve it. When it's rock hard, remove and overwrap it, label and stack it. Come the time you want to cook it, remove all wrappings, put it in its chosen container, and carry on with thawing or cooking or whatever must be done to serve it.

Meanwhile there is in any supermarket an almost endless variety of aluminum-foil pans in all shapes and sizes; they are re-usable if you clean them carefully and make sure that they don't get punctured the first time you serve from them.

BUT such metal pans are not for microwave ovens. Instead, shape-freeze in ceramic ramekins or dishes designed to be used for microwave cookery, and follow the wrapping-to-shape idea above. And you can always thaw or reheat in plastic in these ovens, a thing that could mean disaster in a conventional or convection oven.

Foil is ideal for convection ovens, though. Because of the drying effect of the moving hot air, it's a good safeguard to cover the food in the cooking pan with foil crimped around the rim for part of the reheating or cooking time.

It goes without saying that your wrappings and containers should be just as moisture/vapor-proof as for regular freezing. Still, if you must make-do in a hurry, you can always use the best food-grade clinging plastic film, patting and smoothing it to the newly frozen shape, and using several layers; then put it into a plastic food-storage bag of suitable strength (tucking a written label inside with it), remove as much air as possible from the bag, and seal it. This system is particularly good with an odd-shaped piece of food (small unfrozen but precooked poultry, for example—which of course does *not* contain stuffing); and it helps to prevent the freezer-burn that comes from moist air held inside with the contents.

Perhaps the most valuable container/shape for freezing a 6-serving main dish is the 8-×-8-inch square cake pan. Its 2 inches is deep enough, it can be divided evenly in thirds by two cuts vertically, and then divided further by one horizontal cut across the center. Makes generous helpings.

But the beauty of doing your own convenience foods is the leeway you have in freezing very small portions, collecting them in a fairly large bag, and taking out what you want—instead of taking an ice pick to one end of a quart brick of, say, spaghetti sauce. So freeze it in muffin tins: a large muffin's-worth of sauce should be generous for a normal serving of pasta; if not, make it two.

If the consistency is stiff-ish, freeze any such thing in dollops on a cookie sheet. Whatever its shape when it is frozen hard, first wrap each bit separately (either in clinging plastic food film, or in a flimsy little plastic storage bag), then collect the pieces in best-quality true freezer containers.

Tuck a label inside the large freezer bag, or stick a label on the box with clear plastic tape you can read through. Be sure to say what the measurement is.

These Don't Freeze Well

Rather than note crankiness or poor behavior in freezing as some ingredients come along, we think it's sensible to deal with them in a bunch. So:

These suffer flavor changes:
Garlic, especially if uncooked, gets stronger.
Onion, though, tends to lose its flavor (although being sautéed before its being added to the other ingredients will help it hold).
Sweet green (or the ripe red stage) bell peppers get stronger.
Sage gets bitter; so does some pepper.
Cloves get both stronger and sharp. (A number of spices either give up or over-do when frozen, but short storage will be O.K.) You can always add apple-pie spices before cooking, etc.
Artificial vanilla essence gets truly unpleasant in freezing.
Artificial sweeteners should wait until actual serving time to be added.
Artificial table salt—substituting for a sodium compound—should wait to be added until the food is being served.
Table salt (sodium chloride) fades, and has the added drawback of inhibiting good freezing if used in pronounced quantities (ham, bacon, etc., don't hold long in the freezer, although part of this failure is their fat content).
Fried, especially deep-fried, foods taste stale: not rancid, just tired.

These suffer texture changes:
Hard-cooked egg whites get rubbery and tough. (So don't freeze dishes garnished with chopped egg or egg slices; or chopped-egg sandwich fillings; or stuffed eggs.)
Cooked soft meringue toppings get tough and shrink.
Mayonnaise separates and boiled dressings separate when frozen alone.
Cream sauces or egg-thickened sauces, or wheat-flour-thickened gravies separate—*but there's a whole section on special ingredients for frozen sauces under Thickeners in Chapter 5.*
Lettuce, tomatoes, celery, cucumbers, and similar salad vegetables get limp and watery. (But they hold in gelatin salad, also coming.)
Raw apples and grapes get mushy. Raw apples, bananas, avocados, peaches, and pears get dark without an anti-oxidant treatment (see Chapter 14, "Freezing Fruits," and Chapter 5, "Common Ingredients, etc.").
Old potatoes get grainy and soft in a frozen stew; new ones freeze better. But why not add potatoes when you're reheating the stew?
Green peas are better frozen separately and added to a combination during reheating.
Cooked pasta loses texture, but cooked rice does not: try substituting the rice from time to time.

Cheese-and-crumb toppings get soggy and dull: add when you're preheating the food for serving.

Custards—stirred (also called "boiled" though it's not), or baked or used as fillings—separate or weep.

Soft cake frostings and boiled icings get tacky (butter-and-sugar ones freeze well, however).

FREEZING EGGS

There's a giant *IF* with using frozen eggs, and it has several parts.

Frozen eggs should be used only in long-cooked or long-baked foods. The reason is simple. Uncooked eggs are possibly the favorite growing medium of Salmonellae, which cause severe but usually short-term gastrointestinal illness in the person who eats them. And freezing does not destroy bacteria, it merely slows their growth to a halt. So if they have entered a nest-cracked egg or have been transferred inside from an uncleansed shell when the egg was opened, they will multiply if the eggs are thawed at room temperature or warmer; and they will not be destroyed by heat low and brief enough merely to set eggs delicately.

Be cautious, therefore, of using defrosted eggs in mayonnaise, Hollandaise sauce and its cousins, stirred (or so-called "boiled" custard that of course is not really boiled), quick-scrambled eggs, or omelets. You must satisfy all the safeguards above before you make chocolate mousse or the old-fashioned "snow" puddings.

Instead, think of thawing your eggs to use for fancy cakes or breads, or for any dessert like Indian pudding or rice pudding.

How Much in a Batch?

As you prepare to freeze your eggs, examine each one before adding it to the others in the freezer container. For eggs to be frozen whole—i.e., yolks and white combined gently by stirring—break each egg into a saucer, and look for desirable firm whites and plump, high-standing yolks. For freezing separated eggs, put each white and each yolk in its own saucer before adding either to the batch being frozen.

Eggs to be frozen must be stirred gently in the freezing container after they are counted and added. For whole eggs, combine by stirring. Do the same for whites, taking care lest any bubbles get incorporated (air will dry the whites). Do the same for yolks—*with this added treatment to help prevent them from coagulating during storage:* gently stir in ½ teaspoon salt OR 1 teaspoon granulated sugar, or other natural sweetener, into 6 yolks. And note the sugar/salt on the label, so you'll know if it's for a dessert dish or not.

If you are dealing with many eggs at once, and want to prepare them in bulk even though they will be packaged in small amounts, stir in the total anticoagulant needed for however many times you have multiplied the per-6-yolks proportion, then fill individual freezing molds according to the equivalents given below.

Surely you don't want to whack off a chunk of frozen egg and guess at the resulting measurement: much simpler to have in mind several basic recipes you use a good deal, and package the eggs according to the amounts you'll want. For the rest, freeze in small quantities—bring out the muffin tins and the ice-cube trays again—and use the following measurements:

Equivalent measurements in large fresh eggs:
1 tablespoon stirred egg yolk = 1 egg yolk.
2 tablespoons stirred egg white = 1 egg white.
3 tablespoons mixed whites and yolks = 1 whole egg.
1 cup whole mixed eggs = 5 whole eggs.
1 pint mixed whole eggs = 10 whole eggs.
1 pint stirred whites = 16 whites.
1 pint stirred yolks = 24 yolks.

Thawing Eggs

Frozen eggs must never be thawed by warming of any kind (viz. the increased bacterial load of Salmonellae, for one example, if the fresh eggs were mishandled). Therefore, a 1-pint container may take up to 10 hours to thaw properly in a refrigerator. An alternative would be to thaw the container under *cold running water:* this would cut thawing time to about 3 hours. Better still, defrost in your microwave oven, then heat it or refrigerate after it comes out of the microwave.

It's much simpler to package in smaller amounts, and thaw correctly in correspondingly less time.

Thawed eggs may NEVER be refrozen.

FREEZING DAIRY FOODS

Homogenized Milk

As an emergency ration, sealed 1-quart cartons of homogenized milk may be held for up to 3 months in a freezer; or the milk may be decanted into straight-sided freezer jars or rigid plastic containers, with 1 inch of head-room for pints, $1\frac{1}{2}$ inches of headroom for quarts.

Usually the milk thaws smoothly enough to drink; certainly it does well for sauces or soups or custards.

The fat in milk *not* homogenized separates out as flakes that will not blend again when the milk thaws. It may be used for some cooking purposes, however.

Freezing Cream

Cream must be heavy, with at least 40 percent butterfat, to freeze successfully. It sends an oily film over hot coffee, although this drawback may be minimized if the cream is heated to 175 F/80 C for 10 to 15 minutes, and 3 tablespoons of sugar are added to each 1 pint of cream. Cool quickly, pour into straight-sided freezer jars or rigid plastic containers with tight covers, leaving 1 inch of headroom for each pint.

Thaw in the refrigerator.

Frozen cream whips well.

Cream rosettes: whip heavy cream, as above, with $\frac{1}{4}$ cup confectioner's sugar for each 1 pint of cream. When it peaks, drop it in rosettes onto freezer film laid over a cookie sheet, and freeze on the coldest shelf at Sharp Freeze setting. Check after 8 hours (they should be solid); when they can be handled without losing shape, remove, and wrap each separately in several folds of fresh film (which helps to cushion as well as to prevent freezer burn); pack in rigid boxes or between two paper or foil pie plates taped together to form a hollow container.

Little thawing is necessary: just lay rosettes atop individual servings of pudding, parfait, etc., before carrying to the table.

Sour cream separates when frozen, perhaps because of the butterfat content or because of the commercial souring method. It may be combined with other ingredients and frozen. Spread as a topping, it may be sweetened slightly and "set" by a few minutes in a hot oven (about 450 F/232 C).

Freezing Butter

For best and safest results, freeze it freshly made from sweet pasteurized cream, salted or unsalted (salted has shorter ideal storage, 2 to 3 months). If it's store-bought and in $\frac{1}{4}$-pound portions, overwrap the carton.

If it's in bulk, devise your own portions as to volume/weight; roll each piece like a small cylinder of cookie dough, wrap with plastic film, then store several pieces together in freezer bags. Be sure to label according to the date when frozen and the amount in each portion.

Thaw in the refrigerator.

Freezing Cheese

Cheeses that freeze well are Camembert, Port du Salut, Mozarella, Liederkrantz and their cousins, and Parmesan.

Cheeses with a high fat content (such as Cheddar, Swiss, and American brick, etc.), are best kept at refrigerator temperatures (32 to 40 F/Zero to 4 C). If you have more than you can use soon, though, cut it in ½-pound (or less) pieces, wrap each piece tightly, label, and freeze.

Plain cream cheese (fatty) mixed with cream for dips, etc., will freeze satisfactorily.

If the curds of cottage cheese are *not washed,* it keeps quite well. This means you can freeze homemade cottage cheese, but not the commercial kind. But it, too, may be combined with other ingredients in a gelatin salad for freezing.

Freezing Ice Cream

Either purchased or homemade ice cream keeps its quality up to 2 months in the freezer—although your own recipe, using a rich custard or gelatin base, holds the better of the two. Neither will keep well unless it is carefully wrapped and sealed after every time it is opened.

Store homemade ice cream in good plastic freezer tubs with tight covers; allow 1½ inches of headroom for each 1 pint because of its expansion. Either repack commercial ice cream in good plastic freezer containers, or put the carton/tub in a freezer bag and seal with a tie.

Press a layer of plastic freezer film down on the ice cream in a partially used container to prevent crystals from forming; leave headroom, cover securely.

FREEZING MAIN DISHES

Most bulletins will tell us to leave main dishes not-quite-done with the idea that they'll finish cooking while they reheat. This is too tricky for our purposes. The simplest advice *PFB* has to offer is to have all dishes *fully cooked.* Just be careful in reheating.

You have your own favorite recipes for main dishes. See also the publications dealing in part with convenience dishes listed in the Appendix, under the subsection referring to this chapter.

Roundup of Points That Make a Difference

The TV dinner pitfall is an ordinary hazard: we all start out wanting to create platter meals like the ones in the supermarket, only better. However, the bought ones usually have at least one part that is undercooked deliberately, while its sidebars can stand long reheating. The makers have established why to leave some parts sealed, others partly covered, and one wholly uncovered during the platter's time in the oven. And we haven't.

So cook and package your own favorite things in separate reheating containers or decant them into a saucepan over low heat; deal with each part of your meal on its own merits when you're putting it together—and stand back for the compliments.

Re-Cooking Times

These can be tricky. Probably the best solution to the problem is to start by cooking fully any dish you plan to freeze and reheat, leaving off the cheese-crumb topping until it goes into the oven for the last time; then *thaw the food,* and rely on oven time only for decent reheating.

Later, with experience (and jotting notes on your recipe card) you can figure on the time required for cooking the dish as it comes frozen from the freezer. Of course, your microwaving instructions will have you defrosting and heating with no trouble. However, there are no rules-of-thumb for either microwave or conventional reheating in a regular oven.

Temperature. Use the oven-setting at which the dish was originally cooked; or, if it's a pot pie whose filling is precooked, use the oven-setting required to cook the top crust perfectly.

Time. Start with *less than double* the original cooking time: if it took 30 minutes to bake your pasta-plus-sauce-plus-cheese dish for serving, think first of baking it frozen for 50 minutes. Then check: it should be bubbling around the edges, and the center must be hot—and neither of these is happening. Check in another 15 minutes: nearly ready; add a few minutes more. *And use your pencil thermometer* (the good one you got when you were canning in cans, or were exhausting seafood in jars); or *use your roasting thermometer,* by golly. They'll tell you if the center is hot enough when the sides are bubbling just right.

In the back of your mind will be the feeling that you will end up by using double the original cooking time, but there's no need to be rigid about it. Use the appearance around the edges of the dish, and the color of a topping or a crust—AND a thin and reliable thermometer.

And then there are the foods to be heated loose—dumped from their containers into an ovenproof serving dish (and heated in a medium oven, with minimum stirring; and perhaps a crumb topping added in the last few minutes). Or heated over simmering water in a double boiler. No problems here: you can tell when they're ready to serve.

Length of Freezer Storage for Top Quality

The maximum storage times for convenience foods given by experts is 6 months for cookie dough (more, if they're baked) down to 2 weeks for gelatin salads (which break down over longer hauls). These and the times for more robust dishes mean the length of time that the food is still at its peak and retains best texture, flavor, etc.; it does *not* mean that, if stored longer than for the stated recommendation, the foods have become spoiled or dangerous to eat.

Dairy products' best storage ranges from up to 6 months for butter, down to 3 months for cream, 6 weeks for ice cream, and about 1 month for whipped cream. These spans are approximate, but they're a good yardstick.

Pastries have a longer freezer life than the creams. Hearty soups and stews are good for from 2 to 4 months; their relatively short life comes from the fact that they often contain seasonings that do not stand up well for long freezing (see These Don't Freeze Well, earlier).

Roasted meats are frozen usually cut off the bone, or in serving-size slices, etc. It is important to remove fatty skin and fatty tissue to prolong freezer life (fat impairs freezing). Small game birds, rabbit, or poultry may be precooked as the start of becoming a fricassee or being served in a chafing dish with a special brown sauce; if their cooking is arrested so it can be finished later, the storage time is less than for fully cooked meat dishes, which can hold well up to 4 months.

Large pieces of meat may be packed in a large container, but it is best to separate the pieces by double folds of freezer film, or to wrap each piece in foil or film before adding it to the pack. Such dividers allow you to remove easily only part of what's in a large container, or to spread all the contents out for quicker thawing or reheating.

All small pieces of meat, or sliced meats, keep much better if they are covered with a gravy or a cream sauce to help keep out the air. These sauces are the most difficult aspect of preparing homemade convenience main dishes, and they are worth a separate main section all their own, which follows.

FREEZING PARTS OF A MEAL

Soups

Concentrate soups to save space. From some broths/stocks canned in Chapter 10, take other soups to freeze, *except for* onion soup. The flavors in French Onion Soup will suffer from freezing.

The base for New England Fish Chowder also freezes well, although it does not store so long as stocks or clear soups do. Thaw it in the refrigerator, then stir in hot milk before putting it over heat in a heavy pot.

Baked Goods

USING FILO PASTRY SHEETS

The freezers *PFB* knows best always contain a supply of *filo* (also *phylo,* which is closer to the Greek) pastry sheets, bought sealed in boxes by the pound from Antonio Recchia's Milano Importing Company in Springfield, Massachusetts. We make a point of asking Mr. Recchia for them, because he notifies us when a fresh supply is coming in; then it is up to us to freeze them for the first time. The filo comes in two widths (we use them both) and the sheets are so parchment-thin that they're translucent. They can be used as pastry for meat-and-rice pies, for the famous spinach pie, as "thousand-layer" cases for tiny dessert tarts or for canapes or for main-dish piroshkis.

Basically, each sheet is laid out on a board or in a pan, then dribbled/ painted with melted butter and oil, and another sheet is placed on it, treated the same way. By using half the sheets as a bottom crust, as it were, and putting a filling in the middle, and topping the whole thing with more buttered layers of filo, you have achieved a dish that freezes well, heats splendidly, and always brings pleasure.

The trouble in working with filo is that the sheets dry extremely quickly and start to crumble at the edges even while you are spreading the bottom layers and preparing to deal with the top. So work fast. And take a scrupulously clean tea towel, wring it out in warm water, and lay it on the sheets that are waiting to be oiled/buttered. If you have unused sheets, let them dampen ever so little from the towel, then fold them over and over with waxed paper between the folds; wrap the baton of filo in freezer film, using several thicknesses, then return the filo to its own moisture/ vapor-proof envelope; slide it back into the box; seal the box. Remember to label the box with how much filo remains before you freeze it.

If you have bought fresh filo and frozen it in its unopened box, let it thaw for 2 hours before trying to use it: the folds will break unless they're limber—and, though it's all edible and you can use scraps in the layers— the sheets will tear.

BREADS AND ROLLS

Generally, homemade doughs for yeast breads and rolls do not bake as well as frozen dough from the supermarket does; and they hold well in the freezer only for several weeks.

Unless you must have loaves of sandwich bread on hand for emergencies, save your space for already baked yeast breads made with extra shortening and sugar: party or sweet rolls or holiday breads or individual breakfast pastries.

It is possible to freeze quick-bread dough packaged in the form of rolled-and-cut biscuits, or muffins in baking cups. However, they take much longer to bake when put directly from the freezer into the oven (no thawing here), that the results are often disappointing. They hold well only for several weeks in the freezer. Best for these reasons to whip up batches as needed.

CAKES

Angelfood, sponge, and butter cakes freeze well; fruit cake is the longest freezer without suffering loss of quality.

Icings made from butter and confectioner's sugar freeze well; boiled and soft frostings do not freeze well. Custard fillings do not freeze well.

Wrap closely in freezer film, then put in a freezer bag and seal; it is often good to hold them in pastry boxes to protect them from heavier foods.

Avoid artificial vanilla.

Ideal storage time: up to 2 months.

COOKIES

Baked cookies freeze well, but why bother to wrap against breakage when raw cookie doughs are such a godsend when rolled and frozen in cylinders from which thin slices may be cut off and put in the oven?

Avoid artificial vanilla; also, strong spices can get bitter.

Baked or unbaked, they hold well up to 9 or 10 months.

TWO-CRUST PIES

Do not skimp on the shortening used in your pastry if you intend to freeze your pies.

Freeze pies with pastry *uncooked* for best texture. Use thoroughly chilled cooked fillings for best results. Starch- or egg-yolk-thickened pies lose texture, unless you use mochiko—see Chapter 5—measure for measure instead of wheat flour, or twice the called-for measure of mochiko if you'd use cornstarch.

Thicken cooked fruit fillings more for pies to be frozen than if they are to be cooked and served without freezing.

Freeze pies and pie shells completely *before* wrapping them for freezer storage; this prevents breaking the crust. Line the pie pan with freezer film, shape the rolled pastry to fit and crimp the edge, etc., as you like. When the pastry is fully frozen, lift it carefully from the molding container, wrap carefully, label; store in flat boxes if you have them, to protect the pastry from blows during storage.

Do not slit the tops of pies that are to be frozen before baking. And wait to glaze with milk or beaten egg until you put them in the oven.

Come the time for baking, frozen pastry cooks better in oven-proof glass or in dull tin or darkened aluminum. Preheat the oven to 450 F/232 C, bake the frozen pie for 20 minutes; reduce heat to the normal baking temperature and continue until it is done (about 350 to 375 F/177 to 191 C). It will take longer to finish cooking than if it were thawed before baking.

Pies frozen *after being baked* should be thawed in a slow oven (about 325 F/163 C).

Pies with gelatin-based fillings should be thawed in the refrigerator. (Soft meringue toppings do not freeze well; add and bake them before cooling to serve.)

Freezing fruits for pie fillings are mentioned in Chapter 14. If you plan to freeze specially prepared pie fillings, allow a good 2 cups of the mixture for an 8-inch pie, and 3½ cups for a 10-inch pie. Results are usually better, though, if the berries are frozen loose and stored in rigid containers, to be seasoned and sweetened, etc., just before baking. Remember that custard-based fillings do not freeze satisfactorily.

Other Meal-Makers to Freeze

Thick Tomato Sauce "Stretcher"
About 8 Pints

"Oh, this is good!" Ruth wrote at the top of the recipe that uses leftover roasted meats or limited portions of ground beef. And it is! This basic rule is a starting point. Amounts of ingredients may be varied to taste and/or availability. It is intended to be frozen in convenient pints, but may be canned and processed in a Pressure Canner for the time recommended for the lowest acid ingredient.

She writes: "Menu suggestion for two—1 to 1½ cups cooked meat simmered for 20 to 30 minutes in one pint of this tomato sauce, steamed rice, tossed garden salad, and a crusty bread."

To Make the Sauce:
Wash thoroughly about 8 quarts of ripe tomatoes, remove skins (optional), stems, and cores. Cut crossways, and hand-squeeze most of the seeds and a bit of the juice into one bowl and put tomato pulp into another. (Later strain the juice from the seeds and use it instead of water in canned beef or vegetable soups.)

Dice 2 cups of onions, 1½ to 2 cups celery and 1 de-seeded green pepper; slowly sauté until soft in ¼ cup of vegetable oil in a 6 quart stainless steel kettle. Add 10 to 12 chopped fresh basil leaves (or one tablespoon of dry), one bay leaf, and the tomato pulp. Bring to a simmer over medium heat and, stirring often, cook for 30 minutes or until sauce is of desired thickness. Add ½ cup bottled lemon juice or ⅔ cup white vinegar, 1½ tablespoons salt (optional), 3 tablespoons sugar, and 2 teaspoons Worcestershire sauce. Stir and simmer 5 to 10 minutes longer. Remove the bay leaf.

Fill rigid pint freezer containers, leaving ½ inch of headroom. Seal, label, freeze.

Or pack Hot in hot pint jars, leaving ½ inch of headroom. Cap with disk lid and screwband, and Pressure-process at 10 psig for 20 minutes (the same for ½-pints).

SANDWICHES

Filled sandwiches hold only for a week or so in the freezer; but, frozen, they beat last-minute hassles before a crowd arrives, or the chance of curled edges if stored ahead in the refrigerator.

These fillings *do not freeze well:* any "salad"-type combination that contains chopped egg whites (leathery), mayonnaise or cooked salad dressing (they separate), lettuce or tomato or celery or cucumber (flabby and watery), jam and jelly (weepy).

These fillings are *good for freezing:* cooked egg yolk, peanut butter, minced meat, poultry, or fish with chopped pickle and just enough salad dressing to hold things together; sliced luncheon meat or meat loaf; slivered cheese and chopped olives.

Use bread at least one day old (a bite of thawed perfectly fresh bread can be a wodge), and spread both slices with butter or margarine to prevent the filling from soaking in. Mayonnaise and the like can make frozen bread soggy.

Wrap each cut sandwich closely in plastic freezer film, then package or overwrap in moisture/vapor-proof material. Label and freeze.

Freezing Gelatin Salads

Experiment to arrive at your best combinations before trying these out on guests. Here are some basics to keep in mind:

Use only ¾ of the TOTAL liquid called for in your recipe, whether the salad calls for a dessert gelatin as the vehicle for fruit, minced celery, cucumbers—yes! these fresh salad vegetables keep crisp in a gelatin salad—or you are using unflavored gelatin and say, Tomato Cocktail (in Chapter 12, "Canning Convenience Foods") to carry a medley of cooked vegetables or cooked fish like tuna. This means dissolving gelatin in ½ cup of liquid, and adding only 1 cup more (hot) liquid, to make 1½ cups—when the recipe calls for ½ cup to dissolve, plus 1½ cups added.

Dessert gelatins tend to lose color and get uneven in texture, but they are fine if they are used merely as layers to hold the other ingredients in shape.

Fresh pineapple contains an enzyme that prevents a gel, so it *may not* be one of the fruits used in a fruit-salad mold.

Frozen gelatin salads keep for only 2 weeks in the freezer; then they start to lose quality.

They may be served still nearly frozen on a bed of lettuce. Or they can start thawing for 1 hour before being served.

Whipped cream, whipped softened cream cheese, whipped cottage cheese, a little mayonnaise, sour cream—these all combine well in a gelatin salad to be frozen.

Canned fruit salad, mixed cooked vegetables, flaked fish, even minced left-over chicken or turkey make tasty salads.

Save woe by freezing in the old stand-by square metal 8-×-8-inch cake pan. Line it with freezer film if you are using it only for the frozen mold, and plan to package it merely wrapped (you will unwrap and return it to its original pan for brief thawing and cutting in 6 generous serving portions). This size may be cut smaller, to serve 9, if the salad is to be a fruit dish on the side, rather than a hot-weather mainstay of aspic, vegetables and meat/fish, etc. You can always use a ring mold after your technique is perfected.

Prepare your gelatin base, refrigerate until it becomes syrupy. Meanwhile prepare the fillings: with lime or lemon or raspberry dessert gelatin, use chopped apples, minced celery, raisins; with dessert gelatins use fresh fruit (except for pineapples, and some varieties of grapes can get soft); treat apples with ascorbic acid (see Anti-discoloration Treatments in Chapter 5). Add whipped cream cheese, etc., and perhaps some chopped celery for more texture. Pour some of the nearly set gelatin in the pan, and put it in the freezer for a few minutes to thicken further. Remove it; fold together the solid ingredients and any cream, cream cheese, etc., with the remaining gelatin mixture, and pour all gently into the pan. Refrigerate or freeze until well set, then complete the wrapping in freezer film or moisture/vapor-proof materials. Seal and freeze.

Freezing Dabs and Snippets

The simplest way to deal with this section is to tell what's in the packets stashed in the door shelves of an upright freezer in a kitchen *PFB* knows well, and say what the ingredients were meant for. (There's another freezer, a husky chest affair, down cellar; here are stored the makings for batches of Corn Relish, and a specialty condiment or two, plus the usual large cuts of meat and related staples.) Each packet is labeled with name of the food, the measurement involved, the date, and sometimes the purpose.

You'll note that most of the items are prepared to the point needed for combining them with other ingredients to build a particular dish.

FOR SOUPS

Two-serving portions of the base for Borscht, Dutch Green Pea Soup, and Fish Chowder. Sautéed sorrel, chopped in ½-inch strips, in ½-cup amounts; each little bag will be added to clarified Chicken Broth, to become a serving of *shav* (with sour cream and chopped dill leaves, the latter also on the door-shelf). In summer the shav is chilled in the refrigerator and poured over a cold boiled new potato in its jacket for a hottest-evening-of-the-year supper. There's also sorrel essence to enliven vegetable stews.

The ½-pint freezer jars of Clam Broth (this also was canned, in Chapter 12) will strengthen the character of scalloped oysters (which will be made with a cream sauce using mochiko), or perhaps go into oyster stew (if not clam chowder; or be served hot in its own right as an appetizer). Reduced further, it will help stretch the white clam sauce that goes well on home-made spinach linguini.

HELP FOR MAIN DISHES

There's concentrated Shrimp Stock (again, it was canned as well) to use in a jambalaya or paella, or a deviled seafood filling for filo tarts. Pan juices of chicken, lamb, pork—perhaps for fried rice; or, combined with some of the peeled roasted chestnuts, as a special stuffing for something.

There's enough roast chicken taken off the bone to combine with some chicken broth, and thicken, and drop dumplings in for a chicken fricassee for two plus two. Slices of roast lamb in gravy, each slice in its own flat little bag for easy separating, and all stacked in a box, well labeled. Slices of Baked Liver Loaf; several ramekins of smooth-as-satin chicken liver pâté with wine.

Two divided pairs of shad roe, poached in a court bouillon, and each now wrapped in foil with bacon fat inside for lubrication; come the time to eat them—better soon—they'll be put still wrapped in a 400 F/205 C oven for 15 minutes, then opened and a partly cooked strip of bacon laid on each, to finish heating and cooking.

Some minced Italian capicola, to be used sparingly in a brown-sauce-with-wine for fresh veal kidneys when they're a good buy.

For pasta there's pesto sauce (basil-garlic-pignolias-Parmesan cheese-olive oil), frozen in patties dolloped onto a cookie sheet; then each wrapped separately, and all bagged together. The dab in the small rigid box is stuffing for tortellini pasta. Tucked behind is a chunk of Parmesan cheese.

VEGETABLES

Trimmed kale, blanched with a bit of bacon in the boiling water: this will be served with lemon. So will the several asparagus stalks; but the nearly cooked green beans will be thawed and served cold with a little olive oil.

Every fresh herb from the garden, minced and packed small. But large containers of minced fresh parsley to make taboulleh salad from. The cooked green peppers with stuffing are truly emergency food, as are the grilled tomatoes that have plenty of flavor but little presence.

A whole stuffed cabbage (sausage), cut in wedges and each wrapped in foil for quick heating. Two containers of peas bonne femme, really a vegetable stew of peas and lettuce cooked in a brown sauce. Endive, bought on sale, split and half finished for braising.

DESSERTS

Wedges of cheesecake, each wrapped separately. A handful of small cream-puff shells. Two 1-quart tubs of special boysenberry sherbert; a roll of good cookie dough to slice thin and bake and serve with it.

Fresh lemon juice frozen as ice cubes, each in its flimsy envelope and all gathered in a big freezer bag—well labeled, too, as equaling 1 or 2 lemons. Many twists of microtome-thin strips of lemon peel: start the steel blade of the food processor, put in sugar, then add the lemon zest—and you have grated lemon peel for sponge cake or buttermilk soup or a cold pudding.

That sort of thing . . .

COOK-IN-POUCH FREEZING

For those who do not use microwave ovens to deal with frozen food, there's the still-popular heat-sealer and its especially heavy cooking pouches. Anyone who can read the directions for cooking (certain to be part of the label in a well-run kitchen) can make part of a meal in a pot of boiling water.

The Sealer and How It Works

Because a hot flatiron cannot be guaranteed to make an adequate seal even when wielded with care, you should have a special electric heat-sealer. This costs about as much as a very good automatic toaster; the price is high for just one more fun gadget that takes up space, but it's fair for a sturdy appliance you plan to use often. The leading mail-order firms offer sealers under proprietary names.

The special pouches/bags are included with the sealer, and they can be ordered separately from the outfits that sell the sealers. Usually listed in a range of small sizes, they cost roughly four times as much as conventional freezer bags—which regularly come in much larger sizes to boot.

The preparation and packaging is more complicated than for conventional freezing. Cook-in-pouch foods are either fully cooked (to be reheated), or are at least half-cooked (to be finished), before they are sealed and frozen in relatively small portions.

Filling Pouches and Sealing

Three hands are better than two here. And the sealer can be mounted on a wall or the side of a cupboard; or, as shown in the picture, it can rest on a chopping block to allow the weight of the food to pull the pouch down and help to expel air.

Sealers usually come with a funnel-collar that holds the pouch open wide for filling. If you don't have one, though, make some by cutting the bottoms from small-size aluminum-foil bread pans, then pinching little tucks in the foil to reduce the circumference of the pan to the right size. These handy makeshifts let you and your helper keep several propped-open bags ready for filling.

Filled, each pouch is snapped shut to force out air, and its opening is run under the heated sealing bar.

Cool the sealed pouches in the refrigerator. Then label and lay them flat in the freezer, preferably only one layer deep until they're frozen; pat each bag to distribute the contents to be *uniformly about 1 inch thick.* This even thickness is important: boil-in times are based on it.

What to Freeze in Cook-in Pouches

Unless you have particular needs in mind, your likeliest foods for cook-in-pouch freezing would be programmed extras, sealed in 1-pint bags (about 3 average servings) or 1½-pint bags (5 servings). The mail-order sealer we're most familiar with offers extra 1-quart bags, but a casual consensus from house-holders who actually go in for this type of freezing is that amounts larger than 4 to 5 servings are probably handled more easily over-all if they're packaged conventionally for freezing, and are decanted for regular reheating on the stove-top or in the oven.

These extras can be grouped roughly as main dishes (fully precooked, just to be reheated); side-dish vegetables (partly precooked, to be finished in the boil-in bag); and dessert fruits (raw, to be defrosted by putting the sealed pouch in a bowl of warm water). The manufacturers' pamphlets include many more dishes, but a number of them seem like too much fuss for the benefit, frankly; you'll judge for yourself.

Heat-sealing the special cook-in pouch. Note the flattened headroom in the sealed pouch. (*Photographs by Allan Seymour*)

MAIN DISHES

Thin slices of meat in its gravy, stews, chicken à la king, fillets of fish in a favorite sauce, creamed things—the list goes on. Prepare as for the table: all are fully precooked. Remember that any thickened sauces are best made with the special corn or rice flours described earlier.

Cool the food slightly—just enough so you can handle the filled pouches—and pack it immediately, leaving 2 inches of headroom for both 1-pint and 1½-pint bags (this air-space will flatten away as the bag is held to the sealer).

Boneless meats and fish take 18 minutes' boil-in time for 1-pint bags, 20 minutes for 1½-pint bags.

Casseroles and pastas take 13 minutes' boil-in time for pints, 15 minutes for 1½-pint pouches.

SIDE-DISH VEGETABLES

Of course all your vegetables will be perfectly fresh, and young and tender-crisp; carefully washed and cut/trimmed, etc., as for serving.

Precooking times given below are average for *half*-cooking the individual vegetables—which will then be finished during the boil-in time of 15 minutes for pints, 18 minutes for 1½-pint pouches.

Use only enough water to cover. Don't salt it now because salt and onion flavors tend to disappear in freezing while some herbs get strong or bitter.

When it's half-cooked, the vegetable should be cooled only enough to allow you to handle the filled pouches comfortably. Leave 2 inches of headroom as for Main Dishes. Refrigerate the sealed bags for an hour before freezing them.

Unless you use stick butter or margarine that's easily sliced, slowly melt your butter or whatever, pour it ⅛ to ¼ inch deep in a bread tin; chill quickly until it's solid, and cut squares to insert in the pouches when you fill them.

Asparagus. Choose uniformly slender spears; trim to length to fit your pouches with 2 inches of headroom, or cut small. Cook gently in your usual manner for 5 to 8 minutes, depending on length of the pieces. Drain, cool slightly; pack and seal.

Beans—green/Italian/snap/string/wax. Cook gently for 10 minutes. Drain, cool slightly; pack and seal.

Broccoli. Cook split young spears gently for 5 minutes. Drain, cool slightly; pack and seal.

Carrots. Cook slices gently for 15 minutes. Drain, cool slightly; pack and seal.

Cauliflower. Cook prepared flowerets gently for 5 minutes. Drain, cool slightly; pack and seal.

Corn, whole-kernel. Husk, remove silk, wash. Cut from the cob over a bowl to catch the milk. In its milk—plus only enough water to keep from sticking—cook gently for 3 minutes. Drain, cool slightly; pack and seal.

Peas, green. Cook shelled peas gently for 10 minutes. Drain, cool slightly; pack and seal.

Spinach, etc. Remove stems and any tough midribs, cut large leaves in several pieces. Boil gently for 15 minutes. (Or, for very tender leaves, shake off extra water and steam-sauté in a little oil, covered, for half the full cooking time you use for this method—about 4 minutes.) Drain, cool slightly; pack and seal.

18
JELLIES, JAMS, AND OTHER SWEET THINGS

Traditionally, newcomers to food preservation start by making jam and jellies, and seemingly in no time have branched out to create their own combinations of these prettiest of put-by foods. And it should give them deep satisfaction (and not in the least take the shine off their achievement) to know that they are making use of the fourth major method for preserving food: and this is to decrease the available water that the spoiler micro-organisms need, and thereby prevent their growth. Without perhaps realizing it, jellymakers rely upon the ability of sugar to tie up the water by chemical means. This ability, plus the increased acid of the fruit, the added heat in cooking, *and* the lack of oxygen in the jelly jar, all add up to a virtually unbeatable combination for a safe and attractive product.

Important as sugar is, though, today's feeling for better nutrition leads to reduced sugar in many confections based on old-style recipes—and the best way to reduce this sweetener is to rely on only the natural pectin in the fruit, or to be cagey in adding pectin. Reason: big yields depend on lots of sugar, and jellies and jams made with added commercial pectin usually call for a good deal more sugar to go with it. This chapter, therefore, is organized according to some-sugar/low-sugar/no-sugar procedures, and will include methods for handling the often equal measures of fruit and sugar that our great-grandmothers believed in, then tell how to deal with low-methoxyl pectin and its sparing use of an added sweetener, and finally how to make spreads without any sugar at all.

254

CONVERSIONS FOR JAMS, JELLIES, AND PRESERVES

Do look at the conversions for metrics, with workable roundings-off, and for altitude—both in Chapter 3—and apply them. **Note: this chapter has B–W Bath processing—make your altitude adjustments accordingly.**

Equipment for Jellies, Jams, Etc.

For starters, your regular kitchen utensils will be adequate. Roughly in order of importance, you will need:

6- to 8-quart enameled or stainless steel kettle with a good lid (so it can double as a B–W Bath canner if you do not have one).

A jelly bag for straining juice. You can make a good one from ½ yard of 36-inch wide top grade unbleached muslin, folded so selvage edges are together. Machine-stitch with durable thread down the side seam and across the bottom, leaving top open for filling. Make a stout hem around the opening, through which you'll run a strong cord to tie the bag shut; the cord will hold the filled bag above a wide container to catch the juice. Wash before using to remove any filler in the fabric.

Boiling–Water Bath canner.

Jars/glasses in prime condition, with lids/sealers/gaskets ditto.

If you're a member of the dwindling users of paraffin for sealing jelly—note for *jelly,* because jam is too lumpy to allow a workable coating of wax to be laid on it—there will be fresh blocks of paraffin, lids to protect wax after it has set, plus a small metal pitcher or toy kettle for melting and pouring the paraffin.

Household scales.

Clock with sweep second-hand for close timing.

Minute-timer with warning bell for longer processing periods.

Sieve or food mill for puréeing (better is a food blender and the ultimate is a food processor).

Jelly (syrup) thermometer.

Shallow pans (dishpans are fine).

Ladle.

Long-handled wooden spoon for stirring.

Wide-mouth funnel for filling containers.

Jar-lifter.

Sieve or strainer for de-seeding blackberries or similar fruits.

Colander, for draining.

Large measuring cups, and measuring spoons.

Plenty of clean dry potholders, dishcloths, and towels.

Much paper toweling.

Large trays.

GENERAL PROCEDURES FOR JELLIES, JAMS, AND SWEET PRESERVES

Jellies, jams, preserves, conserves, marmalades, and butters are the six cousins of the fruit world. All have fruit and sugar in common, but differences in texture and fruit-form distinguish one from another.

Jelly. Made from fruit juice, it is clear and tenderly firm. Quiveringly, it holds its shape when turned out of the jar.

Jam. Made from crushed or ground fruit, it almost holds its shape, but is not jelly-firm.

Preserves. These are whole fruits or large pieces of fruit in a thick syrup that sometimes is slightly jellied.

Conserves. These glorified jams are made from a mixture of fruits, usually including citrus. Raisins and nuts also are frequent additions.

Marmalade. This is a tender jelly with small pieces of citrus fruit distributed evenly throughout.

Butters. These are fruit pulps cooked with sugar until thick.

The Four Essential Ingredients

Fruit

This gives each product its special flavor, and provides at least a part of the pectin and acid that combine with added sugar to make successful gels.

Full-flavored, just-ripe fruits are ideal, because their flavor cannot be overpowered by the large amount of sugar needed. However, often up to one-fourth of berries should be *under*ripe, as providing more natural pectin than fully ripe fruit does. But never use overripe fruit: too old to have its proportionate supply of pectin.

Unsweetened frozen fruit makes good jelly and jam.

Pectin

This substance, which combines with added sugar—or other sweeteners, *except* artificial ones—and natural or added acid to produce a gel, is found naturally in most fruits. Pectin content is highest in lightly underripe fruit, and diminishes as the fruit becomes fully ripe; overripe fruit, lacking adequate pectin of its own, is responsible for a good deal of runny jam and jelly. Pectin is concentrated in the skins and cores of the various fruits: this is why many recipes say to use skins and cores in preparing fruit for juicing or pulping.

Most commercial pectin is made from the white pulp under the skin of citrus fruits; only one major brand that we know of—although of course there may be more—is made from apples, and the label on the packet announces the fact. Again with one exception that we know of, pectin marketed widely for home use comes in powdered/granular form in 1¾-ounce packages. The exception is the liquid that used to be marketed in 6-ounce bottles, and since 1980 or so has come in dual 3-ounce foil pouches.

(The low-methoxyl pectin used in reduced-calorie jellies and jams—discussed in the introduction to the second major subsection of this chapter, With Low Sugar/No Sugar—is also made from citrus, though by a method different from that used in widely sold commercial varieties.)

This natural pectin in the fruit can be activated only by cooking—but *cooking quickly,* both in heating the fruit to help start the juice, and later when juice or pulp is boiled together with the sugar. And *too-slow cooking,* or *boiling too long,* can reduce the gelling property of the pectin, whether natural or added.

In the old days, apple juice was added to less pectin-rich juices to make them gel, and this combination still works. Today, though, the readily available commercial pectins take the guesswork out of jellies, jams, and the like.

Testing for pectin content. There are several tests, but the simplest one uses ready-to-hand materials. In a cup, stir together 1 teaspoon cooked fruit juice with 1 tablespoon non-methyl alcohol. No extra pectin is needed if the juice forms one big clot that can be picked up with a fork. If the fruit juice is too low in pectin, it will make several small dabs that do not clump together. DON'T EVER TASTE THE SAMPLES.

HOMEMADE LIQUID PECTIN

Liquid pectin is especially helpful in making peach, pear, strawberry, or other jellies whose fruit is low in pectin.

Four to 6 tablespoons of homemade pectin for every 1 cup of prepared juice should give a good gel: but experiment! These pectins can be frozen or canned for future use. Freeze in small quantities; can in ½-pints or pints, refrigerating after opening. To can, ladle hot into hot jars, leaving ½-inch of headroom; process at a simmering 185 F/85 C for 15 minutes. Remove from canner, cool upright and naturally.

Crab Apple Pectin

2 pounds sliced unpeeled crabapples
3 cups water

Simmer, stirring, for 30 to 40 minutes adding water as needed. Plop into colander lined with one layer of cheesecloth and set over a bowl; press to force the juices. To clear, heat the collected juice and pour through a stout jelly bag that has been moistened in hot water. The result is the pectin you will can, or freeze, or use right away.

Tart Apple Pectin

4 pounds sliced apples with peels and cores
8 cups of water

Simmer, little stirring needed, for 3 minutes. Press apples through a sieve
to remove cores, etc. Return liquid to a heavy kettle to cook briskly, stirring,
until volume is reduced to one-half. Clarify by pouring through a stout jelly
bag that has been moistened. Use, can, or freeze as above.

Acid

None of the fruits will gel or thicken without acid. The acid content of
fruits varies, and, like natural pectin, is *higher in underripe* than in the
fully ripe fruit.

Taste-test for acid content: this is a comparison. If the prepared fruit
juice is not so tart as a mixture of 1 teaspoon lemon juice, 3 tablespoons
water, and ½ teaspoon sugar, your juice needs extra acid to form a successful gel. A rule-of-thumb addition would be 1 tablespoon lemon juice
or homemade citric acid solution (for how to make it, see Acids to Add
for Safety in Chapter 5) to each 1 cup prepared juice.

Sweeteners

Sugar helps the gel to form, is a preserving aid, and increases flavor in
the final product. The sugar called for in the recipes for jellies, jams, and
other preserves is, unless otherwise specified, refined white cane or beet
sugar. Other natural sweeteners—as well as artificial, or non-nutritive
ones—are described at length in Chapter 5, "Common Ingredients and
How to Use Them."

In recipes using *powdered pectin,* light corn syrup may replace ½ the
sugar needed in either jellies or jams. Where *liquid pectin* is used, light
corn syrup may replace up to 2 cups of the sugar.

In recipes *without added pectin,* we suggest substituting no more than
½ the sugar with a mild-flavored honey. In recipes *with added pectin,* we
replace no more than 2 cups of the required sugar with an equal measure
of honey. *Caution:* in small batches (5- or 6-glass yield), no more than 1
cup of the sugar should be replaced by honey.

Pectin/Acid Content of Common Fruits

Group I. These fruits, if not overripe, usually contain enough natural
pectin and acid to gel with only added sugar: apples (sour), blackberries
(sour), crabapples, cranberries, currants, gooseberries, grapes (Eastern
Concord), lemons, loganberries, plums (except Italian), quinces.

Group II. These fruits usually are low in natural acid or pectin, and *may* need added acid or pectin: apples (ripe), blackberries (ripe), cherries (sour), chokecherries, elderberries, grapefruit, bottled grape juice (Eastern Concord, lower than Group I because of previous processing to preserve it), grapes (California), loquats, oranges.

Group III. These fruits *always* need added acid or pectin, or both: apricots, figs, grapes (Western Concord), guavas, peaches, pears, prunes (Italian), raspberries, strawberries.

Steps in Making Cooked Jelly

The recipes that follow are for cooked jellies—that is, ones boiled with sugar and pectin as indicated. (Uncooked jellies are discussed later.)

Always work with the recommended batch. The quantities given are tailored for success: the longer boiling needed for larger amounts can zap the effectiveness of the pectin and the result will be runny and sad.

Preparing Fruit and Extracting Juice

A rough, very rough, rule-of-thumb for estimating how much fruit will be needed to make a particular batch of jelly is: 1 pound of prepared fruit (i.e., washed, stemmed/trimmed/cut as the recipe says to) will make 1 cup of juice.

Plan to process the fruit as soon as possible after it's picked or bought; refrigerate, for no more than 1 day, soft fruits and berries if you can't handle them right away. When you do start, keep at it and work right along.

Pick over the fruit carefully, discarding any that is overripe or has rotten spots. For a successful gel from recipes that have no pectin added, make up the amount called for with one-fourth the total in underripe fruit.

Wash the fruit quickly but thoroughly. Don't let it soak; lift it out of the basin of fresh water, don't pour it with the water into a strainer. The lighter and quicker you are in handling berries, the better. And always use good, clean drinking water for washing your fruit.

Remove the stems and blossom ends of apples and quinces and guavas, but retain their skins and cores. The skins of plums and grapes also contain a good deal of pectin, so keep them too. The stems and pits of cherries and berries need not be removed: the jelly bag will take care of them when the pulp is strained.

TO EXTRACT JUICE

Sparkling clear, firm jelly calls for carefully strained juice. Modern recipes describe the way the juice is to be extracted—simply by crushing; or by short heating, with or without "enough water to keep from sticking"; or by longer cooking with more water added—and these instructions should be followed.

Sometimes, though, you will like the sound of an older recipe that's not explicit about method, so we offer the following ideas as a help in figuring out what to do.

Always start heating the fruit at a fairly high temperature.

To heat ripe soft berries without any water, crush a layer in the bottom of the kettle to start the juice (mashing them with the bottom of a drinking glass, or with a pastry-blender); pile on the remainder and put the kettle on fairly high heat, stirring to mix the contents; reduce heat to moderate and boil gently and stir until all the fruit is soft—5 to 10 minutes.

To heat soft berries that are slightly underripe, Concord and wild grapes, or currants, add no more than ¼ cup water to each 1 cup of prepared fruit. With currants, cook until they are translucent and faded. Add ½ cup of water to chokecherries and wild cherries; add a scant ¼ cup to juicy sour cherries.

To cut-up (but unpitted) plums, add water to *just below the top layer* in the kettle, and cook until soft—about 15 minutes.

To prepared apples, crabapples, quinces and guavas, add water *just to cover,* and cook until soft—20 to 25 minutes.

Strain all crushed raw or cooked fruit through a proper jelly bag that holds at least 6 cups of simmered fruit. Dampen the bag to encourage the juice to start dripping through it; bunch the top together and tie it with strong string. Hang it high enough over a big mixing bowl so the tip of the bag cannot touch the strained juice (a broomstick laid across the tops of two kitchen chairs makes a good height).

Squeezing the jelly bag forces through bits of pulp that will cloud the jelly, but pressing the back of a wooden spoon against the bag will often quicken the flow without clouding the juice.

If there is traffic through the room, with attendant insects and dust, drape a clean sheet over the whole business.

Be fussy about washing the jelly bag after each use and rinsing it well; even a little diluted juice left in the fabric will spoil, and a musty, winey bag will hurt the next juice that's strained in it.

Refrigerate, in a tightly covered sterilized container, any juice left over from measuring for the batch of jelly.

Sugar and Pectin

When you add the sugar depends on the type of commercial pectin you use. Each recipe stipulates the type—*and they are not interchangeable.* Always follow the recipe exactly, because time and quantity variations almost always bring failure.

Powdered pectin is added to the strained juice *before* heating. Heat rapidly, bringing to a full rolling boil—i.e., a boil which cannot be stirred down; *then add the sugar,* bring again to a full rolling boil, and boil for 1 minute.

Liquid pectin (except for homemade pectin, above) is added to the strained juice and sugar *after* the mixture is brought to a full boil. Stir constantly during heating. Add pectin, bring again to a full rolling boil, boil for 1 minute.

WITHOUT ADDED COMMERCIAL PECTIN

Jellies with enough natural pectin (like Basic Apple Jelly) require less sugar per cup of juice than jellies with added *store-bought* pectin do. The longer cooking needed to reach the jelly stage produces the right proportion of sweetness, acid, and pectin.

Testing for Doneness

Because barometric pressure as well as altitude affects the boiling point, make necessary adjustments for heights more than 1000 feet/305 meters above sea level, and for whether the day is close and damp, or clear and dry.

Jelly with added pectin will be done if boiled as the individual instructions for time and quantity specify.

Jelly without added pectin is done when it reaches 8 F (the Celsius comparison is meaningless here) above the boiling point of water—usually, under good conditions at 1000 feet/305 meters or less, this is 220 F/ 104.4 C.

220 F/104.4 C and a full rolling boil in all its glory: the jelly is done. (*USDA photograph*)

If you have no jelly thermometer, use the Sheet Test. Dip a cold metal spoon in the boiling jelly and, holding it 12 to 18 inches above the kettle and out of the steam, turn it so the liquid runs off the side. If a couple of drops form and run together and then tear off the edge of the spoon in a sheet, the jelly is done.

Progress of sheeting test, left to right. (*Drawing by Irving Perkins Associates*)

BUT this test is not for jam: sheeting in the same manner means that the jam has cooked too long, and will be stiffer than the soft texture one is accustomed to.

Or use the Refrigerator Test. Remove the kettle from the heat (so it won't raise Cain while your back is turned) and spoon some hot, hot jelly onto a chilled saucer, and return the saucer to chilling for a minute. Then look at the jelly: if it wrinkles when you push it from the side, and seems generally tender-firm, it is ready to pour and seal.

Jelly at High Altitude

The Sheet Test is likely to be especially helpful if you live at 3500 ft/1067 m or more, because, since jelly boils at a lower temperature than it would in the sea-level zone, the *result* of the boiling is the important thing—not what the temperature was that brought it to the point of sheeting.

Thus, with the gelling point set at 8 F/4.4 C above the boiling point of water at sea level, and you live in Duluth or Thunder Bay, you'd have jelly at 220 F/104.4 C. But if you live in Denver, one mile high, the temperature reading for finished jelly would be around 211 F/99.4 C. These figures work out to an approximate *decrease* of 2 F/1.1 C for each increase of 1000 ft/305 m in altitude.

Quickly, some other equivalents for finished jelly in Fahrenheit and feet (to save space, look up your own Celsius in Chapter 3): 218 F at 2000 ft; 216 F at 3000 ft; 214 F at 4000 ft; 212 F at 5000 ft; 210 F at 7000 ft; 206 F at 8000 ft.

Pouring and Sealing with Paraffin

Old-style topping with melted paraffin is nowhere near so reliable as using the modern two-piece screwband closure that allows a quick finishing bath in boiling water to (1) drive all air out of the headroom, (2) destroy micro-organisms that could have lit on jar or disk cap—or the jelly/jam itself—and not been rendered harmless by the heat, and (3) create a stronger vacuum that ensures good shelf-life despite the micro-climate in the storage area.

Some householders still like paraffin, though—possibly because they have a stock of old-style jelly glasses on hand (but the plastic snap-on lids are no longer made and are hard to find even in a corner of some old variety store).

For these householders, but with hope that they will convert to the modern closures and jars, we say to use only clean, fresh paraffin, and melt it in a double-boiler arrangement, so it doesn't get too hot and it will pour easily. Ladle jelly into jars, leaving $\frac{1}{4}$ inch of headroom. Before topping with paraffin, clean around the inside of the glass where wax will adhere, using grain alcohol (the potable kind, a little will go far if it's just to seal jelly) pressed out on a cotton swab. It's much better for removing smudges of stickiness than even a fresh paper towel dipped in boiling water. Then pour melted wax on jelly, no more than $\frac{1}{8}$ inch deep. Until you can cover the waxed jars with protective lids of some sort, drape a clean tea towel over them to keep dust and spoilers away. If the wax was too hot—a haze would have risen from its surface if it was—the cooled wax will have little breaks in it. With a sterilized needle, prick any air bubbles in the soft, hot wax: the bubbles can cause holes in the wax as it cools—and there goes your seal.

Do follow the $\frac{1}{8}$-inch rule for the wax: a too-thick layer will tend to contract and pull away from the side of the jar. And never try to cover a faulty thin layer by topping it with more wax. Trouble. Put metal or plastic caps on the jars to protect the wax in storage. Lacking these, cover with a double layer of heavy aluminum foil, held tight with thin wire secured by twisting.

Pouring and Sealing with Modern Lids

Keep in mind the virtues of the two-piece screwband lid described above. Ladle in your hot, hot jelly/jam, leaving $\frac{1}{4}$ inch of headroom. Quickly wipe the sealing rim of the jar with fresh paper toweling pinched out in boiling water; take care of any smudges that have climbed up inside the jar, and wipe the threads as well. Put on the prepared disk lid, screw down the band firmly tight, and fill and cap the next hot jar. When there's a canner-load, put your jars in the hot water, add boiling water as needed (but don't pour it on top of the jars) to come 2 inches above the lids. Cover the canner, bring to boiling, and boil convincingly but not violently for 5 minutes only.

Remove jars; let cool upright and naturally. When you hear the soft *plink* that means a strong vacuum has formed, you know you have a hardy seal, designed to withstand less-than-ideal storage. Check all jars for good seals every month or so.

The Quick B–W Bath *versus* General Foods' "Inversion"

General Foods, maker of both liquid pectin (Certo) and powdered (the Sure•Jells), has come out in 1987 with a sealing method for jellies and jams that their press releases announce as a safe, new technique for protecting homemade cooked jams and jellies—and affording this protection in less time, and by strong implication more easily and simply, than the finishing 5-minute Boiling–Water Bath does. Promo pitch aside, the new technique calls for leaving only ⅛ inch of headroom, wiping the jar carefully, clapping on the disk and screwband, and immediately inverting the jar, leaving it to sit upside-down for 5 minutes. It is then turned upright and allowed to continue cooling in the standard, natural fashion.

Unsaid in the news release but voiced by staff responding to telephoned queries to the GF Consumer Center in White Plains, New York, the benefits are that the jam/jelly—being still at a temperature to destroy spoiler micro-organisms—will sterilize the underside of the sealing disk, *and* the little amount of air trapped under the lid.

A vacuum can form if the jars are hot and the contents are about 165 F/74 C. But it won't be a *strong* vacuum, because any amount of air left in the jar will invite growth of mold eventually—even though the jar is technically sealed. While a vacuum formed for us at *PFB* using the "inversion" method, the "inversion" vacuum was not so strong as the vacuum seal on the B–W Bath-treated jars. This fact is a reminder that the *finishing* Boiling–Water Bath was welcomed by food scientists in the South, to counteract heat and humidity of storage in the region; and soon it was adopted for dryer and more temperate climates. At the same time, food scientists determined that 5 minutes in a B–W Bath was adequate (instead of a longer time advocated earlier) to strengthen the seal and drive air from the headroom, and sanitize the surfaces where micro-organisms could have lit.

Presumably the reason for standing the jars on their heads is to hold the hot contents securely against the head and the sealing disk to equal the action of the 5-minute B–W Bath. A further help would be to deal with floating fruit as the medium gels: turned back upright, the contents would shake down by themselves. The same result can be got by giving the jars a twirl several times after they're set aside to cool upright after their bath.

(*Note:* in filling and capping the jars, we at *PFB* must have left the bands a bit loose. After we inverted it, one jar spurted hot, hot jelly over a hand in a rather mean scald. This indirect hazard can also make "inversion" less than foolproof.)

PFB is not gainsaying General Foods just to be tiresome: we, too, used to advocate the quick "inversion" with almost nonexistent headroom—(though never setting the jars upside-down, regarding this practice as harking too far back to old-time ways with preserving)—so we reverse our own recommendations, too.

Postscript: extension food scientists whose work we admire have expressed their worries over the "inversion" technique used at high altitudes, and they are against it.

Post-postscript: General Foods shows fairness in their news release in saying that they will continue to mention the B–W Bath method as an alternative on all their printed materials.

Labeling and Storing

During the 24 hours before you store your jelly, check it for loose texture and faulty seals (see Failures and What to Do in this chapter). Carefully clean away any stickiness from the glasses/jars, giving extra attention to the tops around the closures where outside mold can attack the seal. Label each container with the kind and the date; it's a good idea to indicate the batch number if you made more than one lot of the same sort that day. And if you used a method or recipe different from your usual one, note this fact too: it could help to pinpoint reasons for trouble, as well as for outstanding successes.

Storage in a "cool, dark, dry place" is the same as for canned goods: ideally 32–50 F (Zero–10 C)—certainly not where the contents will freeze (which can break seals), certainly not at normal room temperature (which encourages growth of spoilers). Dark, because the pretty colors can turn brown or fade in the light; otherwise you put them in cartons or wrap them in paper or cover the whole shebang with an old blanket or the like. And dry because humidity can corrode metal caps and lids and lead to broken seals.

Steps for Cooked Jams and Marmalades

Sugar and Pectin

The general proportions for substituting other *non-artificial* sweeteners for called-for sugar are the same for jams and marmalades as for jelly.

With added pectin, the crystalline-type is mixed with the unheated prepared fruit. Liquid pectin is added to the cooked fruit-and-sugar mixture after the kettle is removed from heat. With either form of pectin the cooking time is the same: 1 minute at a full boil.

Without added pectin, the cooking time is increased to a range of from 15 to 40 minutes, depending on the character of the fruit. Jam is more likely to scorch than jelly is, so stir it often during cooking.

If you have worries with long cooking, put your jam kettle in a 300 F/ 149 C oven, as for Butters later.

The so-called Diet jams and jellies are "set" with gelatin and their nonnutritive (artificial) sweetening is not cooked with the fruit, but is added just before pouring and sealing. See the section With Low Sugar/No Sugar, later in this chapter.

Testing for Doneness

Jams, etc., with added pectin will be done when they are boiled according to the individual instructions for time and quantity.

Without added pectin, jam is done when it reaches 7 F above the boiling point of water, usually 219 F at 1000 feet of altitude or below.

No thermometer? Jam is ready when it begins to hold shape in the spoon. Or use the Refrigerator Test for jelly, above.

Pouring, Sealing, Processing, Storing

Remove the kettle from heat, skim carefully, and then stir the jam gently to cool it slightly and thus prevent floating fruit.

Ladle the still-hot jam or marmalade carefully into hot sterile canning jars, leaving ¼ inch of headroom in ½-pint jars; wipe the sealing rims carefully with a clean paper towel dipped in boiling water; cap and process in a Boiling–Water Bath (212 F/100 C) for 5 minutes. Complete seals if necessary. Cool upright and naturally, twirling the containers several times in the first 30 minutes of sitting: doing this distributes any floating fruit well through the contents. Hours later, when the contents have cooled, clean the containers, removing screwbands if you used this two-piece closure, and carefully wiping the threads and around the seal. Label; store in a dry, cool, dark place.

Special note: Because the general methods for making Preserves, etc., and Butters are described as part of the recipes, and are quite brief, the procedures for them are given with the specific instructions at the end of the next major section, With Old-style Use of Sugar.

Steps for Freezer (Uncooked) Jelly and Jam

The general handling of these jellies differs from that for conventional ones in several ways. Not being sterilized through boiling, they must be stored in the freezer to prevent spoilage—although freshly made or defrosted jelly will keep well for up to 3 weeks in the refrigerator. Also, because their natural pectin is not activated by boiling, pectin must be added; and it is added after the sugar, regardless of whether it is liquid or powdered. And finally, the jellies must be packed in sterilized freezer-proof jars, with headroom to allow for expansion; and must be sealed with sterilized tight-fitting lids, *not* with paraffin.

Filled and capped, the containers must stand at room temperature until the jelly is set—which can take up to 24 hours—before going into the freezer.

The juice for this jelly is made from *unheated* fresh fruit. However, it can be made with frozen juice *not heat-extracted or sweetened;* or with juice from berries that have been frozen, *without added sugar,* for making jelly later on. It *cannot be made successfully from canned fruit juices,* because there the natural pectin has been impaired by the heat of processing.

People seem to make more freezer jams than jellies perhaps because there is more leeway to a jam's consistency than there is for jelly. Certainly the lovely garden-fresh flavor of berries and fruits is more pronounced in jams. Like their jelly counterparts, opened freezer jams must be refrigerated.

The folders that come with the containers of commercial powdered or liquid pectin have a number of good jams and jellies; and the proportions have been worked out after much testing. We recommend them.

Failures and What to Do

In theory, we'd all have perfect jellies, jams, marmalades, preserves, conserves, and fruit butters if we used prime ingredients, if we measured carefully, and if we followed procedures conscientiously. But things can go wrong, even when we mean to be careful. Therefore, here is a rundown of the symptoms and causes of common failures, listed now (rather than at the end of all the recipes, as other publications do), so you can keep them in mind as you work along.

We'll start by saying that you shouldn't stash away any of these preserving kettle products until they have stood handy by for 24 hours. Aside from allowing you a wonderful gloat, this day of grace before storing them will let you check the seals in time to do them over again if you find any poor ones. With paraffin it is not enough merely to add another layer of melted wax: you must remove the wafer of old wax, wipe the inside lip of the glass with a scrupulously clean paper towel twisted out in boiling water, and immediately pour on a ⅛-inch layer of fresh and very hot wax, rolling the glass slightly to make the wax climb the sides. (If you have grain alcohol, apply it sparingly around the inside of the glass, using a cotton swab to clean off any stickiness; then pour on new wax.)

Now for the problems that can be dealt with *safely*—provided that the seal is intact, and that there is no mold or fermentation in the contents.

Too stiff, tough. Too much pectin in proportion, or cooking no-added-pectin products too long; sliced citrus rinds in marmalades not precooked before added to syrup. Nothing can be done for pectin-added things, and it's not feasible to do the others over with more liquid. They're still probably tastier than store-bought.

Too-soft jelly. Tilt the containers: if you can see the contents shift, the jelly is too soft. This condition can be caused by cooking too long (as when the batch was too big and so was boiled beyond the ideal time limit); or by cooking too slowly; or by too much sugar; or by too little sugar or pectin or acid; or by not cooking long enough. Sometimes you can salvage such jelly by cooking it over; not always—but it's worth a try. Work with only 4 cups of jelly at one time:

Without added pectin: bring 4 cups of the jelly to boiling and boil it hard for 2 minutes, then test it for signs of gelling. Let it try to sheet from a cold spoon, or (having removed the kettle from heat) chill a dab of it; if it shows signs of improving, boil a minute or so longer until it tests done. Then take it off the heat, skim, pour into hot sterilized containers, and seal.

With added powdered pectin: for each 4-cup batch of jelly, measure 4 teaspoons of powdered pectin and ¼ cup water into the bottom of the kettle; heat the pectin and water to boiling, stirring to keep it from scorching. Add the jelly and ¼ cup of sugar, bring quickly to a full rolling boil for 30 seconds, stirring constantly. Remove from heat, skim, pour into hot sterilized containers, seal.

With added liquid pectin: bring 4 cups of jelly to a boil quickly. Immediately stir in 2 tablespoons of lemon juice, ¾ cup of sugar, and 2 tablespoons of liquid pectin. Bring it back to a full rolling boil, and boil it hard for 1 minute, stirring constantly. Remove the kettle from heat, skim, pour into hot sterilized containers, and seal.

Runny jam. Jam isn't supposed to be as firm as jelly, so if it's only a little bit looser than you'd like it to be, don't bother to remake it. If it's really thin, though, try one of the remedies for too-soft jelly. If a test batch won't turn out right, make sure all the seals are intact and that storage is good—cool, dark, and dry—and mark the remaining jars to be used as a sweet topping for ice creams, puddings, pancakes, etc.

Runny conserves and butters. Often simply cooking them over again will help; try a small batch. Is your storage too warm?

"Weeping" jelly. This is the partial separation of liquid from the other ingredients, and it can come from too much acid or from gelling too fast— or from storage that's warm. So check the seals, make sure there's no mold or fermentation, and move it to a cool, dark, dry place; this should help keep it from getting worse. Such jelly is still usable: decant it just before serving and mop up the juice with clean paper towels.

Mold. Imperfect seals, unsterilized jars and lids; wax (if used) not clean or fresh; warm and damp storage—take your choice of causes. But the mold on these *high-sugar* products does not offer the danger that it would do if it was on a lower-acid product like canned fruit, tomatoes, etc., or a low-sugar confection. Therefore, it may be scraped off so long as it does not have "feeler roots" deep into the product. Dig out at least ½ inch beyond any sign of mold; if the jar is contaminated more than one-fourth its depth, throw out the contents.

Fermentation. The stuff has spoiled. Heave it.

WITH OLD-STYLE USE OF SUGAR

These recipes are indeed old-style. What could be more so than proportions or methods followed since sugar became a staple for the common man as well as for the rich? For tastes more hurried, there will be the confections made with that relatively new thing: store-bought pectin, which increased the size of the batches by asking for so much more sweetening.

Some of the following "receipts," to use the pleasant old term, are translated into today's methods. Ingredients for all are given for small, easy-to-work-with batches; especially with jellies, you get best results if you handle no more than 3 to 6 cups of juice at a time.

Random and handy: 1 cup juice + 1 cup sugar = 1¼ cups jelly. Overboiling will reduce the jelling ability of the natural gel in the juice.

I. With No Added Pectin

Strawberry Jam with No Added Pectin
Four to five ½-pint jars

4 cups prepared crushed berries (about 2¼ to 2½ quarts fresh)
4 cups sugar

Sort and wash and crush berries. Bring to boiling over medium-high heat. Measure pulp into a kettle, add an equal amount of sugar; over medium-high heat bring to boiling and, stirring constantly, cook until jam begins to thicken (but not so much as jelly does at this stage). Remove from heat, skim, and stir a couple of minutes, then ladle into clean, hot ½-pint jars, leaving ¼ inch of headroom. Cap with two-piece screwband lids. Process in a Boiling–Water Bath for 5 minutes after the canner returns to a full boil. Remove jars. Cool.

Red Currant Jelly without Added Pectin
5 to 6 medium glasses

4 cups currant juice (about 2½ quarts currants)
3½ cups sugar

Pick over the currants, discarding overripe or spoiled ones; wash quickly but carefully, and drain off excess water. Measure the washed currants into a large kettle, and *add no more than ¼ as much water as currants*. Over moderate heat, cook the currants until they are soft and translucent, stirring as needed to ensure that they cook evenly—about 10 minutes. Strain the currants with their juice through a damp jelly bag; do not squeeze, lest the juice become cloudy (this classic jelly should always be sparkling clear and jewel-like). Measure 4 cups of juice into a large kettle, bring to boiling, and boil briskly for 5 minutes. Add the sugar, stirring to dissolve it, and boil rapidly until the jelly sheets from a cold spoon or the temperature reaches 8 F/4.4 C above the boiling point of water in your kitchen. Remove from heat and skim off the foam; pour immediately into hot ½-pint jars, leaving ¼ inch of headroom; cap, process in a 5-minute B–W Bath.

Blackberry Jam
About four ½-pint jars

4 cups crushed prepared blackberries (3 quarts fresh if to be seeded)
4 cups sugar

You may need only about 2½ quarts if you are not going to remove seeds. But the exact amount is not vital, because you will be adding an amount of sugar equal to the measurement of prepared berry pulp.

Sort and wash berries, remove stems and caps; crush well. Put berries into an enameled or stainless steel kettle. If you will seed the berries, add ½ cup water; stir, place over high heat, and bring to boiling, stirring well. When berries are soft, put them through a food mill, and—if you want absolutely *no seeds*—press again through a fine-meshed wire sieve. Measure purée into the rinsed-out kettle, add an equal amount of sugar, stirring to dissolve. Bring to boiling, stirring as the mixture thickens (test on an icy saucer). Remove from heat, skim, stir for a minute or two. Ladle into clean, hot ½-pint jars, leaving ¼ inch of headroom; cap with two-piece screwband lids, turning until firmly tight. Process in a Boiling–Water Bath for 5 minutes, after water has returned to a full boil. Remove jars; cool upright and naturally.

Apple (or Quince) Jelly with No Added Pectin
Four ½-pint jars

3 pounds tart red apples (or quinces)
3 cups water
3 cups sugar
2 tablespoons lemon juice (optional)

Four cups of prepared juice are needed.

To prepare juice, use ¼ *underripe* and ¾ fully ripe apples. Wash; remove blemishes, stems, and blossom ends; do not peel or core. Cut apples in small pieces, add water; cover and bring to boiling over high heat. Reduce heat and simmer for 20 to 25 minutes, or until apples are soft. Put apples through a moistened jelly bag. Measure 4 cups of juice into an enameled or stainless steel kettle. Add sugar; strain lemon juice, and add, stirring well. Bring to boiling over high heat to 8 F/4.4 C above the boiling point of water, or until two drops of jelly merge and tear from a spoon. Remove the kettle from heat, skim the jelly quickly, and immediately pour it into hot ½-pint jars, leaving ¼ inch headroom. Cap with two-piece screwband lids. Process in a B–W Bath for 5 minutes. Remove jars; cool upright and naturally.

Crabapple Jelly with No Added Pectin
Five ½-pint jars

3¼ pounds sound fruit, ¼ (13 oz.) underripe and ¾ (39 oz.) just ripe
3 cups water
4 cups sugar

Four cups of prepared juice are needed.

Sort; wash; remove blemishes and stem and blossom ends—*don't pare,* because you want the pectin lying near the skin. Do not core (pulp will be strained). Cut apples small, add water, cover the kettle, and bring to boiling on high heat, reduce heat and simmer for 25 minutes until fruit is soft. Put pulp through a moistened jelly bag. Measure juice into a large enamelware kettle, add sugar, stirring to dissolve. Boil over high heat to 8 F/4.4 C above boiling point of water, or until two drops of jelly merge and tear off the edge of a spoon. Remove from heat, skim, and pour into sterilized ½-pint jars; leave ¼ inch headroom, cap with two-piece screwband lids. Finish with a 5-minute B–W Bath. Cool upright and naturally.

Concord (or Wild) Grape Jelly with No Added Pectin
About five ½-pints

For best results use Eastern Concord or wild grapes (the latter have a flavor especially good with meats and game), and they should be *slightly underripe* for a natural pectin content higher than in fully ripe fruit. Holding the juice overnight in a cool place, and then straining again, will remove the crunchy little slivers of tartrate crystals that form in grape juice.

4 cups grape juice (3½ to 4 pounds of grapes)
1 firm apple
3 cups sugar

Wash and stem the grapes, put them in a large kettle and crush. Wash the apple and cut it in eighths *without peeling or coring,* and add it with ½ cup of water (to prevent sticking). Bring all quickly to a boil, stirring, then reduce the heat and let the fruit cook gently until it is soft—about 10 minutes. Turn the pulp and juice into a damp jelly bag and drain well without squeezing.

Measure 4 cups of juice into a large kettle, stir in the sugar, and boil quickly to the jelly stage; pour immediately into hot ½-pint jars, leaving ¼ inch of headroom. Cap; process for 5 minutes in a B–W Bath. Cool upright and naturally.

The following three jams are from the good Louisiana State University CES booklet *Jellies, Jams and Preserves* (No. 1568). After fruit and sugar are combined over high heat, stir gently but continuously to keep from scorching. When the jam tears off the spoon, remove from heat and let stand 3 to 5 minutes, again stirring gently to dissolve froth and foam. Pour into hot jars, leaving ¼ inch of headroom; simmer (180 F/82 C) in a Hot–Water Bath for 15 minutes. Remove jars; let cool upright and naturally.

Peach Jam

4 pounds peach pulp
2 pounds sugar
1 cup peach juice

Remove peach skins, then put fruit through a food chopper. Weigh, add the peach juice, bring slowly to a boil, add sugar, and finish according to method above.

For spiced jam: to this rule add a spice bag containing 3 inches of cinnamon stick, broken, 1 teaspoon whole allspice, 2 teaspoons whole cloves, and 1 teaspoon of minced fresh gingerroot. Remove the spice bag before filling jars.

Plum Jam

2 pounds plum pulp
1½ pounds sugar

Wash and halve plums, and place in a covered vessel with just enough water to keep the fruit from burning. Cook, stirring, until the seeds slip easily, then press through a colander. Weigh ingredients and continue according to method above.

Pear Jam

2 pounds pear pulp
1½ pounds sugar

Wash the pears, slice without peeling, place in a covered vessel with just enough water to keep from burning and cook slowly until soft. Press through a sieve or colander to remove seeds and skin. Weigh ingredients and continue according to method above.

II. With Added Pectin

Grape Jelly
Eight to nine ½-pints

---

4 cups grape juice (3½ to 4 pounds grapes)
7 cups sugar
3-ounce pouch liquid pectin

Sort, wash, and stem ripe Concord (or wild) grapes; crush, add ½ cup of water, and bring to a boil. Reduce heat and simmer for about 10 minutes. Turn into a damp jelly bag and drain well; do not squeeze. Hold the juice overnight in a cool place, then strain through 2 thicknesses of damp cheese-cloth to remove the tartrate crystals that form in grape products. Measure 4 cups of juice into a large kettle, add the sugar and mix well; bring quickly to a full boil that cannot be stirred down. Add the pouch of pectin, bring again to a full rolling boil, and boil hard for 1 minute. Remove from the heat, quickly skim off the foam, and pour the jelly into hot ½-pint jars, leaving ¼ inch of headroom. Cap with two-piece screwband lids. Process 5 minutes in a B–W Bath; cool upright and naturally.

Helen Ruth's Sand Plum Jelly
With pectin, nine ½-pint jars

The cherry-sized sand plum of the American Southwest is kin to the beach plum, that favorite for preserves from the sandy coasts of the Northeast up into the Canadian Maritimes. The sand plum is ripe in early June; the season for beach plums starts around the middle of August; the sand plum is a lovely pink when ripe, the beach plum is purple for conserve later in the month but is picked red for jelly. Both varieties gel better if at least one-fourth the amount of fruit is not quite ripe, thus having more natural pectin.

4 pounds sand plums, 3 pounds ripe and 1 pound underripe
1 cup water
1 package powdered pectin (1¾ ounces)
7 cups sugar

Wash and pick over the plums; do *not* pit or peel. Crush them in the bottom of a large enameled kettle with the 1 cup water, bring to a boil, and simmer for 15 minutes. Crush again with a vegetable masher as the fruit softens. Strain through a jelly bag: add a little water to bring the measure up to 5 cups of juice. Return juice to the kettle, reserving 1 cup in which to mix the pectin; combine pectin mixture with juice and bring to a full boil, stirring constantly. Add the sugar, continue stirring, and boil hard for 2 minutes. Remove from heat, skim, and immediately pour into hot ½-pint canning jars, leaving ¼ inch of headroom. Cap with two-piece screwband lids. Give a 5-minute B–W Bath.

Freezer (Uncooked) Berry Jellies with Powdered Pectin
Six 8-ounce jars

3 cups prepared juice (about 2 to 2½ quarts of fresh strawberries, blackberries or red raspberries)
6 cups sugar
1 package powdered pectin (1¾ ounces)
¾ cup water

Crush the berries and strain them through a damp jelly bag or four layers
of damp cheesecloth; squeeze gently if necessary. Add the sugar to the measured 3 cups of juice; stir well and let stand for 10 minutes (a few sugar crystals
may remain, but they will dissolve in the time it takes the jelly to set). In a
small saucepan stir together the ¾ cup water and the pectin; bring the mixture
to the boil and boil hard for 1 minute, stirring constantly. Remove from heat
and add it to the sweetened juice; continue stirring for 3 minutes. Then pour
into sterilized jars that are freezable and have tight-fitting lids or two-piece
screwband lids, leaving ½ inch of headroom; seal. Let stand at room temperature until set—up to 24 hours—then freeze.

"Best-Ever" (Frozen) Strawberry Jam
Five to six 8-ounce jars

2 cups prepared fruit (about 1 quart ripe strawberries)
4 cups sugar
¾ cup water
1 box powdered fruit pectin (1¾ ounces)

Thoroughly crush, one layer at a time, about 1 quart of fully ripe strawberries. Measure 2 cups of crushed berries into a large bowl. Add the sugar
to the fruit, mix well, and let stand for 10 minutes; a few sugar crystals may
remain but they'll dissolve as the jam sets. Mix water and pectin in a small
saucepan, bring the mixture to a boil and boil for 1 minute, stirring constantly.
Remove from heat and stir the pectin into the fruit; continue stirring for 3
minutes. Ladle quickly into sterilized freezable jars, leaving ½ inch of headroom. Seal immediately with sterilized tight-fitting lids. Let jars stand at room
temperature until the jam is set—which may take up to 24 hours—then freeze.

Tomato Jelly
Five ½-pint jars

1¾ cups home-canned tomato juice
½ cup strained fresh lemon juice
2 teaspoons Tabasco sauce
4 cups sugar
1 three-ounce pouch liquid pectin

Combine all ingredients except pectin. Stir over high heat until mixture reaches a full, rolling boil. Stir in pectin and bring again to a full, rolling boil. Boil 1 minute, stirring constantly. Remove from heat, skim, and pour into hot ½-pint jars, leaving ¼ inch of headroom. Process 5 minutes in a B–W Bath. Cool upright and naturally.

Pyracantha (Firethorn) Jelly
Five ½-pint jars

The red-orange pomes of this spiky hedge plant are a favorite for jelly-making in the American Southwest.

3 generous quarts Pyracantha berries
3 cups water
Juice of 1 grapefruit
Juice of 1 lemon
1 box powdered pectin (1¾ ounces)
4½ cups sugar
¼ teaspoon salt

Sort, pick over the berries, and put them in a large enameled kettle with the 3 cups of water. Boil for 20 minutes, add the citrus juices, bring briefly to boiling again, and pour all into a dampened jelly bag to strain slowly into a crockery bowl. The result should be 3½ cups of juice. In a bowl, combine the pectin and 1 cup of the juice, then pour the mixture into the preserving kettle with the remaining 2½ cups of juice. Bring to a hard boil, add the sugar and the salt, and bring to a full rolling boil; stir and boil 3 minutes. Remove from heat, skim quickly, and pour into hot ½-pint jars, leaving ¼ inch of headroom. Cap, process 5 minutes in a B–W Bath. Cool upright and naturally.

III. Marmalades, Butters, and Preserves

Ginger Squash Marmalade
About 4 pint jars

This is a paraphrase of a very old Scottish rule.

4 pounds prepared winter squash (or pumpkin)
4 pounds sugar
2 lemons, grated and juiced
2 ounces crystallized ginger
½ teaspoon ground ginger
3 cups water

(Recipe continued next page)

Peel the squash, scrape out the strings and seeds from the center; save the seeds. Slice and cut the peeled squash to make 4 pounds of $\frac{1}{4}$-inch cubes. Put the squash in a crockery bowl with the sugar, gingers, grated peel and juice of the lemons. Meanwhile break the seeds, put them in a pan with the water, and boil gently for 30 minutes; strain, and add $2\frac{1}{2}$ cups of the liquid to the squash mixture. Cover and refrigerate overnight. Put into an enameled kettle, and over medium heat, bring to a slow boil, stirring constantly. (Here you can finish it in a 325 F/163 C oven, or let it putter along with a heat-reducing pad under the kettle.) Cook and stir until the squash is clear and will set— 45 to 60 minutes. Ladle immediately into clean hot jars, leaving $\frac{1}{2}$ inch of headroom for pints, $\frac{1}{4}$ inch for $\frac{1}{2}$-pints. Adjust lids; process in a Boiling–Water Bath for 5 minutes for pints or $\frac{1}{2}$-pints. Remove; cool upright and naturally.

Green Tomato Marmalade
2 pints

- 2 quarts sliced, small, green tomatoes
- $\frac{1}{2}$ teaspoon salt
- 4 lemons, peeled (save the rind)
- 4 cups sugar

Combine tomatoes and salt in a stainless steel kettle; chop lemon rind fine and add. Cover with water and boil 10 minutes. Drain well. Slice the peeled lemons very thin, discarding seeds but reserving all juice. Add lemon slices and juice and the sugar to the tomato mixture. Stir over moderate heat until sugar melts. Bring to boiling, reduce heat, and simmer until thick—about 45 minutes. Stir frequently. Pack in hot jars and process 5 minutes in a B–W Bath. Cool upright.

This classic from long ago is especially good served with meat.

Classic Orange Marmalade
Five to six $\frac{1}{2}$-pint jars

The Scots make highly prized marmalades, among them this one from Mildred Wallace, which is characterized by a darker color and slightly bitter flavor compared with the most popular supermarket brands in the United States. The precooking prevents the peel from becoming tough when it is boiled with the sugar.

- 2 pounds Seville oranges (or other bitter variety left whole)
- 2 large lemons, whole
- 8 cups water (about)
- 8 cups (4 pounds) sugar

Wash the oranges and lemons well, removing any stem "buttons," and put the clean washed whole oranges and lemons in a large kettle with enough water to cover them; put the lid on the kettle and bring to a boil, then simmer until a slender fork will easily pierce the fruit—about 1½ hours. Remove the fruit to cool, saving the liquid; when they are cool, cut them in half the long way, then cut the halves in *very thin* slices (your knife must be sharp!), and take out and save the pips. Return the pips to the juice in the kettle and boil for about 10 minutes (this contributes to the bitter flavor). Strain the juice and return it to the kettle. Add the fruit slices and heat to boiling. Add the sugar, stirring until it dissolves, and continue cooking at a fast boil—stirring only enough to prevent scorching—until it starts to thicken and its temperature reaches 9 F (1 degree more than for jelly) above the boiling point of water in your kitchen. Remove from heat, skim off any foam, pour at once into hot sterilized ½-pint jars with two-piece screwband lids, leaving ½ inch of headroom. Adjust lids; process in a Boiling–Water Bath (212 F/100 C) for 5 minutes. Cool upright.

Spicy Carrot Marmalade for Game
Four to five ½-pint jars

4½ cups coarsely ground raw carrots
3¼ cups sugar
About ¼ cup fresh lemon juice
¾ teaspoon finely grated zest of lemon
½ teaspoon ground ginger
½ teaspoon ground cloves
½ teaspoon ground cinnamon
½ teaspoon salt

Trim and scrape carrots; run them through the coarse knife of a food grinder. Put them in a heavy enameled kettle; add the sugar, spices, and salt. Warm or roll on a hard surface two lemons (to make them easier to juice), but do *not* cut: the next step is to grate off the thin yellow skin of the rind, taking care not to get any of the bitter white portion. Add the grated zest to the ingredients in the kettle. Halve and juice the lemons; add the juice to the kettle. Over low heat, and stirring, bring the mixture to a very slow boil, and cook for about 30 minutes or until it is thick. Pour into hot ½-pint jars, leaving ¼ inch of headroom; cap with clean hot lids. Process in a Boiling–Water Bath for 5 minutes. Remove; complete seals if necessary. Cool upright and naturally.

Yellow Tomato Marmalade
Three ½-pint jars

3¼ cups coarsely chopped, peeled, ripe, yellow plum-type tomatoes
¼ cup fresh lemon juice
Grated rind of 1 large lemon
6 cups sugar
2 pouches liquid pectin (6 ounces)

(Recipe continued next page)

Place chopped tomatoes in small pan and set over low heat and cover. *Do not add any water.* Bring to boiling, reduce heat, and simmer about 10 minutes, stirring frequently. Remove from heat and measure out 3 cups of the tomatoes and liquid. In a large kettle, combine the 3 cups tomatoes with the lemon juice, grated rind, and sugar. Stir over moderate heat until boiling. Boil hard 1 minute. Turn off heat and add pectin. Stir for 5 minutes. Skim as needed. Pack in hot ½-pint jars; process 5 minutes in a B–W Bath.

Steps in Making Butters

Butters are nice old-fashioned spreads and they're good with meats. Their virtues are that they take about ½ as much sugar as jams from the same fruits (½ cup sugar to each 1 cup of fruit pulp), and they can be made with the sound portions of windfall and cull fruits that you'd probably not bother with for jelly or jam. Their one drawback is that they require very long cooking—and careful cooking at that, because they stick and scorch if you turn your back.

Butters are made from most fruits or fruit mixtures. Probably apple is the best-known ingredient, but apricots, crabapples, grapes, peaches, pears, plums, and quinces also make good butters.

TO PREPARE FRUIT

Use prime ripe fruit or good parts of windfalls or culls. Wash thoroughly and prepare as follows:

Apples. Quarter and add ½ as much water or cider (or part water and part cider) as fruit.

Apricots. Pit, crush, add ¼ as much water as fruit.

Crabapples. Quarter, cut out stems and blossom ends, and add ½ as much water as fruit.

Grapes. Remove stems, crush grapes, and cook in own juice.

Peaches. Dip in boiling water to loosen skins; peel, pit, crush, and cook in their own juice.

Pears. Remove stems and blossom ends. Quarter and add ½ as much water as fruit.

Plums. Crush and cook in their own juice. The pits will strain out.

Quinces. Remove stems and blossom ends. Fruit is hard, so cut in small pieces and add ½ as much water as fruit.

MAKING THE PULP

Cook the fruits prepared as above until their pulp is soft. Watch it—it may stick on.

Put the cooked fruit through a colander to rid it of the skins and pits, then press the pulp through a food mill or sieve to get out all fibers.

SUGAR AND COOKING

Usually ½ cup of sugar to each 1 cup of fruit pulp makes a fine butter. It's easiest to use at one time not more than 4 cups of fruit pulp, plus the added sugar.

Let the sugar dissolve in the pulp over low heat, then bring the mixture to a boil and cook until thick, stirring often to prevent scorching.

When the butter is thick enough to round slightly in a spoon and shows a glossiness or sheen, pack while still hot into hot, sterilized ½-pint or pint canning jars, leaving ½ inch of headroom. Adjust lids and process the jars in a Boiling–Water Bath (212 F/100 C) for 10 minutes. Remove jars, complete seals if necessary. Cool and store.

Alternative cooking method: butters stick so easily when they are cooking on the stove top that it's a real chore to keep them from scorching. Some cooks put about ¾ of the hot uncooked purée in a large, uncovered, heatproof crockery dish or enameled roasting pan and cook it in a 300 F/149 C oven until it thickens. As the volume shrinks and there is room in the dish, add the other ¼ of the purée. When the butter is thick but still moist on top, ladle it quickly into containers and process.

OPTIONAL SPICES

Any spices are added as the butter begins to thicken. For 1 gallon of pulp use 1 teaspoon ground cinnamon, ½ teaspoon ground allspice, and ½ teaspoon cloves. Ginger is nice in pear butter—1 to 2 teaspoons to 1 gallon of pulp. For smaller quantities of pulp reduce measures of spices proportionately. If the butter is to be light in color, tie whole (not ground) spices loosely in a cloth bag and remove the bag at the end of the cooking.

Apple Butter
About 6 pints

---

5 pounds juicy tart apples, 12 to 15 (Winesap, Northern Spy, Jonathan)
1 cup apple cider (or water)
About 2½ cups sugar, or to taste
½ teaspoon ground cinnamon
¼ teaspoon ground cloves
¼ teaspoon ground nutmeg

Remove blemishes, core, and cut apples in eighths. Put apples and cider in a heavy enameled kettle and cook over medium heat until soft; stir to prevent sticking. Remove from heat, and when the pulp is cool enough to handle, put it through a food mill or sieve. For every 1 cup pulp add ¼ cup white or brown sugar. Return to the heavy kettle and bring it to a low boil until sugar is melted.

Now choose: (1) cook quickly over medium-high heat to a brisk boil, stirring constantly. Or (2) put the kettle in a preheated 300 F/149 C oven, where you need stir only occasionally to keep a caramelized skin from forming on top. The quick stove-top boil produces brighter color, but scorches the purée if you turn your back; the oven makes darker color and takes longer, but doesn't scorch (any caramelized skin can be rolled off easily).

By either method, cook the purée until it mounds slightly on the spoon and has a sheen to it. On a jelly thermometer it will be a bit under 220 F/104 C at sea level. Ladle into clean hot jars, leaving ¼ inch of headroom for ½-pint jars, ½ inch for pints. Adjust lids, process in a Boiling–Water Bath for 5 minutes. Remove jars; complete seals if necessary.

Variation: omit cinnamon and cloves; instead use ½ teaspoon scraped zest of lemon and ½ teaspoon freshly grated nutmeg.

Blender/food processor: instead of putting cooked fruit through a food mill or sieve to remove skins before adding sweetener, purée with skins in small amounts at the highest speed of the blender, then strain the sauce into the heavy kettle, adding sweetener and seasonings as liked.

In a food processor with the steel blade in place, purée the cooked apples and their skins, small amounts at a time; the result will be smoother—and runnier—than with the blender, but you may want to sieve the purée. (Also using the steel blade, you can save time by whirring thin shavings of fresh lemon peel with 1 cup of the sugar.)

Creating "All-Fruit" Spreads

Big on the market at this writing are several confections made without any added sweetening: all the sugar is natural fructose from the fruit. Here is a chance to experiment (especially since these spreads cost an arm and a leg to buy).

Use a filler fruit able to provide the consistency: ripe peaches, peeled plums, sieved berries, apples—even canned fruits. Frozen pineapple juice concentrate would help: if you fear that it's likely to act like fresh pineapple, whose tricky enzymes prevents any sort of jel from forming, dump it into a small pot, add ¼ cup water, and bring it to boiling; watch it carefully so it doesn't scorch. Concentrated grape juice would be worth a trial. Be sure to save and purée all skins likely to have good pectin in them. And you can always add some homemade apple pectin to help the gel. Orange zest is a charmer with blackberries; try a breath of cloves instead of cinnamon with blueberries.

When you think it's done, can it very hot in ½-pint jars, leaving ¼ inch of headroom; process for 5 minutes in a B–W Bath. Cool upright and naturally.

Steps in Making Preserves

Wash the fruit and remove stem and blossom parts. Peel peaches, pears, pineapples, quinces, and tomatoes. Shred pineapple, discard the core. Cut slits in tomatoes and gently squeeze out the seeds, cut large tomatoes in quarters, leave small ones whole. Pears and quinces are thinly sliced after halving and coring. Take the pits from sour cherries. Of course strawberries and raspberries are left whole.

To cook, carefully follow the specific recipe. Generally, dry sugar is added to the soft fruits to start the juice flowing. There should be enough juice to cook the fruit. Hard fruits are cooked in a sugar-and-water syrup. The recipe will tell you how long to cook each of the preserves.

Ladle hot preserves into hot sterilized canning jars, leaving ¼ inch of headroom; wipe the mouths of the containers carefully with a clean cloth wrung out in boiling water; adjust lids, and process in a Boiling–Water or simmering Bath for 10 minutes; complete seals if necessary. Cool, clean the jars, label, and store.

Cantaloupe Preserves
About 1½ pints

1 pound cantaloupe flesh
¾ pound sugar
1 lemon

Cut cantaloupe into sections. Remove seeds. Cut more tender portion of pulp for use fresh. Remove rind. Cut firm portion of pulp into uniform pieces. Add sugar to melon in alternate layers of melon and sugar. Let stand 24 hours. Add the juice of one lemon. Bring to boil and boil quickly until the fruit is clear and tender. Place fruit in shallow trays. If syrup is too thin continue cooking until the desired consistency is reached. Pour hot syrup over fruit and allow to stand overnight so fruit will plump. Pack cold in sterilized jars, seal, and process at simmering temperature for 30 minutes.

Blueberry Conserve
Six ½-pint jars

1 orange
1 lemon
3 cups water
5 cups sugar
½ cup dark seedless raisins
6 cups blueberries, stemmed and washed

With vegetable peeler, remove each outer rind of the orange and the lemon—cutting so thinly that none of the white underlayer comes with it; chop fine. Remove and chop the pulp, discarding any seeds. Bring the water and sugar to a boil in large stainless steel kettle, and add orange, lemon, and raisins; simmer for about 5 minutes. Add the blueberries, and cook over moderate heat until the mixture thickens—about 30 minutes—stirring often to prevent sticking. Pour boiling hot into hot ½-pint jars leaving ¼ inch of headroom. Adjust the lids and process in a Boiling–Water Bath for 10 minutes. Remove and complete seals if necessary.

This is especially good as topping for ice cream or plain white or yellow cake.

Louisiana Pear Conserve
7 or 8 pints

5 pounds firm pears—ideally Kieffer—cut small
10 cups sugar
1 pound seedless raisins
Rind of 2 oranges
Juice of 3 oranges
Juice of 1 lemon

Peel pears, remove core and midvein, cut in small pieces. Stir pears and sugar together in a large enamelware bowl; let stand overnight. Next day, in a large stainless steel or enameled kettle, combine juices, raisins, chopped orange rind. Bring to gentle boil, and cook, stirring, until thick (about 30 to 35 minutes). Ladle promptly into hot jars, remove any air bubbles from contents, wipe jars' sealing rims; cap, and process 15 minutes in a simmering Hot–Water Bath at 180 F/82 C. Remove jars; cool upright and naturally.

Seven-Minute Strawberry Preserves
About 1½ pints

A classic from Louisiana State University Extension

4 cups perfect strawberries
3 cups sugar
Boiling water

Wash and hull perfect berries, place in a colander. In a large flat-bottomed pan (so the water will fall out in a sheet, not in a harsh stream as from a teakettle) bring 3 quarts of water to boiling, pour water over berries in the colander. Put heated berries promptly into a stainless steel or enamelware stewpot, fold into the hot berries 1½ cups sugar, slowly bring to a boil, and boil gently 4 minutes. Off heat, add remaining 1½ cups sugar, shake the pan to distribute and dissolve the sugar. Bring to a gentle boil, and boil 3 minutes. Skim off any foam, lift berries out and into a shallow dripping pan, pour syrup over them to cover, and let berries rest, covered, for 24 hours to become plump. Next day, pack berries and syrup in hot jars, leaving ¼ inch of headroom. Remove any air bubbles from contents, wipe sealing rim, cap with prepared disk and screwband firmly tight. Process in a Hot–Water Bath, simmering at 180 F/82 C for 20 minutes for pints or ½-pints. Remove jars; let cool upright and naturally.

Sweet Cherry (or Other) Preserves
2 pints

4 cups pitted sweet cherries, tightly packed
3 cups sugar

In a 4-quart saucepan crush the cherries lightly to start the juice flow. Boil cherries and their juice about 10 minutes—or until fruit is tender. Add sugar to the cherries, stir well, and boil for 5 minutes more. Now cover the kettle and let the cherries stand for 2 minutes while they absorb more of the sugar. Stir the hot preserves to prevent floating fruit, then pour into hot sterilized jars, leaving ¼ inch of headroom; process for 5 minutes in a B–W Bath (212 F/100 C).

WITH LOW SUGAR/NO SUGAR

Many things are called "diet" these days: it depends, really, on what one is trying to stay away from—sucrose, sodium, cholesterol, the list can go on. In general, though—and certainly in context here—the word indicates that the amount of white cane or beet sugar has been reduced, by using low-methoxyl pectin, even below the amounts used in proportion to fruit juice/pulp when no pectin was added at all. Then from low-methoxyl pectin, they continue down till the end is Zero sugar in the spreads using artificial sweeteners.

Time Out for Low-Methoxyl Pectin

Low-methoxyl pectin must be accompanied by added calcium, but it can make cooked fruit juice or pulp gel with only a fraction of the sugar called for by the universally available pectins.

The following procedures are a meld of the instructions that accompanied bulk low-methoxyl pectin *PFB* bought, along with the di-calcium phosphate to boost its power to gel, from two Northeast suppliers; to the directions are added what could be called side comments from the company that distributes pectin of all types to the conglomerates that put it eventually in shopping carts in grocery stores across North America.

The operative ingredient in low-methoxyl confections is calcium. It doesn't matter technically what kind of calcium it is so long as it is food-grade—monocalcium, di-calcium or tri-calcium phosphate—or a calcium compounded with the less-pleasing chloride—*and* that the amount of free calcium in each of the compounds has been taken into account. Actually, though, we found that di-calcium phosphate made jelly more nearly like the old thing we were used to.

Idiosyncrasies to Watch For

First: low-methoxyl pectin will form globs like half-cooked tapioca if you give it the slightest chance. So, if you are using it with a granulated nutritive sweetener of some sort (as against an *artificial* sweetener, either liquid or granular), mix the pectin with the sugar, which then acts as a wetting agent.

Then, di-calcium phosphate is put into what is termed a solution. But it really doesn't dissolve at all—it just hangs there in suspension; and, though it can be made ahead and kept in the refrigerator, its storage jar must be shaken hard each time you put the measuring spoon into it.

And add the calcium at the end: *do not add it too early and cook it*. This is the warning of Dianne Leipold of Hercules, Inc., who supplies the people who in turn supplied us. One of the main causes for low-methoxyl jelly/jam to go wrong, Dr. Leipold said, is allowing the calcium to cook, and cooking it results in grainy jelly that releases its water. Calcium can be added even when the kettle is taken from the stove, and be stirred into the sweetened fruit-and-pectin just before the boiling mixture is ladled into ½-pint jars. Process 5 minutes in a B–W Bath.

Using Low-Methoxyl Pectin with Some Sugar

You can increase this batch to 4 cups of juice, or even more. However, we recommend that you start with 1-cup batches until you get the consistency you like. After all, your water may be hard (contain calcium): this will affect the stiffness of your jelly/jam. The day may be muggy. All the things that could go awry with other jelly-making could play hob here. The proportions:

   1 cup prepared fruit juice (prepare it as for any jelly) OR
   1 cup fruit pulp (seeded, skins removed, etc., as for any jam)
   ¼ cup (4 tablespoons) sugar OR 3 tablespoons fructose
   ½ teaspoon granular low-methoxyl pectin
   1 teaspoon di-calcium phosphate *solution*

And you make the solution by adding ¼ teaspoon granular di-calcium phosphate to ½ cup boiled and cooled water, and storing it in a sterilized, tightly closed container in the refrigerator. This amount of solution will "do" 24 cups of prepared juice or prepared fruit pulp, because at the end of cooking you add 1 teaspoon of the solution for each 1 cup of fruit.

Procedure

Prepare fruit; boil jars and hold them and their two-piece screwband lids in the hot water. In a completely dry container, and with a dry spoon, add ½ teaspoon granular low-methoxyl pectin to the ¼ cup sugar and mix thoroughly. Bring the fruit to boiling over moderate heat, add the sugar-pectin in one swoop, and stir well to dissolve it. When the sugar and pectin are dissolved in the hot fruit, add the 1 teaspoon of (well-shaken) calcium solution, stir, and fill hot jars to within ⅛ inch of the top. Wipe rims, adjust lids, and screw them down tightly. Stand the jars upright out of drafts and let them cool naturally. Jars may be stored in the pantry until opened; then they go in the refrigerator.

"Diet" (Sugar-free) Jams and Jellies

These are perishable and therefore must be stored in the refrigerator. Unopened, they will keep 3 to 4 weeks in the fridge, so make only small batches at a time. (Gelatin sweets "weep" when frozen unless they have added solids; sometimes the gelatin toughens in the separation. So don't freeze: freezing is for pectin-added jams that take sugar to help form a gel.)

For refrigerator storage, headroom is ⅛ inch in ½-pint containers, maximum ¼ inch in pint-size jars.

Generally, jams have a more expectable consistency than jellies do. Fruit "butters"—which are cooked down for a fairly long time—do not need gelatin to give them body: just add some sweetener, perhaps a spice or zest of orange or lemon.

Procedure

Wash jars and boil for 15 minutes to sterilize; hold in hot water until used. Prepare new disk lids at manufacturer's directions; wash and boil used lids; hold in hot water. Use unblemished ripe fruit, wash, core; peel as necessary.

FOR JAMS

Put ½-inch layer of berries or chopped fruit in a stainless steel or enameled kettle, mash with pastry blender, then add succeeding layers and crush. Add ½ cup water, set over medium heat, bring to simmer; increase heat, bring to boiling, boil 1 minute. Off heat, add gelatin softened in ½ cup water, stir, bring just to boiling. Add lemon juice or spice flavors, stir in sweetener. Pour into sterilized containers, leaving appropriate headroom. Cap, seal, cool, refrigerate.

FOR JELLIES

Use bottled juice or fresh cider. Or prepare and cook fruit as for jam. Pour blackberry or raspberry pulp through cheesecloth to remove most seeds, then strain through jelly bag. "Let down" juice by diluting with boiling water to equal ½ the volume of juice (or to taste); experiment— without the traditional load of sugar to lessen its impact, the undiluted juice is likely to be "too much of a good thing." Add softened gelatin, any flavorings, boil one minute; stir in sweetener. Pour into sterilized jars, cap, seal, cool, refrigerate.

Unopened, will keep for up to 4 weeks in refrigerator storage.

Sugarless Jellies and Jams with Gelatin

Ruth Hertzberg's comments about these confections should introduce the following recipes. First, she says, it is just possible that the amount of gelatin used for a dessert may be "just a whisker" too stiff for smooth spreadability. And stirring finished jam may break it up too much to be spreadable the next time you want it.

It's possible to freeze these, but they can weep.

Without the usual sugar as a preservative, these are perishable, so there is little use in making much at a time (but the temptation is great for a sweet-starved diabetic).

Finally, Ruth recommends that next to no headroom be left: ⅛ inch is plenty for refrigeration. For freezing, though, allow ½ inch for expansion.

Diet Peach Jam
Four ½-pint jars

---

   2 tablespoons unflavored gelatin
   ¼ cup cold water
  3½ pounds ripe peaches
   8 teaspoons lemon juice
   4 tablespoons liquid artificial sweetener

In a small bowl soften the gelatin in the ¼ cup of water. Wash and peel peaches, remove pits and cut in chunks. In a 4-quart enameled or stainless steel kettle over medium heat, bring the cut peaches to a simmer until soft. Remove from heat and crush lightly. Measure and add enough water to make 4 cups of pulp. Return to the kettle and add softened gelatin and lemon juice, stirring well to dissolve the gelatin. Over medium heat, bring to a boil and boil 1 minute, stirring all the while. Remove from heat, skim, and stir in well the liquid artificial sweetener, and pour immediately into hot sterilized ½-pint jars, leaving ⅛ inch of headroom (½ inch if to be frozen, to allow for expansion). Cap tightly with two-piece screwband lids.

When cool and set, store the jars of jam in the refrigerator for use within one month. Each 1 tablespoon of jam has 10 calories.

Apple Butter
About 30 ounces

  2½ pounds juicy, tart apples, 5 or 6 large (Winesap, Northern Spy, Jonathan)
   ⅓ to ½ cup apple cider (water will do, cider is better)
  Scant ¼ teaspoon ground cinnamon
  Pinch each of ground nutmeg and cloves
  Sweetener to equal ¾ cup sucrose, or to taste (added last thing)

Remove any blemishes, core, cut apples in eighths (do not peel). In a heavy steel or enameled pot, combine apples and cider; cook, stirring often, until quite soft. Off heat, put through a sieve or food mill to remove noticeable skins. (Blender at highest speed will incorporate most of the skins, but the pulp will be runnier.) Return to rinsed heavy pot, add flavorings except for the sweetener, and cook down, stirring constantly lest it scorch, until the purée mounds on a spoon and is glossy. (A 325 F oven, stirring often, is a safeguard against scorching.) Off heat, stir in sweetener, ladle into hot sterilized jars/glasses, cap, and seal.

Frozen Diet Raspberry Jam
Two to three ½-pint jars

1 quart fresh red raspberries
1 tablespoon lemon juice
1 package powdered fruit pectin (1¾ ounces)
3 to 4 teaspoons liquid artificial sweetener

Crush raspberries thoroughly; put half (or more) through a food mill to remove some seeds, if you like. Crush the berries directly into a large stainless steel or enameled kettle. Add the lemon juice and the powdered pectin, stirring until the pectin is thoroughly dissolved. Over medium heat, bring the mixture to boiling, and boil hard for 1 minute. Remove from heat, add the sweetener, and stir for 2 minutes. Ladle into hot sterilized ½-pint glass freezer jars, leaving ⅛ inch of headroom if to be refrigerated, ½ inch if to be frozen. Cap at once with sterilized two-piece screwband lids; *do not use paraffin*. Let the jars stand at room temperature for a day or so, until the jam is set. Freeze.

Store in the refrigerator after opening and thawing. Each 1 tablespoon of jam has 7 calories.

Sugarless Apple Jelly
Two 8-ounce jars

2 cups unsweetened apple juice
4 teaspoons unflavored gelatin
1½ tablespoons lemon juice
2 tablespoons liquid artificial sweetener, to equal 1 cup sugar (read the label, especially if substituting a dry artificial sweetening agent)
1 or 2 drops red food coloring if you like

Soften the gelatin in ½ cup of the apple juice. Meanwhile heat the remaining 1½ cups juice to boiling; remove from heat and stir in the softened gelatin until it is dissolved. Add the lemon juice, the food coloring if you like it. Return to heat and bring to the boil; stir in the sweetener, then pour into hot sterilized ½-pint jars that have two-piece screwband lids, leaving ⅛ inch of headroom; seal. Store in the refrigerator when cool, and use within 3 to 4 weeks.

Blackberry Jam
About 16 ounces

Substitute 2 cups crushed blackberries in the recipe for Strawberry Jam; increase lemon juice to 1½ tablespoons. Use same gelatin and sweetener. Follow Strawberry method.

Remove most of the seeds by putting *heated* crushed berries through a food mill before adding gelatin, etc. You'll need an extra cup or so of fresh berries if you plan to seed the jam.

Diet Grape Jam
Four ½-pint jars

4 teaspoons unflavored gelatin
¼ cup water
3 pounds Concord or other juicy grapes
2 tablespoons liquid artificial sweetener

In a small bowl soften the gelatin in the cold water. Wash and stem the grapes, then press them through a food mill or coarse sieve to remove skins and seeds. Measure the pulp and add enough water to make 4 cups, then put it into a 4-quart kettle and add softened gelatin; stir well to dissolve it. Over medium heat, bring to a boil and boil for 1 minute, stirring all the while to prevent scorching. Remove from heat, skim off any foam, and stir in well the liquid artificial sweetener. Pour into hot sterilized ½-pint jars, leaving ⅛ inch of headroom if to be refrigerated, ½ inch if to be frozen. Cap tightly with sterile two-piece screwband lids.

When cool and set, store in the refrigerator for use within one month. Each 1 tablespoon of jam has 12 calories.

Sugarless Strawberry Jam
Two 8-ounce jars

2 cups crushed strawberries (about 1 quart whole berries)
4 teaspoons unflavored gelatin
1½ tablespoons lemon juice (particularly if blackberries are substituted)
2 tablespoons liquid artificial sweetener, to equal 1 cup sugar (read the label, especially if substituting a dry artificial sweetening agent)

Soften the gelatin in ½ cup of the juice from crushed berries. Meanwhile heat the remaining 1½ cups of crushed berries to boiling; remove from heat and add the softened gelatin, stirring until it is dissolved. Add the lemon juice and the artificial sweetener. Return to heat and bring to the boil, then pour into hot sterilized jars that have two-piece screwband lids, leaving ⅛ inch of headroom; seal. Store in the refrigerator when cool, and use within 3 to 4 weeks.

Grape Jelly from Frozen Concentrate
About 26 fluid ounces

---

1 six-ounce can frozen unsweetened concentrated grape juice
1½ cups water
2½ teaspoons unflavored gelatin
Artificial/non-nutritive sweetener to equal 2¼ cups sucrose (added last thing)

Thaw grape juice; soften gelatin in ½ cup of the water. Combine juice, water (remaining 1 cup), bring to boiling; stir in gelatin, and boil 1 minute. Off heat; stir in sweetener. Pour into hot sterilized glasses/jars, cap, seal.

19
PICKLES, RELISHES, AND
OTHER SPICY THINGS

By increasing acidity, we preserve what at first might seem to be unlikely foods to be prized so highly. Yet pickles and relishes and condiments have been regarded as treats since the first gastronome reported them at a pharaoh's feast, and praised the inventiveness of the cook.

Acid is the key to their safety as food, and to their charm. The acid either is added to the vegetables or they are induced to create acid by undergoing fermentation. In either case, the result is put into a clean jar with a safe closure—today, the two-piece screwband lid—and is given a simmering *Hot*–Water Bath between 180 F/82 C and 185 F/85 C for 10 or 15 minutes (individual recipes will say). The bath is simmering lest it turn canned sauerkraut into dull cooked cabbage, or ruin fresh cucumbers, or soften mixed vegetables.

Avoid boiling for a long time any vinegar solution that is to be used as the canning liquid for pickles, because the acetic acid in vinegar is rather volatile, and it will lose its ability to keep stored pickles safe. (Remember how the acid added to canned tomatoes called for twice as much 5 percent vinegar as 5 percent lemon juice, just because the acetic acid was less stable than citric?)

Pickles and relishes are first cousins. Their major difference is that vegetables and/or fruits for relishes are chopped before being put with the vinegar, and those for pickles are left whole or cut to fairly large size.

Equipment for Pickles and Relishes

The characteristic to keep in mind is the interaction of the vinegar and salt with metals: use enameled, earthenware, or glass containers to hold or cook these mixtures—*never use anything that's galvanized,* or copper, brass, iron, or aluminum.

Water–Bath canner (or stockpot with rack).
6- to 8-quart enameled kettle for short brining and cooking pickles.
Jars in prime condition, with lids/sealers/gaskets ditto.
Minute-timer with warning bell to time processing periods.
Shallow pans (dishpans are fine).
Ladle or dipper.
Long-handled wooden or stainless spoon for stirring.
Wide-mouth funnel for filling containers.
Jar-lifter.
Colander for draining.
Large measuring cups, and measuring spoons.
Squares of cheesecloth to hold spices.
Plenty of clean dry potholders, dishcloths, and towels.
Household scales.
Stoneware crocks.

Essential Ingredients

The Produce Itself

Any firm-fleshed vegetable or fruit may be used. Fresh, prime ingredients are basic. Move them quickly from garden or orchard to pickling solution. They lose moisture so quickly that even one day at room temperature may lead to hollow-centered or shriveled pickles.

Perfect pickles need perfect fruits or vegetables to start with. The *blossom ends* of cucumbers must be removed (since any enzymes located there can cause pickles to soften while brining), but do leave $\frac{1}{4}$ inch or so of *stem.*

Salt

Use only plain, pure salt, either coarsely or finely ground, without additives. The 5-pound bags of canning/pickling salt from the supermarket are ideal. Avoid like the plague any salt designated as "solar" or "sea" (even though the latter may be food-grade in a gourmet shop, it contains minerals that could play hob with food during the fermenting or curing process).

Do not use table salt. Although pure, the additives in it to keep it free-running in damp weather make the pickling liquids cloudy; the iodine in iodized salt darkens the pickles.

Do not use the so-called rock salt or other salts that are used to clear ice from roads and sidewalks: they are not food grade.

Salt, as used in brining pickles, is a preservative. A 10-percent brine, about the strongest used in food preservations, is 1 pound/about $1\frac{1}{2}$ cups salt dissolved in each 1 gallon of liquid. Old-time recipes often call for a brine "that will float an egg"; translate this to "10-percent brine."

Brine draws the moisture and natural sugars from foods and forms lactic acid to keep them from spoiling.

Juices drawn from the food dilute the brine, weakening the original salt solution.

Vinegar

Use a high-grade cider or white distilled vinegar of 4 to 6 percent acidity (40 to 60 grain). Avoid vinegars of unknown acidity or your own home-made wine vinegar. The latter develops "mother" that clouds the pickling liquid. Use white vinegar if you want really light pickles.

Warning: NEVER reduce the proportion of vinegar called for if, when you make the recipe again, you have decided that the pickle was too tart. Instead, deal with the unwanted tartness by adding sugar—$\frac{1}{4}$ cup white sugar for every 4 cups of vinegar called for in the recipe—and you will not have upset the tested-for-safety ratio of vinegar to low-acid ingredients. If it's still too puckering, get yourself another recipe.

Sweeteners

Use white sugar unless the recipe calls for brown. Brown makes a darker pickle. Sometimes a cook in the northern United States or in Canada may use maple sugar or syrup in her pickles for its flavor—but this is feasible only if she has lots of it to spare. (See Sweeteners in Chapter 5.)

Spices

Buy fresh spices for each pickling season. Spices deteriorate and lose their pungency in heat and humidity, so they should be kept in airtight containers in a cool place.

Additives for Crisping Pickles

To enhance the crispness of various cucumber and rind pickles, old cookbooks sometimes called for a relatively short treatment with *slaked lime* (calcium hydroxide) or *alum* (see also Firming Agents in Chapter 5). Such chemicals are *not necessary* for good-textured products if the ingredients are perfect—well grown, unblemished, and perfectly fresh—and are handled carefully according to directions.

However, a company called Dacus, Inc, from Tupelo, Mississippi, offers a coordinated range of pickling materials, and *PFB* feels that, if you truly want to try these chemical additives, it's best to use one of the Dacus mixtures ("Mrs. Wage's") because the balance has been tested, and there would be no danger of trouble from the traditional beginner's notion that if some is good, a lot more is better.

A happier crisper-upper also mentioned in heirloom recipes is a natural one that, so far as we can discover, is safe to use without restriction. One bygone rule we've seen says to cover the bottom of the crock with washed grape leaves and put a layer of them on top of the pickles; and a Southern homemaker says scuppernong leaves are best. Another uses cherry leaves.

Water

Water used in making pickles should of course be of drinking quality (because otherwise contaminants can increase the bacterial load that leads to spoilage). Also, water with above-average calcium content can shrivel pickles, and iron compounds can make them darker than we like.

See Water in Chapter 5 for ways to rid water of some excess minerals, and for a discussion of how "hard" or "soft" water affects texture.

Methods
Long-Brine

Vegetables such as cucumbers are washed and dropped into a heavy salt solution (plus sometimes vinegar and spice) and left in a cool place to cure for 2 to 4 weeks. Scum *must be removed* from the brine each day. Following this the pickles are packed loosely in clean jars and covered with the same or freshly made brine and processed in a Simmering Hot–Water Bath (180–190 F/85–88 C).

Short-Brine

Vegetables are left overnight in a brine to crisp up. The next day they are packed in jars, covered with a pickling solution and finished in the Simmering Hot–Water Bath (212 F/100 C) for a suitable time.

Complete Precooking

Complete precooking is the rule for relishes and similar cut-up pickle mixtures in a sweet-sour liquid. Packed hot in regular canning jars, these products then have a short water bath.

Pickle, etc., Troubles, and What to Do

If you find any imperfect seals during the 24 hours between processing and storing your pickles, relishes, and sauces, you can dump the contents into the preserving kettle, bring to boiling, pack into hot canning jars, cap with new closures, and process again in the water bath for the required time. (Of course if only one seal is imperfect, it's easier to pop that jar in the refrigerator and eat the food within the next week.)

The interim day before storing in a cool, dry, dark place is the only time that these foods can be salvaged by repacking in sterilized containers and processing over again from scratch.

If, after these foods are checked and put in the storage area, you find any of the following, DESTROY THE CONTENTS SO THAT THEY CANNOT BE EATEN BY PEOPLE OR ANIMALS; then deal with the containers as described in Chapter 6, "Canning Methods."

- Broken seal.
- Seepage around the seal, even though it seems firmly seated.
- Mold, even a fleck, in the contents or around the seal or on the underside of the lid.
- Gassiness (small bubbles) in the contents.
- Spurting liquid, pressure from inside as the jar is opened.
- Mushy or slippery pickles.
- Cloudy or yeasty liquid.
- Off-odor, disagreeable smell, mustiness.

If this sounds strict, it's meant to. Our most shiver-producing bedside reading is not the latest Elmore Leonard, but the cumulative statistics on outbreaks of botulism in the United States, published regularly by the Centers for Disease Control of the U.S. Public Health Service. Surprisingly to a layman, condiments—including home-canned tomato relish, chili sauce, and pickles—have been found to contain *C. botulinum* toxin. Maybe the product was not truly pickled, because the recipe was altered—perhaps by cutting down on the vinegar to reduce tartness, rather than by increasing the sweetener to achieve a result more bland. Or, as has been described in Chapter 8, "Canning Tomatoes," unclean handling or faulty processing allowed spoilage that reduced the acidity of the food, and thus in turn the botulinum spores could grow.

In addition to the sloppy treatment noted above, warm storage conditions contribute to such spoilage. As does the old "open-kettle" canning method now in disrepute.

Often, low-acid vegetables are spoiled by the scum that naturally forms on top of the fermentation brine; the scum should be removed faithfully. And it is not only the top layer of pickles that is affected, for the scum (which contains wild yeasts, molds, and bacteria) can weaken the acidity of the brine.

Also, hard water that contains a great deal of calcium salts can counteract some of the acid, or keep acid from forming well enough during brining, and thus interfere with the process that is meant to make certain pickles safe with an otherwise adequate water bath.

And of course "knife out" air bubbles before capping.

Warning about measurements: The critical ingredient is the vinegar in pickled products or tart relishes, and this is easy to measure. But proportions can be thrown off by how chunky ingredients are measured: for example, ¾-inch cubes of vegetables should be measured by the quart—not by 4 separate level cupfuls—and be rounded: this rounding compensates for the wasted space in the container. On the other hand, shredded cabbage is pressed down to the rim of the measuring cup (but not down to the cup-mark slightly below the rim).

Not Perfect, but Edible

If jars have good seals, if there are none of the signs of spoilage noted above, and if the storage has been properly cool, you can have less-than-perfect pickles that are still O.K. to eat.

Hollow pickles. The cucumbers just developed queerly on the vine; you can spot these odd ones when you wash them: usually they float. So use them chopped in relishes. Or they stood around more than 24 hours after being picked. If you can't get around to doing them the day you get them, refrigerate.

Shriveled pickles. This can come from plunging the cucumbers into a solution of salt, vinegar or sugar that's too strong for them to absorb in one session (here's the reason why some recipes handle pickles in stages). Or they've cooked too fast in a sugar-vinegar solution. Or the water used was too hard (see also Water, in Chapter 5).

Darkened pickles. Iron in hard water, or loose ground spices.

Bleached-looking pickles. With no signs of spoilage present, this could mean that jars were exposed to light during storage. Wrap the jars in paper or put them in closed cartons if the place they're stored is not dark.

CONVERSIONS FOR PICKLES AND RELISHES

Do look at the conversions for metrics, with workable roundings-off, and for altitude—both in Chapter 3—and apply them. **Note: this chapter has B–W Bath processing—make your altitude adjustments accordingly.**

Pickles and Relishes

Sweet Pickle Chips
4 to 5 pints

These are so delicious and so easy that you'll want to make several separate batches as the cucumbers come along.

(Recipe continued next page)

4 pounds pickling cucumbers (3 to 4 inches long)

Brining solution:
 1 quart distilled white vinegar
 3 tablespoons salt
 1 tablespoon mustard seed
 ½ cup sugar

Canning syrup:
 1⅔ cups distilled white vinegar
 3 cups sugar
 1 tablespoon whole allspice
 2¼ teaspoons celery seed

Wash the cucumbers, remove any blemishes, nip off the stems and blossom ends, and cut them crossways in ¼-inch-thick slices. In a large enameled or stainless steel kettle, mix together the ingredients for the *brining solution;* add the cut cucumbers. Cover and simmer until the cucumbers change color from bright to dull green (about 5 to 7 minutes).

Meanwhile have ready the *canning syrup* ingredients heated to the boil in an enameled kettle. Drain the cucumber slices and pack them, while still piping hot, in hot 1-pint canning jars, and cover them with very hot syrup, leaving ½ inch of headroom. Remove air bubbles, and adjust lids. Pack and add the hot syrup to one jar at a time, returning the syrup kettle to low heat between filling and capping each jar, so the syrup doesn't cool. Process filled and capped jars in a Hot–Water Bath (185 F/85 C) for 10 minutes. Remove jars and complete seals if necessary.

Little Cucumber Crock Pickles

This is an old-time rule, producing small, crisp, whole pickles with good flavor. They take 4 to 5 weeks to make; and if they're put in brine as they come along in season, and kept in a cool place, they should last well into winter.

1 gallon cider vinegar (regular 5 percent)
½ cup sugar (or 1 teaspoon powdered straight saccharin)
1 cup whole mustard seed
1 cup pickling salt (pure, no fillers) plus salt to add later
Optional: 4 fresh dill heads; or more, if you like stronger dill
3- 4-inch pickling cucumbers (about 10 pounds total, or a scant peck)

Thoroughly scrub a 5-gallon earthenware crock with hot water and soap, rinse well, then scald with boiling water; be energetic about it, because any residue of fat or milk from a previous use will ruin the pickles. In the crock mix together the vinegar, sweetening, mustard seed, and salt; lay dill heads on the bottom if you like them. Keep the crock in a constantly cool place (40 to maximum 50 F/4 to maximum 10 C).

As they're gathered, wash the little cucumbers well, rub off the blossom ends (where enzymes are concentrated), and drop the cucumbers into the brine. Push newly harvested ones toward the bottom of the container as it fills, so the last ones in will not be the first ones out. Hold the pickles beneath the brine with a weighted plate (a pint jar filled with water weighs enough), and cover the crock with a layer of clean cheesecloth or muslin.

If you put all the cucumbers in at the same time, after three days add 1 cup more pickling salt, laid on the plate where it will dissolve slowly downward (the extra salt counteracts weakening of the brine as the natural juice is drawn from the cucumbers). One week later, put $\frac{1}{4}$ cup more salt on the submerged plate; and continue adding $\frac{1}{4}$ cup salt in this manner each week until the pickles are ready. At the end of a month, test by cutting a pickle crossways: if it is firm, and clear throughout with no white center, the pickles are ready to eat.

If you harvest your cucumbers piecemeal over a period of, say, two weeks, lay $\frac{1}{2}$ cup pickling salt on the plate when the crock is half full, and add another $\frac{1}{2}$ cup salt when the crock is filled; thereafter, add $\frac{1}{4}$ cup salt each week until the pickles are ready.

A gray-white film will appear on the surface of the brine after the cucumbers have been in the pickling solution a few days: skim it off, and keep removing it as it forms. The film is to be expected as a natural part of the brining process, but if allowed to stay on the pickles it will hurt the acidity of the pickling solution, and your pickles will spoil.

When there's no more film, and your pickles test evenly clear to the center, start enjoying them. Always replace the weighted plate after taking any out, and cover the crock to keep the contents clean.

Canning. If the conditions for storing your crock of pickles are not good, or if you foresee that you can't eat them all within their storage life of several months—can them.

Take all the pickles from the brine, and fit them vertically in clean pint or quart jars, leaving $\frac{3}{4}$ inch of headroom. From the pickling solution remove any dill heads (and the mustard seed, if you like); bring the solution to boiling, and pour it over the pickles, leaving $\frac{1}{2}$ inch of headroom. Remove trapped air with the blade of a plastic knife, adjust lids with their clean fresh rubbers or sealers, and process at 185 F/85 C—10 minutes for pints, 15 minutes for quarts. Remove jars; complete seals if necessary.

("Short-Form") Bread-and-Butter Pickles
12 pints

---

    6 quarts of thinly sliced pickling cucumbers (about fifteen 6-inch)
    6 medium onions, thinly sliced
    $\frac{1}{2}$ cup pickling salt
    $1\frac{1}{2}$ quarts white vinegar
    $4\frac{1}{2}$ cups sugar
    $\frac{1}{2}$ cup whole mustard seed
    1 tablespoon celery seed   *(Recipe continued next page)*

Wash cucumbers, remove stem and blossom ends, slice thin on the broad blade of a food grater, or on the thin-slicing disc of a food processor. Peel onions, slice thin. Put sliced vegetables in an enameled, crockery or stainless steel bowl, combine them with the ½ cup of pickling salt; let stand 3 hours, then drain well but do not rinse. Meanwhile combine the vinegar, sugar, mustard, and celery seed in a large stainless steel or enameled kettle and bring to a boil. When boiling, add the cucumber and onion slices; over medium heat, bring up to a low boil, and pack immediately in clean, hot 1-pint jars, leaving a good ½ inch of headroom. Adjust clean, hot lids, and process in a Hot–Water Bath (185 F/85 C) for 10 minutes. Remove; cool upright and naturally.

Good because they're simple. The slicing disk of a food processor is grand help here. Allow several weeks for the flavor to develop. Also may be made with crisp young zucchini.

Watermelon Pickles

4 pints

8 cups prepared watermelon rind
½ cup pickling salt
4 cups cold water
4 teaspoons whole cloves
4 cups sugar
2 cups white vinegar
2 cups water

Choose thick rind. Trim from it all dark skin and remains of pink flesh; cut in 1-inch cubes. Dissolve salt in cold water, pour it over rind cubes to cover (add more water if needed); let stand 5 to 6 hours. Drain; rinse well. Cover with fresh water and cook until barely tender—no more than 10 minutes (err on the side of crispness); drain. Combine sugar, vinegar, and water; add cloves tied in a cloth bag; and bring to boiling; reduce heat and simmer for 5 minutes. Pour over rind cubes, let stand overnight. Bring all to boiling and cook until rind is translucent *but not at all mushy*—about 10 minutes. Remove spice bag, pack cubes in hot sterilized pint jars; add boiling syrup, leaving ½ inch of headroom; adjust lids. Process in a 185 F/85 C water bath for 10 minutes. Remove jars and complete seals if necessary.

Pickled Mixed Vegetables

Fine timber for antipasto, and based on a mix-as-you-like recipe from the University of California, Davis, the combinations may be freewheeling but the method and handling are not. These are much like Dilly Beans, where seasonings are put in the bottom of each clean jar, the prepared food is packed raw (and tight), and a half-vinegar-half-water canning liquid, heated, is poured over all; processing is a simmering bath.

Prepare 1½- to 2-inch pieces of bell peppers, both red and green; cauliflower, celery cut on a slant, small green cherry tomatoes, onions, slices of zucchini and carrots, eggplant strips; perhaps some not-too-savage hot peppers cut small, and some reconstituted sun-dried tomatoes.

As a rough estimate, plan for 8 to 10 cups of the canning liquid to deal with 12 pints of vegetables. Therefore, heat together equal amounts of 5 percent white vinegar and of water—4 to 5 cups of each. In each clean jar put ½ clove garlic (pints) or 1 to 2 in quarts; ½ teaspoon salt in pints, 1 teaspoon in quarts; peppercorns as liked. Pack mixed vegetables tight, leaving a good ½ inch of headroom; cover with hot vinegar-water combination, leaving ½ inch of headroom. Bubble each jar with a plastic spatula to remove trapped air, wipe sealing rim of jar, put on prepared disk lid and screw the band firmly tight. Put in a warming Hot–Water Bath, add hot water as needed to cover jars. When simmering temperature of 185 F/85 C is reached, time for 20 minutes. Remove jars; cool upright and naturally. Allow several weeks for flavors to develop.

48-Hour Sour Pickles
8 pints

```
32 cucumbers, about 4 inches long
 1 cup salt
 4 quarts water
 4 cups vinegar
 1 cup sugar
 ½ tablespoon whole cloves
 ½ tablespoon mustard seed
 ½ tablespoon celery seed
 ½ tablespoon peppercorns
```

Prepare a large clean crock or enamelware kettle. Dissolve the salt in the water. Wash cucumbers, put in the crock/kettle, cover with the salt brine. Place cover over the crock/kettle, let stand 24 hours, then drain.

Hold in cold fresh water 20 minutes, drain. If rinse water is too salty, freshen again. Meanwhile combine vinegar, sugar, spices tied in a bag; bring to boiling, boil 5 minutes; remove from heat. Wash thoroughly the crock/kettle. Cut cucumbers lengthwise in halves or quarters; put into clean kettle. Bring the vinegar/sugar solution back to a boil. Remove spice bag and pour boiling hot syrup over cucumbers. Cover crock/kettle; let stand 24 hours.

Drain syrup off cucumbers, pack cucumbers upright in clean hot jars. Fill jars with boiling syrup leaving ½ inch of headroom. Top, and process 10 minutes in a 180 F/82 C water bath for 10 minutes. Remove jars.

Sweet variation: for 48-Hour Sweet Pickles follow the method above, using 32 cucumbers, 4 inches long; brine of 1 cup salt dissolved in 4 quarts water; a syrup of 4 cups vinegar and 3 cups sugar boiled together with 2 tablespoons of mixed pickled spices tied in a bag.

Small-batch Freezer Pickles
3 pints

---

2 rounded quarts of thinly sliced pickling cucumbers, *not* peeled
2 large yellow onions, peeled and thinly sliced
2 tablespoons pickling salt
1 cup white vinegar
1½ cups white sugar

Scrub cucumbers, remove stem and blossom ends, slice thin on the broad blade of a vegetable grater, or with the slicing disk of a food processor. Peel onions, slice like the cucumbers. Put cucumbers and onions in a large crockery bowl, sprinkle with the salt, let stand for 2 to 3 hours. Meanwhile combine vinegar and sugar, bring to a boil. Drain the vegetables but do *not* rinse, and pack into 1-pint freezer containers (either plastic or straight-sided can/freeze jars), leaving ¾ inch of headroom. Pour in the hot syrup, leaving ½ inch of headroom. Seal, cool, freeze.

Use within one week after thawing.

Note that pickling spices are not used here, because freezing reduces the flavor of spices.

These pickles are a fine way to use up cucumbers that aren't enough for a batch of canned pickles—provided you have a special spot in your freezer for oddments.

Sweet-and-Sour Spiced Crabapples
5 pints

---

3 pounds firm ripe crabapples (about)
3 cups cider vinegar
3 cups water
2¼ cups sugar
3 dozen whole cloves
4 to 6 three-inch sticks of cinnamon
6 short blades of mace

To prepare the crabapples, rub the fuzz from the blossom ends, but leave stems on; wash well, then prick the skins with a large darning needle to keep the fruit from bursting while cooking. Tie the spices loosely in a square of muslin or double thickness of cheesecloth, and put the bag in a large enameled kettle with the vinegar, water, and sugar; bring to boiling and boil together for 3 minutes. Add the crabapples and simmer until just tender—not mushy. (Test after 15 minutes by poking one deeply with a darning needle: there should be a little resistance.) Discard the spice bag, and pack the crabapples immediately in hot pint jars and cover them with the very hot syrup in which they were cooked, leaving ½ inch of headroom. Adjust the lids and process the jars in a Hot–Water Bath (185 F/85 C) for 10 minutes. Remove; complete seals if necessary.

Green Tomato Pickles
6 pints

7½ pounds green tomatoes (about 30 medium)
6 good-sized onions
¾ cup pickling salt
1 tablespoon celery seed
1 tablespoon whole allspice
1 tablespoon mustard seed
1 tablespoon whole cloves
1 tablespoon dry mustard
1 tablespoon peppercorns
½ lemon
2 sweet red peppers
2½ cups brown sugar
3 cups vinegar (approx. 5 percent acidity)

Wash tomatoes well, cut off blossom ends, blemishes, and stems. Slice thin crossways. Peel and slice onions in thin rings. Sprinkle salt over alternate layers of sliced tomatoes and onions in an earthenware dish, and let stand in a cool place overnight. Drain off the brine, rinse the vegetables thoroughly in cold water, and drain well. Slice the lemon thinly and remove the seeds; wash the peppers well, remove stems and seeds, slice thinly crossways. Tie all the spices loosely in muslin or a double layer of cheesecloth, add the spice bag and the sugar to the vinegar in a large enamelware kettle; bring to a boil. Add the tomatoes, onions, lemon, and peppers. Cook for 30 minutes after the mixture returns to a boil, stirring gently to prevent scorching. Remove the spice bag, pack the pickles in hot jars, and cover with boiling-hot liquid, leaving ½ inch of headroom. Adjust lids. Process in a 185 F/85 C bath for 10 minutes. Remove jars; complete seals if necessary.

Quick Dill Pickles
7 pints

---

30 to 40 medium pickling-type cucumbers, 5 inches long
¾ cup sugar
¾ cup pickling salt
1 quart vinegar
1 quart water
7 fresh dill heads
(Optional: 7 garlic cloves)
(Optional: 3 tablespoons mixed whole pickling spices)

Mix together the sugar, salt, vinegar, and water and bring to a boil. Tie optional pickling spices loosely in a thin white cloth and boil in the vinegar mixture for about 10 minutes; remove and discard. Scrub cucumbers, remove stems and blossom ends; cut lengthwise in halves or quarters, not longer than the shoulder-height of the jar. Put 1 whole head of fresh dill in each clean hot jar. Pack the jars with cut cucumbers upright, then tuck in a clove of garlic if you like it. Pour in the boiling vinegar mixture, leaving ½ inch of headroom. Adjust lids. Process in a 185 F/85 C water bath for 10 minutes. Remove jars; complete seals if necessary.

Sweet Mustard Pickle
4 quarts

　　1 quart small green tomatoes, quartered
　　1 quart *tiny* unpeeled cucumbers (about 2 inches)
　　1 quart unpeeled *medium* cucumbers in 1-inch chunks
　　1 quart tiny pickling onions
　　1 small head cauliflower, broken into flowerets
　　3 green peppers, seeded and diced
　　2 cups green beans, cut in 1-inch slices
　　1 cup salt
　　1 cup flour
　　⅓ cup dry mustard
　　2 teaspoons turmeric
　　2 cups sugar
　　2 quarts vinegar

Combine vegetables and sprinkle with salt. Cover with cold water and let stand overnight. Place over moderate heat and bring just to boiling point, then drain thoroughly. Combine remaining ingredients smoothly. Stir over moderate heat until smooth and thick. Add well-drained vegetables and bring *just to boiling point:* they should never be overcooked and mushy. Ladle hot into hot canning jars, allow ½ inch of headroom, and process in a Hot–Water Bath (185 F/85 C) for 10 minutes for pints or quarts.

Pickled Mushrooms in Oil
Four ½-pints

　　6 cups small button mushrooms
　　½ cup lemon juice or
　1¼ teaspoons citric acid
　　1 quart water
　　1 cup vinegar (5 percent)
　　1 cup water
　　2 teaspoons salt
　　½ teaspoon dried oregano
　　3 bay leaves
　　½ teaspoon dried sweet basil
　　2 cloves garlic
　1½ cups olive oil, or ½ cup olive oil and 1 cup salad oil

As with all products that are pickled in oil, the main ingredient—in this case the mushrooms—must take up enough acid to become truly pickled *before* the oil is added to the mixture and the jar is capped. If the oil by mistake were added too early, it would inhibit the mushrooms' ability to take up the acid that pickles them (this acquired acidity makes them safe to be canned).

In a stainless steel or enameled kettle bring to a boil ½ cup lemon juice OR 1¼ teaspoons of citric acid crystals in 4 cups water. Add mushrooms, return to the boil, simmer for 5 minutes; drain. In a crockery or steel bowl, combine the 1 cup water and the 1 cup vinegar; add the precooked mushrooms; cover and let stand 12 hours or overnight.

Meanwhile, mix the salt and the crumbled oregano, bayleaf, and basil in a small dish. In the bottom of each small jar, place ½ clove of garlic and ¼ of the salt/herb mixture. Pack mushrooms into the jars leaving a good ¼ inch of headroom; pour oil over the mushrooms to cover using a plastic spatula to clear out any air bubbles. Wipe sealing rim, clean and dry. Put on prepared disk lid, turn band firmly tight; process in a Hot–Water Bath at 180 F/82 C for 20 minutes. Remove jars; cool upright and naturally.

"Dilly" Green Beans
7 pint jars

---

    4 pounds table-perfect whole green beans
 1¾ teaspoons crushed dried *hot* red pepper
 3½ teaspoons dried dill seed, *or* 7 fresh dill heads
    7 cloves of fresh garlic, peeled
    5 cups vinegar
    5 cups water
    ½ cup less 1 tablespoon pickling (non-iodized) salt

Wash beans thoroughly, remove stems and tips, and cut them as much as possible in uniform lengths to allow them to stand upright in 1-pint canning jars, coming to the shoulder of the jar. Have jars clean and very hot, and lids and sealers ready in scalding water. In each jar place 1 dill head *or* ½ teaspoon dill seed, add 1 garlic clove and ¼ teaspoon crushed hot red pepper. Pack beans upright in jars, leaving 1 inch of headroom. Heat together the water, vinegar, and salt; when the mixture boils, pour it over the beans, filling each jar to ½ inch from the top. Run a plastic knife down and around to remove trapped air, adjust lids, and process in a 185 F/85 C bath for 10 minutes after the water in the canner returns to simmer. Remove jars, complete seals if necessary.

The beans are almost garden-crisp, but the high acidity of the vinegar allows this B–W Bath processing to be safe for a low-acid food.

If you substitute ground cayenne pepper for the crushed hot red pepper, *halve the amount of cayenne:* use only ⅛ teaspoon cayenne to each jar.

Wait at least two weeks for these beans to develop their flavor.

Relishes and Sauces

Pickled Hot Peppers or Bell Peppers
8 pints

This recipe comes from the University of California publication No. 4080.

4 quarts peppers
4 cups vinegar
4 cups water
4 teaspoons salt
Olive oil (optional)

Wash peppers thoroughly. Remove core, seeds, and stems of large peppers. Cut as desired, or leave whole after coring. The small, hot peppers may be left whole with stems intact. Make 2 small slits in whole peppers.

Mix vinegar and water; heat to 150–160 F/66–71 C about to the simmering point. Since it is rather volatile, vinegar should *not* boil a long time. Pack peppers rather tightly into jars. Pour hot vinegar and water over the peppers to ½ inch of jar rim. If oil is desired, add vinegar to only ¾ inch of jar top. Add olive oil to come ½ inch from top. The peppers will be coated with oil when they pass through the oil layer as you use them. Add salt to taste, seal, and process 15 minutes in simmering (180–185 F/82–85 C) Hot–Water Bath.

Beet Relish
About 5 pints

Unusual, because it's made from raw, not precooked, beets. And it's a handy way to use large beets that are on the woody side. A food processor makes quick work of what would otherwise be a splashy preparation.

4 cups coarsely ground fresh beets—about 2 pounds before peeled
6 cups coarsely ground green cabbage—3-pound head before coring
2 cups coarsely ground onions—about 1 pound
2 cups cider vinegar
2 cups sugar
2 teaspoons salt
2 tablespoons freshly grated horseradish (or bottled)

Peel and cut beets lengthwise in eighths or finer, feed them upright onto the grating disk of a food processor. Empty processor bowl, grate slender wedges of cabbage; grate quartered onions. Combine vegetables in an enameled or stainless steel kettle, add vinegar, sugar, salt, and horseradish. Bring to a boil over medium heat, cook and stir until thick—about 25 minutes. Remove from heat, ladle into clean hot 1-pint jars, leaving ½ inch of headroom. Adjust lids and process at 185 F/85 C for 20 minutes. Remove; complete seals if necessary.

Piccalilli
4 pints

---

6 medium-size green tomatoes
6 sweet red peppers, seeded
6 medium onions, peeled
1 small cabbage
¼ cup salt
2 cups vinegar
2½ cups light brown sugar
2 tablespoons mixed pickling spices

Put vegetables through the food grinder, using a coarse knife. Sprinkle with the salt, cover, and let stand overnight. Drain; then cover with fresh water, and drain again. When thoroughly drained, put into a large kettle and add vinegar and sugar. Tie spices in a small cloth bag and add. Bring to boiling, then reduce heat and simmer about 20 minutes, stirring frequently. Remove the spice bag and turn the hot piccalilli into hot jars, leaving ½ inch of head-room; adjust lids and process at 185 F/85 C for 10 minutes.

Margaret Hawes's Zucchini Relish
5 to 6 pints

---

10 cups finely chopped zucchini (if small, leave in the seeds; if over 8 inches, remove seeds)
4 large onions
4 green bell peppers, seeded
4 red bell peppers, seeded
½ cup salt
2½ cups white vinegar
4 cups white sugar
2 tablespoons cornstarch
1 teaspoon ground nutmeg
1 teaspoon turmeric
2 teaspoons celery seed
½ teaspoon ground black pepper

Wash and peel zucchini, removing stems and blossom ends; remove seeds if squash is cut in large chunks for grinding. Peel and quarter onions; seed and quarter the bell peppers. Put vegetables through the food grinder, using a coarse knife. (With a food processor, use the shredding disk: the steel blade can make these ingredients lose too much texture.) Put ground vegetables in a crockery or stainless steel bowl, stir in the salt; keep the vegetables in the resulting brine by holding them down with a weighted plate. Let vegetables stand overnight. The next day, drain off the brine and rinse vegetables with cold water; drain again, and squeeze well by hand. Mix cornstarch with the sugar and four other dry seasonings, add all to the cold vinegar, blending well.

Over medium heat, bring to boiling, stirring well to prevent lumping. When sugar is melted and the syrup is clear, add the vegetables; simmer 30 minutes, stirring often. Pour into clean very hot jars, leaving ½ inch of headroom; adjust sterilized lids, and process for 10 minutes in a 185 F/85 C water bath. Remove; complete seals if necessary.

Corn Relish
4 pints

---

        4 cups corn kernels (about 9 ears' worth)
        1 cup diced sweet green peppers
        1 cup diced sweet red peppers
        1 cup finely chopped celery
        ½ cup minced onion
        1½ cups vinegar
        ¾ cup sugar
        2 teaspoons salt
        1½ teaspoons dry mustard
        1 teaspoon celery seed
        ¼ teaspoon Tabasco sauce
        ½ teaspoon turmeric, for color (optional)
        2 tablespoons flour, for thickening (optional)

Prepare corn by boiling husked ears for 5 minutes, cooling, and cutting from cob (do not scrape). In an enameled kettle combine peppers, celery, onion, vinegar, sugar, salt, celery seed, and Tabasco sauce; boil 5 minutes, stirring occasionally. Dip out ½ cup hot liquid, mix it with dry mustard and turmeric, and return it to the kettle. Add the corn. (If you want the relish slightly thickened, blend the 2 tablespoons flour with ¼ cup cold water and add to the kettle when you put in the corn.) Boil for 5 minutes, stirring extra well if the relish has been thickened, so it won't stick or scorch. Immediately fill clean hot pint jars within ½ inch of the top, adjust lids, and process at 185 F/85 C for 15 minutes. You can use frozen whole-kernel corn that's been thawed slowly: three 10-ounce packages will equal 4 cups of fresh kernels.

Indian Chutney
3 pints

The rule for this fine Calcutta-style chutney was given us by Frances Bond, who lived twenty years in India before moving to Vermont, and she has tailored it for ingredients easy for the North American housewife to come by. It's ideal with budget-stretching curries or pilau, with hot or cold meats, and it makes a delightful present packed in decorative ½-pint canning jars.

For best results, the fruit—whether apples, peaches or pears—should be firm varieties, or slightly underripe. The fruit, raisins, and crystallized ginger are added after the syrup ingredients have cooked together for 30 minutes, to let them keep their identity in the finished product: they should be tender but recognizable in the syrup, which is thick and a rich brown in color. The chutney improves after a couple of months in sealed jars.

Juice, pulp, and peel of 1 lemon, finely chopped
2 cups cider vinegar
2½ cups dark brown sugar (1 pound)
1 clove garlic, minced
Pinch of cayenne pepper (⅛ teaspoon)
Pinch of chili powder (⅛ teaspoon)
1½ teaspoons salt
5½ cups coarsely chopped firm apples, peeled and cored (about 3 pounds), *or* peaches or pears
¾ cup crystallized ginger—cut small but not minced (about 3 ounces)
1½ cups raisins, preferably seeded (½ pound)

Caught without fresh fruit in a chutney-making mood, Mrs. Bond substituted 5½ cups of coarsely chopped canned, drained pears, but added them in the last 10 minutes of cooking. Results: heavenly.

Ground ginger contributes only flavor without texture, and reconstituted dry cracked ginger is usually woody, so don't substitute with either of them. Minced, peeled fresh gingerroot is a logical substitute, but ¼ cup, prepared this way, should do.

Chop the lemon, removing seeds and saving the juice (a blender is good here), and put it in an open, heavy enameled kettle with the sugar, vinegar, minced garlic, salt, cayenne pepper, and chili powder. Boil the mixture over medium heat for 30 minutes, stirring occasionally. Meanwhile prepare the apples (or peaches or pears), and add them to the syrup with the raisins and ginger. Boil all slowly, stirring to prevent sticking and scorching, until the fruit is tender but not mushy and the syrup is thick—about 30 to 45 minutes longer. Ladle the boiling-hot chutney into hot pint or ½-pint jars, leaving ¼ inch of headroom in ½-pints, ½ inch of headroom in pints. Remove trapped air, wipe sealing rim; cap. Process in a Boiling–Water Bath for 10 minutes. Remove; cool upright and naturally.

Chili Sauce
5 to 6 pints

Because of extra acid from the vinegar, this also can be finished safely in a B–W Bath.

4 quarts chopped ripe tomatoes (about 9 to 10 pounds)
5 large onions, peeled and chopped small
4 sweet red peppers, seeded and chopped
2 cups cider vinegar
1 cup brown sugar, packed firmly
2½ tablespoons salt

(Ingredients continued next page)

1 stick of cinnamon 3 inches long, broken in pieces
1 tablespoon mustard seed
2 teaspoons celery seed
1½ teaspoons ground ginger
1 teaspoon ground nutmeg
1 teaspoon peppercorns

Peel, core, and chop tomatoes; peel and chop onions; seed and chop peppers. Put them in a heavy enameled kettle, add the vinegar, sugar, salt; tie the spices in a double thickness of muslin or four thicknesses of cheesecloth (extra density of cloth will hold the ground spices better) and add the bag to the ingredients in the pot. Bring the mixture quickly to boiling, stirring so it won't scorch, then put it in a 300 F/149 C oven to cook slowly, stirring occasionally, until the sauce is thick—from 2 to 3 hours. It wants to be a little thicker than ketchup but not so thick as jam. It will scorch if it's not watched and stirred, especially toward the end. Remove the spice bag and pack Hot in hot canning jars, leaving ½ inch of headroom. Adjust lids and process in a Boiling–Water Bath (212 F/100 C) for 15 minutes.

For "hotter" sauce, substitute 1 teaspoon crushed dried *hot* red pepper pods for the peppercorns. This is one of those recipes whose seasonings can be tinkered with according to the family's taste.

Mincemeat

Mincemeat
5 pints

---

1 pound boiled lean beef
½ pound beef suet
2½ cups seeded raisins
¼ pound chopped citron
3 cups coarsely chopped apples
2 cups dried currants
2¼ cups light brown sugar
3 tablespoons light molasses
2 cups sweet cider
¾ teaspoon ground cinnamon
¾ teaspoon ground mace
¾ teaspoon ground cloves
¼ teaspoon ground nutmeg
¼ teaspoon ground allspice
¼ teaspoon salt
1 cup brandy

Put beef, suet, and raisins through the food grinder, using a coarse knife. Put citron and apples through the grinder. Combine all in a heavy kettle and add remaining ingredients in order, *except the brandy*. Bring to boiling, stirring constantly. Reduce heat and simmer about 1 hour, stirring frequently. *The mixture will scorch easily, so use a heatproof pad under the kettle.* Remove from heat and stir in brandy. Ladle hot into hot pint canning jars, allowing ½ inch headroom, and process at 10 pounds pressure (240 F/116 C) for 20 minutes. Makes enough for five 9-inch pies.

Green Tomato Mincemeat
8 pints

- 3 quarts prepared green tomatoes
- 3 quarts prepared apples
- 1 cup ground suet
- 1 pound seedless raisins
- 2 tablespoons grated orange rind
- 2 tablespoons grated lemon rind
- 5 cups well-packed light brown sugar
- ¾ cup vinegar
- ½ cup fresh lemon juice
- ½ cup water
- 1 tablespoon ground cinnamon
- ¼ teaspoon ground cloves
- ¼ teaspoon ground allspice
- 2 teaspoons salt

Wash the tomatoes, remove stem and blossom ends, and chop fine with a chef's knife or with the coarse blade of a food grinder; wash, peel, and core the apples; chop like the tomatoes. (For a food processor, cut tomatoes and apples in fairly small chunks to drop through the feed tube, and use the shredding disk.) Put suet through a finer blade of the food grinder or mince it by hand somewhat smaller than the tomatoes. Combine tomatoes, apples, and suet with all the other ingredients in a large enameled or stainless steel kettle over medium heat and bring just to a boil, stirring frequently. Move the kettle to a 300 F/149 C oven until dark and thick, stirring occasionally—about 2½ hours.

Toward the end of the cooking time, wash and scald at least eight 1-pint canning jars, and prepare their lids; hold in scalding water. Quickly ladle boiling-hot mincemeat into jars, leaving a good ½ inch of headroom, and cap; process in a finishing B–W Bath (212 F/100 C) for 15 minutes. Remove jars from the canner and complete seals if necessary. Makes 8 pints, enough for eight 9-inch pies.

Dietary note: instead of the suet, cut in small pieces 1 tablespoon of butter or corn oil margarine for each 1 pint of mincemeat, and press the pieces into the filling when you build your pies.

This mincemeat is excellent used for small holiday tarts: see no-fail Pie Pastry in Chapter 23, "Putting By Presents for Christmas."

20
CURING WITH SALT AND SMOKE

In the preceding chapter we dealt with one type of salting: brining cucumbers to "pickle" them by fermentation. Here we'll start the first—and longer—of our two main sections with Salting, breaking it down into treating vegetables and then meats. The other major section, Smoking, will give the *Why/How* of smoking meats, and, as an example of the treatment for fish, coho salmon from the Great Lakes.

CONVERSIONS FOR CURING

Do look at the conversions for metrics (with their workable roundings-off) and for altitude—both in Chapter 3—and apply them.

SALTING

We don't discuss curing two sorts of food: (1) the kind that cannot stand up to the taste of salt—fruit is obvious in this case; and (2) extremely perishable high-protein foods whose flavor, even though enhanced by a little salt, would be ruined by the process of heavy salting—organ meats are an example. (So is fish roe; but what, then, about caviar? Best leave this to the experts.)

Still, there are many cured foods that must have most of their salt washed out before they can be cooked and eaten. Or they were salted so lightly that they must be refrigerated; or, if they really are to be put by, they must be canned or frozen. In the next chapter, "Drying," there are instructions for meat (Jerky) and for cod.

What Salting Does

A concentrated brine—which is salt + juice drawn from the food by the salt (called "dry-salting"), or salt + water if juice is limited or not easily extracted (called "brining")—cuts down the activity of spoilage micro-organisms in direct relation to the strength of the solution. The following general proportions give the idea, with percentages reflecting ratios by weight of salt to water, not sophisticated salinometer readings.

Note: the salt used in the instructions is granulated, food-grade, regular pickling and canning salt—don't use gourmet sea salt and never use "solar" salt evaporated in open basins and unrefined, and never salt with iodine or "free-flowing" additives. The *bulk* of salt changes with its cut— the photo in Chapter 5 shows how—but in the following examples it's simpler to consider 1 pound of salt equaling 1½ cups (12 fluid ounces) in volume.

A 5 percent solution (1 pound of salt to 19 pints of juice/water) *reduces* the growth of most bacteria.

A 10 percent solution (1 pound of salt to 9 pints of juice/water) *prevents* the growth of *most bacteria*.

A solution from 15 percent (1 pound of salt to 5½ pints of juice/water) to 20 percent (1 pound of salt to 4 pints of juice/water) *prevents* the growth of *salt-tolerant bacteria*.

The amounts of salt given in the individual instructions are designed to give the necessary protection to the food being cured, provided that any further safeguards are followed as well. Sometimes a brine is added to make sure that enough liquid is present to carry out the curing process, because you can't add plain water without diluting the strength of the salt required to treat the particular food satisfactorily.

Because salt draws moisture from plant and animal tissues in proportion to its concentration, heavy salting is often a preliminary step in drying or smoking high-protein foods.

Equipment for Curing With Salt

Especially for vegetables:
Large stoneware crocks or jars (5-gallon size is good here), OR—
The biggest wide-mouth canning jars you can get—or ask the high-school cafeteria or your friendly neighborhood snack bar for empty gallon jars (wide-top) that their mayonnaise or pickles came in, OR—

Sound, unchipped enamelware canner (if you can spare it) with lid.

Vegetable grater with a coarse blade; large old-style wooden potato-masher.

Safe storage area at 65 to 70 F/18 to 21 C for fermenting vegetables; plus cooler—about 38 F/3 C—storage for longer term.

Especially for meats and fish:

Large stoneware crocks (10-gallon or larger), OR—

Wooden kegs or small barrels—new, or thoroughly scrubbed and scalded used ones (before curing in them, though, fill them with water to swell the staves tight together, so the containers won't leak when they're holding food).

Moisture/vapor-proof wrappings; plus stockinet—tubular cotton-knit—for holding the wrap tight to the meat after it's packaged.

Safe, cold storage area (ideally 36 to 38 F/2 to 3 C) for curing meats and fish—and for longer-term storage of meats and vegetables in their curing solutions.

For both vegetables and meats, etc.:

Cutting-boards and stainless steel knives (see Chapter 10, "Canning Meats").

Large enameled or glass/pottery pans or bowls for preparing the curing mixtures.

Big wooden spoons, etc., for mixing and stirring.

China or untreated hardwood covers that fit down inside each curing container: an expendable plate, a sawed round, etc.

Weights for these covers, to hold the food under the curing brine—a canning jar filled with sand is good; but nothing of limestone or iron, which mess up the curing solutions.

Plenty of clean muslin (old sheets do beautifully) or double-weight cheesecloth.

Glass measuring cups in 1-cup and 4-cup sizes.

Scale in pounds (up to 25 is plenty, with ¼- and ½-pound gradations).

Good-sized working space, particularly for dealing with meats.

SALTING VEGETABLES

Unless you're fermenting vegetables—as for sour cabbage (Sauerkraut), etc., below—there's only one reason for salting them: you have no other way to put them by, so you either salt your vegetables now or do without vegetables later.

Dry-Salting to Preserve Vegetables

Corn, green/snap/string/wax beans, greens, even cabbage and Chinese cabbage and a number of root vegetables may be dry-salted.

Salted Sweet Corn

Select sweet corn in the milk stage as you'd choose it for serving in season as corn-on-the-cob. Husk, remove the silk, and steam it for 10 to 15 minutes or until the milk is set. Cut it from the cob about ⅔ the depth of the kernels, and weigh it. Mix 4 parts of cut corn with 1 part salt—1 pound of pure pickling salt for each 4 pounds of corn; or 1 cup of salt to 4 cups of cut corn if you don't have scales.

Pack the corn-salt mixture in a crock to within about 4 inches of the top, cover with muslin sheeting or a double thickness of cheesecloth, and hold the whole business down with a clean plate or board on which you place a weight. If there isn't enough juice in 24 hours to cover the corn, add a salt solution in the proportions of 3 tablespoons salt to each 1 cup of cold water; replace the weighted plate to submerge the corn.

Store the crock in a safe, cool place (about 38 F/3 C). The corn will be cured in from 3 to 5 weeks. Remove meal-sized amounts by dipping out corn and juice with a glass or china cup (don't use metal). Change the cloth as it becomes soiled, and always replace the weighted plate. Keep the crock in cool storage.

To cook the corn, freshen it (soak in cold water a short time, drain, and repeat) until a kernel tastes sweet. Simmer until tender in just enough water to prevent scorching; serve with butter or cream and seasoning to taste.

Salted Green/Snap/String/Wax Beans

Use only young, tender, crisp beans. Wash, remove tips and tails; cut in 2-inch pieces, or french them. Steam-blanch 10 minutes and cool. Weigh the beans, and measure 1 pound (1½ cups) of pure pickling salt for every 4 pounds of beans. Sprinkle a layer of salt in the bottom of a crock, add a layer of beans; repeat until the crock is filled to within 4 inches of the top or until the beans are used; top with a layer of salt. Cover with clean muslin sheeting or doubled cheesecloth and hold down with a weighted plate. If not enough brine has formed in 24 hours to cover the beans, eke it out with a solution in the proportions of 3 tablespoons of salt to each 1 cup of cold water.

Proceed as for Salted Sweet Corn, above.

Salted Dandelion (or Other) Greens

Green salads were a rarity with New England hill folk in the early nineteenth century; nor did they go in for leaf vegetables much, except for dandelions in early spring and beet or turnip tops from their gardens in late summer.

Sometimes they salted their greens according to the 1-to-4 rule. Nowadays we'd go them one better, though, and steam-blanch the washed, tender leaves until they wilt—from 6 to 10 minutes, depending on the size of the leaves. Cool the greens, weigh them, and layer them in a crock with 1 pound (1½ cups) of pure pickling salt for every 4 pounds of greens. Proceed as for Salted Sweet Corn and Salted Green Beans, above.

To cook, rinse well and freshen in cold water for several hours, rinse again, drain, and simmer gently in the water adhering to them. Season with small dice of salt pork cooked with them, or serve with vinegar.

Salted Rutabagas (or White Turnips)

Use young, crisp vegetables without any woodiness. Peel; cut in ½-inch cubes. Steam-blanch from 8 to 12 minutes, depending on size of the pieces; cool. Weigh the prepared turnips and proceed with the 1-to-4 rule—1 pound (1½ cups) of pure pickling salt for each 4 pounds of turnips—and handle thereafter like Salted Sweet Corn, above.

To cook, rinse and freshen for several hours in cold water, rinse again; then simmer until tender in just enough water to keep from scorching. Mash if you like and serve with butter and seasoning to taste.

Salted Cabbage

Remove bruised outer leaves; quarter, cut out the core. Shred as you would for cole slaw. Steam-blanch for 6 to 10 minutes until wilted. Cool, weigh; follow the 1-to-4 rule for Salted Greens above, and continue with the cure.

To cook, rinse and freshen for several hours, rinse again; then simmer until tender in just enough water to prevent scorching. Season during cooking with 2 teaspoons of vinegar and ¼ teaspoon caraway; or drain and return to low heat for 3 minutes with crumbled precooked sausage or small dice of salt pork; or serve with butter and seasoning to taste.

Old-Style Dry-Salting to Ferment Vegetables

Most often fermented are cabbage (Sauerkraut) and Chinese cabbage, and rutabagas or white turnips. Generally speaking, the sweeter vegetables make a more flavorful product, while firmer ones provide better texture. Don't relegate tough, old, woody vegetables to the souring crock—use the best young, juicy ones you can get.

If you feel like experimenting with a small batch (5 pounds, say, in a 1-gallon jar, or less in a smaller container) you could add with the salt the traditional German touches of caraway or dill; or try a bay leaf or two, or some favorite whole pickling spice, or some onion rings, or even a few garlic cloves, peeled (but whole, so you can fish them out before serving).

Some rules advocate starting fermentation with a weak brine, but this procedure offers a loophole for too low a concentration of salt, and the likelihood of mushy food or even of spoilage instead of the desired acidity. Unless you're an old hand with sauerkraut and its relatives, you'll do well to stick to dry salting here.

As with vegetables preserved with salt earlier, you should never mix a fresh batch with one already fermenting.

Produce to be soured *is not blanched:* you want to encourage the micro-organisms that cause fermentation.

For fermenting you use $\frac{1}{10}$ the amount of salt you needed for the pre-serving just described. This means $2\frac{1}{2}$ percent of pure pickling salt by weight of the prepared food: 10 ounces (15 tablespoons or a scant 1 cup) of salt to 25 pounds of vegetables; 4 ounces (6 level tablespoons) of salt for 10 pounds of vegetables; 2 ounces (3 level tablespoons) of salt for 5 pounds of prepared food. This ratio of salt turns the sugar in the vegetables to lactic acid, and the desired souring occurs.

The vegetables should be kept between 68 to 72 F/20 to 22 C during the fermenting period, which takes from 10 days to 4 weeks, depending on the vegetable being processed. Temperatures below 68 F/20 C will slow down fermentation; above 72 F/22 C, and you court spoilage.

As a rough estimate, allow 5 pounds of prepared vegetables for each 1 gallon of container capacity, with the crock/jar holding a slightly greater weight of dense food that's cut fine. The instructions below use 10-pound batches, but you may want to deal with 25 or 30 pounds of cabbage or turnips at a time, using a 5-gallon crock.

Keep all souring vegetables covered with a clean cloth and weighted below the brine during fermentation. A top-quality vegetable should re-lease enough juice to form a covering brine in around 24 hours; if it hasn't, bring the level above the food by adding a weak brine in the proportions of $1\frac{1}{2}$ teaspoons of pickling salt for each 1 cup of cold water.

By the second day a scum will form on the top of the brine. Remove it by skimming carefully; then replace the scummy cloth with a sterile one, and wash and scald the plate/board before putting it back and weight-ing it.

Take care of this scum every day, and provide a sterile cloth and plate every day; otherwise the scum will weaken the acid you want, and the food will turn mushy and dark. If the brine gets slimy from too much warmth it's best not to tinker with it: do the simplest thing and decant the batch on the compost pile—and wait until cooler weather to start over again.

Fermentation will be continuing as long as bubbles rise to the top of the brine. When they stop, remove the cloth and weighted plate, wipe around the inside of the headroom; cover the vegetable with a freshly scalded plate/board, and put a close-fitting lid on the container. Then store the whole thing in a cool place at about 38 F/3 C.

Sauerkraut (Fermented Cabbage) with Today's Tools

Quarter each cabbage, cut out the core; shred fine and weigh. Using $2\frac{1}{2}$ percent of pickling salt by weight—6 tablespoons to each 10 pounds of shredded cabbage—pack the container with alternate layers of salt and cabbage, tamping every two layers of cabbage to get rid of trapped air and to start the juice flow. You don't need to get tough with it: just tap it gently with a clean wooden potato-masher or the bottom of a small jar. Top with a layer of salt.

Carefully cover cabbage well with a layer of tough saran plastic pressed against the top of the food and tucked down at the sides. Top this with a 5-gallon freezer bag partly filled with water so that it plops into every possible cranny, and keeps air from getting to the cabbage. For the cure, 68–72 F/20–22 C is the range for good fermenting. Good temperature, clean handling, keeping air away—these make for good sauerkraut.

When the fermenting has stopped in about 2 weeks or so, the sauerkraut will be a clear, pale gold in color and pleasantly tart in flavor. It's a good idea to lay a clean plate on it to keep it below the brine's surface; at any rate cover the container with a close-fitting lid. Store in a cool place and use as needed.

If your storage isn't around 38 F/3 C, you'd better can it (pp. 143–4).

Chinese Cabbage Sauerkraut

Follow the method for Sauerkraut. The result usually has more flavor than regular fermented cabbage does, thanks to more, and sweeter, natural juice.

Sour White Turnip (Sauer Rüben)

Peel and quarter young rutabagas or white turnips (rutabagas are usually firmer and juicier than turnips). Shred fine or chop with medium knife of a food grinder, catching stray juice in a bowl placed underneath. Pack with layers of salt as for Sauerkraut, but do not tamp down—there should be juice aplenty without tamping, and it's enough to press down on the topmost layer to settle the pack.

Proceed in every way as for Sauerkraut.

Sour Rutabagas

Handle like White Turnips.

Souring Other Lower-Acid Vegetables

Even though correct fermentation raises the acidity (lowers the *pH* rating) of lower-acid raw vegetables, *unless they are heat-processed for storage* they cannot be regarded as safe from spoilage or from growth of certain dangerous heat-sensitive bacteria.

So, because you should can them anyway for safe storage, it doesn't make much sense to go through the business of fermenting them as a preamble to putting them by for serving much later as accompaniments to meat or whatever.

SALTING MEAT

The four keys to successful salt-curing of any meat are (1) strictly fresh meat to start with, properly handled and chilled, (2) sanitation, (3) temperature control, and (4) salt content. The same quality, cleanliness, and care required for canning or freezing meat obtain in the procedures described below, and we give in the specific instructions the exact proportions of salt required to do each job.

However, temperature control demands special emphasis here. The meat must be kept chilled—held as constantly as possible at 38 F—before curing; this is why country-dwellers wait for winter weather to slaughter hogs and beef for their own tables. Once in the cure, meat should be held 36–38 F/2–3 C; for the largest pieces this means a thermometer inserted to the center of the meatiest part.

Below 36 F/2 C, salt penetrates the tissues too slowly. If the temperature of the storage area drops below freezing and stays there for several days, *increase the days of salting time by the number of freezing days*.

Above 38 F/3 C, the chances of spoilage increase geometrically with each degree of rise in temperature, and the cure changes from a clear, fresh liquid to a stringy-textured goo. It is the rare modern home that has natural storage constantly cool enough for curing meat right. Indeed, failure to ensure good temperature control is the main cause of unsuccessful curing in town and country alike.

In general, home-frozen meats do not cure well: even when defrosted completely, their texture has been changed too much to allow the cure to penetrate the tissues uniformly.

The term "pickle" is used in some manuals to designate a sweetened brine that contains some sugar as well as salt; it is *not* the solution with added vinegar that is described for pickles. "Sugar cure" usually means adding $\frac{1}{4}$ as much sweetener as there is salt in the mixture; this amount of sugar is important as food for benign flavor-producing bacteria during long cures.

Salt-curing of meats is almost always followed by exposure to smoke in order to dry the surface of the meat, to add flavor, and to discourage attacks by insects. Smoking procedures are described in detail in the section following this one.

"Saltpeter" and Nitrates/Nitrites

For generations, householders—and, back in less technologically sophisticated times, commercial processors too—used saltpeter in the cures
for many meats and meat products. What is still called saltpeter is either
potassium nitrate or sodium nitrate (this latter often termed "Chilean"
nitrate). If you buy the substance at a drugstore you are likely to get the
compound with potassium, and note that it is labeled as a diuretic. If you
buy it at a farm-supply store, it will probably be sodium nitrate. Many
store-bought curing mixtures already made up—some even containing
spices and simulated hickory-smoke flavoring—contain both a nitrate *and*
sodium nitr*ite*.

Saltpeter has been used for centuries as a means of intensifying and
holding the red color considered so appetizing in ham and allied pork
products, and in corned beef, etc. Nitrites also help to prevent the growth
of *C. botulinum*.

Storing Cured Meat

Their heavy concentration of salt protects Corned Beef and Salt Pork for
several months if the brine in which they're held is kept below 38 F/3 C.

Freezer storage of sausage and cured meats is relatively limited: after
more than 2 to 4 months at Zero F/−18 C, the salt in the fat causes it
to become rancid.

Warning: A home-cured ham is not the same as a commercially processed
one that has been "tenderized," etc. *Home-cured pork is still RAW.*

Salting Beef

Because they lack what producers and butchers call "finish," veal or calf
meat shouldn't be used to make corned or dried beef. The product is
disappointing.

Corned Beef

Use the tougher cuts and those with considerable fat. Bone, and cut them
to uniform thickness and size.

To cure 25 pounds of beef, pack it first in pickling salt, allowing 2 to
3 pounds of salt (3 to 4½ cups) for the 25 pounds of meat. Spread a generous
layer of coarse pickling salt in the bottom of a clean, sterilized crock or
barrel. Pack in it a layer of meat that you've rubbed well with the salt;
sprinkle more salt over the meat. Repeat the layers of meat and salt until
all the meat is used or the crock is filled to within a couple of inches
below the top.

Let the packed meat stand in the salt for 24 hours, then cover it with a solution of 1 gallon of water in which you've dissolved 1 pound (2 cups) of sugar, ½ ounce (about 1 tablespoon) of baking soda, and 1 ounce (about 2 tablespoons) of saltpeter.

Put a weighted plate on the meat to hold every speck of it below the surface of the brine; cover the crock/barrel; and in a cool place—not more than 38 F/3 C—let the meat cure in the brine from 4 to 6 weeks.

The brine can become stringy and gummy ("ropy," in some descriptions) if the temperature rises above 38 F/3 C and the sugar ferments. The baking soda helps retard the fermentation. But watch it: if the brine starts to get ropy, take out the meat and wash it well in warm water. Clean and sterilize the container. Repack the meat with a fresh sugar-water-etc. solution (above), to which you now add 1½ pounds (2¼ cups) of pickling salt; this salt replaces the original 2 pounds of dry salt used to pack the meat.

To store it, keep it refrigerated in the brine; or remove it from the brine, wash away the salt from the surface, and can or freeze it.

Dried (Chipped) Beef

Dried beef—which has about 48 percent water when produced commercially—is made from whole muscles or muscles cut lengthwise. Select boneless, heavy, lean-muscled cuts—rounds are best—and cure as for Corned Beef (above) *except* that you add an extra ¼ pound of sugar (½ cup) for each 25 pounds of meat.

The curing is completed in 4 to 6 weeks, depending on size of the pieces and the flavor desired. After it has cured satisfactorily, remove the meat, wash it, and hang it in a cool place to air-dry for 24 hours.

Then it is smoked at 100 to 120 F/38 to 49 C for 70 to 80 hours (see Smoking)—or until it is quite dry.

To store, wrap large pieces in paper and stockinette (tubular, small-mesh material, which holds the wrap close to the meat) and hang them in a cool (below 50 F/10 C), dry, dark, insect-free room; certainly refrigerate small pieces. Plan to use all the dried beef before spring.

Salting Pork

All parts of the pig may be cured by salting. Some—such as the fat salt pork for baked beans, chowders, etc.—are used as they come from the salting process. The choice hams, bacon and, perhaps, loins are carried one step further and are smoked after being cured.

Have the meat thoroughly chilled, and hold it as closely as possible to 38 F/3 C during the process of curing: salt penetrates less well in tissues below 36 F/2 C, and spoilage occurs with increasing speed in meat at temperatures above 38 F/3.3 C.

"Pumping"—i.e., forcibly injecting a strong curing solution into certain parts of a large piece of meat—is not included in the instructions below because we're leery of it: much safer to allow safe curing time than to try to speed the process by localized "spot" applications of the cure.

Allow 25 days as the minimum curing time for dry-salted pork, with some of the larger pieces with bone taking longer. Allow at least 28 days for sweet Pickled (Brined) pork, and more for the larger pieces. The days-per-pound are given for each cut cured by each method.

Before smoking or storing large pieces containing bone, run a skewer up through the meat along the bone, withdraw the skewer and sniff it. If the odor is sweet and wholesome, fine—proceed with the smoking or storing; but if there's any "off" taint, any whiff of spoilage, destroy the entire piece of meat, because it is unsafe to eat.

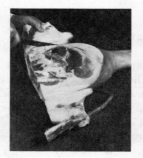

Basic cuts from a side of pork. Left to right: separating picnic shoulder and shoulder butt; trimming the ham; separating spareribs from the bacon strips. (*USDA photograph*)

Dry-Salting Large Pieces (Hams and Shoulders)

For each 25 pounds of hams and shoulders mix together thoroughly 2 pounds (about 3 cups) of coarse-fine pickling salt, $\frac{1}{2}$ pound (about 1 cup) sugar—and $\frac{1}{2}$ ounce (about 1 tablespoon) of saltpeter. Rub half the mixture in well on all surfaces of the meat. Poke it generously into the shank ends along the bone (you can even make a fairly long internal slit with a slender boning knife inserted at the shank, and push the mixture up into it: this is better than relying on "pumping" a strong solution to such areas where the salt must penetrate deeply). Plan to leave an $\frac{1}{8}$-inch layer of the mixture on the ham face (the big cut end), with a thinner coating on the rest of the ham and on the shoulders.

Fit the salt-coated meat into a clean sterilized barrel or crock, taking care lest the coating fall off. Cover with a loose-fitting lid or cheesecloth and let cure in a cold place, 36 to 38 F/2 to 3 C.

One week later, remove the meat, re-coat it with the remaining half of the curing mixture, and pack it again in the barrel/crock.

Curing time: at least 25 days. Allow 2 to 3 days for each 1 pound of ham or shoulder, being sure to leave them in the curing container even after all surface salt is absorbed.

Then smoke them.

Dry-Salting Thin Cuts (Bacon, "Fat Back," Loin, Etc.)

For each 25 pounds of thin cuts of pork, mix together thoroughly 1 pound (about 1½ cups) of pickling salt, ¼ pound (½ cup) of brown or white sugar, and 1½ teaspoons of saltpeter.

Coat the cuts, using all the mixture. Pack the meat carefully in a sterilized crock or barrel, and cover it with a loose-fitting lid or layer of cheesecloth, and let it stand at 36 to 38 F/2 to 3 C for the *minimum* total curing time of 25 days; allowing 1½ days per pound. Thin cuts do not require an interim salting—that's why you used all the mixture in the first place. And leave them in the crock even after the surface salt has been absorbed.

All but the "fat back" (Salt Pork) is then smoked. Wrap the Salt Pork in moisture/vapor-proof material; refrigerate what is intended for immediate use, and freeze the rest.

"Sweet Pickle" Salting Large Pieces (Brining Hams, Etc.)

Curing hams and shoulders in brine is slower than the dry-salting treatment just described, and therefore is well suited to colder regions of the country.

Pack the well-chilled (38 F/3 C) hams and shoulders in a sterilized crock or barrel. For every 25 pounds of meat, prepare a solution of 2 pounds (about 3 cups) of pickling salt, ½ pound (1 cup) of sugar, ½ ounce (about 1 tablespoon) of saltpeter and 4½ quarts of water. Dissolve all thoroughly, and pour over the meat, covering every bit of it: even a small piece that rises above the solution can carry spoilage down into meat submerged. Put a weighted plate or board over the meat to keep it below the brine, and cover the barrel/crock. Hold the storage temperature to 38 F/3 C.

After 1 week, remove the meat, stir the curing mixture, and return the meat to the crock/barrel, making sure that every bit of it is weighted down below the surface of the brine.

Remove, repack, and cover with the stirred brine at the end of the second and fourth weeks.

If at any time during the cure you find that the brine has soured or become ropy and syrupy, remove the meat, scrub it well, and clean and scald the barrel/crock. Chill the container thoroughly, and return the meat, covering it with a fresh, cold curing solution made like your original brine, except that you *increase the water to 5½ quarts.*

Curing time: a minimum of 28 days; allow 3½ to 4 days for each 1 pound of ham or shoulder.

"Sweet Pickle" Salting Small Pieces (Bacon, Loin, "Fat Back")

Pack the pieces in a sterilized crock or barrel, and cover with a brine like that for large pieces, except in a milder form: use 6 quarts of water, rather than 4½ quarts. Proceed as for hams and shoulders, keeping the pieces well submerged, and overhauling the contents as above at the end of the first, second, and third weeks.

Curing time: a minimum of 15 days for a 10-pound piece of bacon, allowing 1½ days per pound; but 21 days for heavier pieces of bacon, or for the thicker loins.

Pork that is not to be smoked may be left in the brine until it is to be used—but it will be quite salty.

Heavy salting for heavy cuts; then into the barrel/crock to cure. (*USDA photograph*)

SMOKING

Without intending either to pun or to discuss the pro/con of this traditional finishing process for many cured meats and a few cured fish, we feel duty bound to note that smoking any food is under fire nowadays from some critics.

However, as we indicated in the individual salting instructions above, meats may be left in brine or dry salt until they're ready to be used. Or remove them from the cure, scrub them well to remove surface salt, and hang them in a cool dry, well-ventilated place for from several days to a week to let them dry out a bit before storage.

We do not recommend using so-called "liquid smoke" or "smoke salt" in place of bona fide smoking. Either smoke your meat or call it a day at the end of the salt cure.

What "Cold-Smoking" Does

We're not concerned here with what is known as "hot-smoking"—which in effect is cooking in a slow, smoky barbecue for several hours, thus making the food partially or wholly table-ready at the end of the smoking period.

What we'll do is hold the food in a mild smoke at never more than 120 F/49 C, and usually from around 70 to 90 F/21 to 32 C, for several days to color and flavor the tissues, help retard rancidity and, in many cases, increase dryness—the actual length of time depending on the type of food.

The food is then stored in a cool, dry place, or is frozen, to await future preparation for the table.

Making the Smoke

Use only hardwood chips for the fire—never one of the evergreen conifers, whose resinous smoke can give a creosote-y taste, or other softwoods. Among the most popular woods are maple, apple, and hickory.

Or use corncobs. These should be the thoroughly dried cobs from popcorn or flint corn that has dried on the ear: cobs saved from a feast of sweet corn-on-the-cob aren't the same thing at all. Look in the Yellow Pages for a handler of hardwood sawdust or shavings; such a dealer often has chopped cobs to use as a tumbling medium for polishing. Two bushels of cut corncobs can produce 72 hours of smoke, or enough to do a whole ham in a small smoke-box.

Avoid chemical kindlers. Small, dry hardwood laid teepee-fashion over crumpled pieces of milk cartons catch well, and form a good base for the fire. Get your fire well established and burning clean, but do not have it hot; keep it slow, just puttering along evenly so the meat is in no danger of cooking. Hang a thermometer beside the food closest to the fire: fish, which is so highly perishable (even when lightly salted for smoking), should be smoked at 40 to 60 F/4 to 16 C, and then for a relatively short time compared to the temperature for meats.

The fire can be made and held in any sort of iron or tin brazier suitable for the size of the smokehouse or box.

If you use sawdust or fine chips or chopped corncobs, the smoke might also be maintained well enough by using an electric hot plate to fire a tin pie pan that's filled with the smoke-making material. Set the hot plate on High to start the pan of stuff smoldering, then reduce the heat to Medium or Low. Experiment.

SMOKING MEAT

Because bacteria in meat grow faster between 70 and 100 F/21 to 38 C, you should smoke meat in fairly cold weather, in late fall or early spring, when temperatures are between 30 and 50 F/ − 1 and 10 C during the day. However, really cold weather, down to Zero, is not for the beginner.

Smoking should be as sustained as is reasonably possible, simply because you want to get it over with and get the meat cooled and wrapped in moisture/vapor-proof material and stored in a cool, dry place (or frozen). But it won't suffer from the hiatus if you can't smoke at night: the weather will probably keep it cold enough without freezing so you can leave it in the smokehouse and just start your smoke-maker again in the morning.

If you have a sudden sharp drop in temperature, though, you had better bring inside to cool storage any meat that shows danger of freezing without the warmth of the smoke. Resume counting the total smoking time when the smokehouse is operating again.

Preparing the Meat for the Smokehouse

Remove the meat from the salting crock, scrub off surface salt, using a brush and fresh lukewarm water. Then hang the meat in a cool, airy place for long enough to get the outside of it truly dry—up to 24 hours.

Run several thicknesses of food-grade cord or a stainless steel wire through each piece of meat several inches below one end; tie the string or double-twist the wire to form a loop that will hold the weight of the meat. Hams are hung from the shank (small) end.

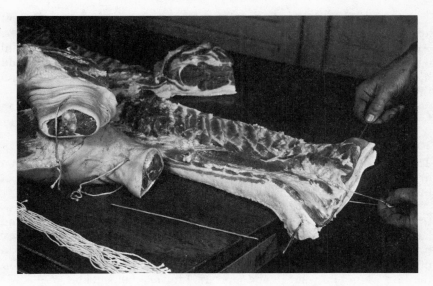

Food-grade twine is run through each piece of meat to prepare it for hanging in the smokehouse. (*USDA photograph*)

Small Homemade Smoke-Boxes

The USDA bulletin on processing pork, cited earlier, contains plans for a full-fledged smokehouse, and also a description of a barrel smoke-box. Here are our variations of the barrel.

A Barrel

You can get the smoking parts of half a 200-pound pig in a 55-gallon steel barrel that you make into a "smokehouse." This means that one ham, one shoulder, and one side of bacon cut in pieces can be smoked at the same time.

Wooden barrels large enough to do the job are (1) hard to come by these days, and (2) their staves shrink when dried out (as they'd be after several days' worth of warm smoke) and open. So use a metal barrel with one head removed. If it's had oil in it, set the residue of oil on fire and let it burn out; then scour the drum thoroughly inside and out with plenty of detergent and water; rinse; scald the inside, and let it dry in the air.

OUTSIDE ON THE GROUND

In the bottom of the barrel cut a hole large enough to take the end of an elbow for whatever size of stovepipe you want to use (see the sketch). Set the barrel on a mound of earth—with earth banked high enough around it to hold it firm and steady—and dig a trench from it down to a fire-pit at least 10 feet away, and inclining at an angle of something like 30 degrees. Via the trench either connect the barrel to the pit with stovepipe, or build a box-like conduit (stovepiping is easier to remove and clean). You should have the length in order to cool the smoke on its way to the meat, and the pitch to encourage the draft.

Put a cover of close-fitted boards over the fire-pit, arranged so it can be tilted to increase the draft when necessary.

ON THE PORCH, OR IN A GARAGE

The electric-plate/pan-of-sawdust arrangement should be used only in a dry place with fire-retardant material underneath it. This can be sheet metal, or a concrete floor.

Set your barrel on supports—cinder-blocks or trestles of some sort—to hold the elbow well away from the floor. Connect the stovepipe, and lead it from the barrel to the electric smoke-making unit. Make a wooden box, lined with fire-retardant material, to house the hot plate and the pan of sawdust, cut adequate slits for regulating air intake; and merely lift off the box when you want to add more fuel for making smoke.

SMOKING IN THE BARREL

Get the fire or smoke-maker well established and producing evenly before hanging the meat to smoke. There should be good ventilation from the top in order to carry off moisture the first day (to keep the fire from getting too hot, though, reduce the air intake at the bottom of the fire-pit as much as you can without letting the fire go out).

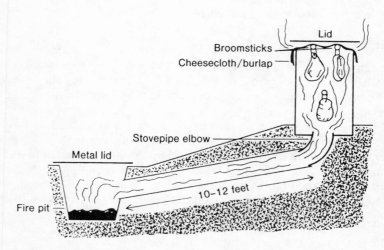

Smoke barrel made from a 55-gallon oil drum. (*Drawing by Irving Perkins Associates*)

Hang the ham, shoulder, and chunks of bacon (or comparable sizes of beef pieces) from broom-handles or stainless steel rods—not galvanized, not brass, not copper—laid across the top of the barrel. Stagger the meat so that none of it touches other pieces or the side of the barrel; suspend smaller pieces on longer loops of strong steel wire so they drop below the large pieces. Hang your thermometer.

Over the whole business lay a flat, round wooden cover slightly bigger than the barrel's top. It will be held up from the rim by the thickness of the supporting rods. If this isn't enough clearance at the beginning, or if the draft seems to be faltering, prop the lid higher with several cross-pieces of wood laid parallel to the supporting rods.

Close down the ventilation on the second day by draping a piece of clean burlap or several thicknesses of cheesecloth over the supporting rods *under* the lid. The cloth will also protect the meat inside from debris, or from insects attracted to it if the smoke stops. Weight the lid down over the cloth with a good-sized rock to keep it in place.

HOW LONG?

Smokiness—color and flavor of the meat—is a matter for individual taste. If it's oversmoked, the meat is likely to be too pungent, especially on the outside. And you can always put the meat back for more smoking if the flavor isn't enough for you.

So try out your system in a small way. Give a shoulder of cured pork, say, 45 to 55 hours of smoking; take it out and slice into it—you may want to give it a few hours more: 60 hours is about average for a smoked shoulder.

The average ham takes about 72 hours of total smoking time.

Bacon, being a thinner piece of meat, is usually smoked enough in a total of 48 hours.

BUT ALL THESE TIMES ARE APPROXIMATE—they're mentioned merely as guides.

Smoking Fish

The following procedure deals with coho salmon and related species.

Pre-Smoking Preparation

Dress, scrub, and fillet your salmon, taking special care to remove the dark lateral line of flesh that is capable of harboring unusually large numbers of spoilers; hold the pieces as close to 32 F/Zero C as possible. In your largest kettle or tub—enameled, ceramic, or wooden, *never* one that can corrode—prepare enough ice-cold brine to cover your fish, made in the proportion of 3 cups pure pickling salt dissolved in each 1 gallon of fresh drinking water. Depending on the thickness of the fillets, hold the fish in this 30–40 F/ – 1–4 C brine for 1 to 2 hours, during which time diffused blood will be drawn out, the oil in the tissues will be sealed in to a large extent, and the flesh will be chilled so much that the following dry-salt cure will not penetrate too rapidly.

Remove the fillets, drain, and scrub away debris. Using pickling salt in the proportion of 3½ pounds (5¼ cups) for each 10 pounds of fish, dredge the pieces completely in salt and pack them in a large non-corrodible container with plenty of salt between the layers. Put the pieces skin-side down, except for the top layer; cover the top layer with salt. Keep the container as cool as possible, and hold the fish in it for 3 hours.

Remove fish, rinse well. Air-dry in single layers away from sun or heat for 1 to 3 hours until a thin shiny "skin," or pellicle, forms on the surface. The fish is now ready to smoke.

Cold-Smoking the Fish

Many beginners are confused by the term "hot-smoking," which is a sort of long-distance barbecue in which the flesh reaches an internal temperature of up to 180 F/82 C after which it is eaten within a couple of days— as with any cooked food—or is frozen. We are not speaking here of this type of smoke-cooking.

Build your regular hardwood fire; after it is burning well, smother it with fine hardwood chips or sawdust to produce a very dense smoke with little heat—the temperature inside the smoke chamber ideally should never exceed 70 F/21 C in order to inhibit growth of bacteria in this highly perishable food. Tend the fire night and day: smoking fish is a continuous process.

After the end of 4 full days of smoking, sample a piece of fish to see if its color, flavor and texture are what you want. If not, smoke it 24 hours more, and test again. When it is smoked to your satisfaction, air-dry the pieces in a cool place for several hours. Then package the fillets individually in plastic wrap and store at 32–40 F/Zero–4 C for up to 3 months.

Freezing will cause salt in the tissues to deposit on the surface of the fillets. We do *not* recommend that smoked fish be canned at home.

21
DRYING

The purpose of drying is to take out enough water from the material so that spoilage organisms are not able to grow and multiply during storage: to be remembered as one of the six factors in preservation. The amount of remaining moisture that is tolerable for safety varies according to whether the food is strong-acid or low-acid raw material, or whether it has been treated with a high concentration of salt—and, to some degree, with the type of storage.

Although the terms "drying," "dehydrating," and "evaporating" are often used casually as meaning the same thing, the USDA Research Service's fine multivolume *Agriculture Handbook No. 8,* which tells the composition of raw, processed, and prepared foods of all sorts, lists as dehydrated those foods containing only 2.5 to 4 percent water—the other 96 + percent having been removed by highly sophisticated processes that we can't hope to equal at home. It lists as dried those foods still containing roughly 10 to 20 percent water (the amount depending on whether they're vegetables or fruits). We can take out all but this much moisture with the equipment and methods described in this section—and we'll call it *drying*.

General Procedures in Drying

Dry only food that is in prime condition and perfectly fresh, just as you choose it for any method of putting by; and handle it quickly.

Be scrupulously clean at every step. A number of the micro-organisms that cause food poisoning, ranging from the Salmonellae to *C. botulinum* and including molds and fungi, contaminate the food because they are in the soil or on the surfaces of our workplaces or even in the air around us.

The procedures described hereafter do not undertake to *sterilize* food. However, a moisture content of less than 35 percent can greatly slow the growth of micro-organisms.

Before drying starts and after the food is pared/cored/sliced or whatever, much of it will be given some sort of treatment to preserve color, prevent decomposition, and safeguard nutrients (in general, though, drying is hard on some of the vitamins). Depending on the type of food, these treatments are: for fruit, coating with an anti-oxidant or sulfuring; for vegetables, blanching in boiling water or steam to stop enzymatic action; and, in the case of meat or fish, salting.

Throughout the drying process the food must be protected from airborne spoilers and from vermin—and simply from poor handling. Regardless of where it is being dried, it will lie on only food-grade materials and it will be shielded from insects. After it tests dry, any unevenness in moisture content will be equalized by conditioning; insect eggs, if there are any, will be destroyed by pasteurizing the food; it will be stored in food-grade containers, safe from infestations or dampness or temperatures too warm.

Equipment for Drying

Keep everything simple, even rudimentary, in the beginning: aside from saving money it's a lot more fun in this hypertechnical age to return to elementals.

Trays First

Shallow *wooden* trays are necessary whether you dry outdoors in sun or shade, or indoors in a dryer or an oven. *Never* use aluminum, copper, or galvanized metal, or wire with a plastic coating that is not food-grade. The trays should have slatted, perforated, or woven bottoms to let the air get at the underside of the food. Don't make them of green wood— which weeps and warps; and don't use pine, which imparts a resinous taste to the food; and don't use oak or redwood, which can stain the food. The simplest frames to make would be those cut from wooden crates that produce comes in: saw the crates in several sections horizontally, rather as you'd split a biscuit.

Each 1 square foot of tray space will dry around 1½ to 2½ pounds of prepared food.

Loaded trays shouldn't be too large to handle easily, and they should be uniform in size so they stack evenly. The flimsier the construction, the smaller they should be; but even well-built ones for sun-drying are better if they're not more than 2 feet by 2 feet.

However, since you can have an emergency that means you will need to finish off in an oven or dryer a batch you've started outdoors, it makes sense to have the trays smaller, and rectangular. Make the trays narrow enough for clearance when you slide them inside an oven, and 3 to 4 inches shorter than the oven or dryer is from front to back: you'll want to stagger the trays to allow air to zigzag its drying way up and over each tray as it rises from the intake at the bottom to the venting at the top.

Consider having the trays 1 to 2 inches deep, 12 to 16 inches wide, 16 to 20 inches long—but first having found the inner dimensions of the oven or dryer (less the fore-and-aft leeway for staggering the trays).

Don't use metal screening for the bottom. Steer clear of fiberglass mesh: minute splinters of fiberglass can be freed easily and impregnate the food. Vinyl-coated screen in beguiling ¼-inch and ½-inch mesh looks like the answer at first glance, *BUT is it food-grade?* And what will it do at 140 F/60 C, the average heat in a dryer—melt? peel?

Any cloth netting will do if its mesh isn't larger than ½-inch; nylon net is the easiest to keep clean. Two layers of cheesecloth work, as does mosquito net, etc.—but they're hard to clean without getting frazzled. Old clean sheets let less air up through, but they're stouter. (In a pinch you can dry food on sheets laid flat in the direct sun.) When cutting cloth for tray bottoms, allow 2 inches more all around so you can fold it over itself on the outside of the frame; then staple it in place.

We've seen good trays with bottoms of twine or strong cord strung back and forth and then cross-hatched the other way. Draw the twine tight and flat, staple each loop to the outside of the frame-strip before you turn around and go back, keeping the strands ½ inch apart. *Do not use hay-baler twine,* a conscientious reader pointed out to us: this is now treated with a pesticide, so it's bad for food.

Strong, serviceable bottoms are made by nailing ½-inch wood strips to the bottom of the frame ½ inch apart; the strips run in only one direction. More finished—but worth it, because they're smooth and easy to clean— are ¼- or ½-inch hardwood dowels; these are nailed inside the frame with small box nails driven through from the outside, and they also go in only in one direction.

One thickness of cheesecloth laid over bottoms will keep sugar-rich food from sticking to them while it dries; so will a thin coating of oil. Even a few recent publications suggest mineral oil for lubricating the trays—it doesn't impart flavor and doesn't get rancid—but use any fresh, low-flavored vegetable oil. You'll be scrubbing your trays anyway, regardless of what oil you use.

CONVERSIONS FOR DRYING

Do look at the conversions for metrics, with their workable roundings-off, and for altitude—both in Chapter 3—and apply them.

Two Good Homemade Dryers

Buy a dryer if you can, but if you are at all handy and enjoy doing for yourself, *PFB* recommends a make-it-yourself dryer to set on a table. It is described in full in Circular 855, *How to Build a Portable Electric Food Dehydrator,* by Dale E. Kirk, Agricultural Engineer, Oregon State University, Corvallis; but directions—and diagrams—for building it are also contained in USDA *H&G Bulletin 217, Drying Foods at Home* (1977).

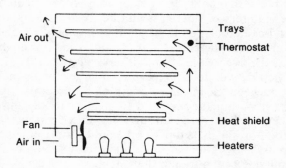

A homemade dryer. (*Drawing by Irving Perkins Associates*)

The dryer offers around 8½ square feet of tray surface, and thereby will handle around 18 pounds of fresh fruit or vegetables. Basically, it is a plywood box that holds five screen trays above the heat source, which is nine 75-watt light bulbs; the heat is dispersed by a shield and forced upward through the trays of food by an 8-inch household fan.

A smaller, simpler, and more passive version is for small-scale drying outdoors with plenty of sunshine. You can put together a dryer that looks, and acts, much as a cold-frame does (see sketch).

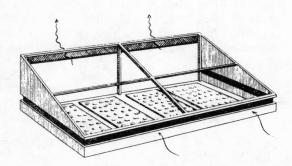

The cold-frame dryer. (*Drawing by Norman Rogers*)

The tilted glass panel—one or more pieces of storm sash are fine—intensifies the heat from the sun, and this rise in temperature inside lowers the relative humidity correspondingly, so that drying occurs faster than is possible outside the dryer. The ample screened venting allows circulation of air.

This dryer is not effective on overcast days.

PROTECTIVE COVERINGS

Food dried in the open, whether outdoors or in a warm room, needs protection from insects and airborne gurry. Simplest to use is a strong nylon netting, as for mosquitoes; nylon because it's easier to keep clean. Many people cut the covering 2 inches larger all around than the tray it's intended for, bend it over, and thumbtack the overlap to the sides of the tray. Or sometimes it's easier to stretch over several trays laid side by side. Or use a maverick window screen laid on top of the trays.

Food that's drying outdoors must be protected from dew at night—unless it is brought inside outright. So stack the trays under a shelter and cover the stack with a big carton, or drape the stack with a clean old sheet.

MISCELLANEOUS FURNISHINGS

Trestles, racks, benches. No set sizes or types for these, so just know where you can get bricks or wood blocks for raising the first course of trays off the ground; scrap lumber for building rough benches or racks to hang drying food from; smaller stuff to use as spacers.

Sulfuring box. We're going to suggest sulfuring in certain instances, and we'll tell how to make and use a sulfuring box in a minute.

Auxiliary heaters. Easier to list what *not* to use: no small stove burning flammable material—wood, oil, coal, etc.—that sits inside any portion of the dryer-box. Never any sort of front-blowing electric heater laid on its back to blow hot air upward. Avoid electric hot plates.

The wiring of all electrical heating and blowing units must meet all safety criteria.

Electric fan. To boost the natural draft in an indoor dryer or to augment a cross-draft when drying in an open room or outdoors. It needn't be large; it should be directable, and *it must have a safety grill covering the blades.*

Thermometers. Even with a dryer or oven having a thermostat, you'll need a food thermometer—a roasting, candy or dairy type will do—to check on the heat of food being processed; plus the most inexpensive kind of oven thermometer to move around between the top and bottom trays to keep track of the varying temperatures.

Scales. Not vital but a great help is scales that go up to 25 pounds/11.34 kilograms, with quarter- and half-pound gradations; use it for judging water-loss by weight, per-pound treatments before drying.

Blanching kettle. Your stockpot or Boiling–Water Bath canner will do.

Assorted kitchen utensils. Dishpan, colander, crockery or enameled bowls; stainless sharp knives for cutting and paring; apple-corer and a melon-ball scoop; non-wood cutting-board; vegetable slicer or a coarse shredder; spoons—some wooden, at least one slotted; also plenty of clean towels and paper toweling, and an extra packet of cheesecloth.

Materials for storing. Several large covered crocks for conditioning dried food before storing—or strong cartons, moisture-proofed with a lining of plastic sheeting; plastic or paper bags (not big) for packaging dried food in small quantities; mouse-proof, sealable containers for the packages. And cool, dark, dry storage when you're done.

THE DRYING METHODS

Basically, drying food at home combines sustained mild heat with moving air to accomplish its purpose. This means (1) heat adequate to extract moisture, but moderate enough so that it doesn't cook the material; and (2) currents of air dry enough to absorb the released moisture and carry it off. These conditions can occur outdoors naturally, or they can be reproduced indoors in dryers.

Open-Air/Sun

Successful outdoor drying is possible only in sun-drenched regions with prolonged low humidity, where foods are exposed for perhaps only a minor portion of their total drying time to direct sun, and are partially shaded by a roof of some sort from the fierce rays at midday. An open, south-facing veranda is a favorite place for drying in many parts of the American Southwest.

Where to Sun-Dry

Hereout, "sun-drying" will mean outdoors in open air, the food exposed to sun but not in full sun at all times of the day, lest it "case-harden"— that is, cook the outside to form a crust that prevents the inside from drying well.

In North America, the interior of California and the high country of Southwestern states possess the ideal climate for sun-drying: predictably long periods of hot sun and low humidity. Next come the wide Plains east of the Rockies in the United States and Canada, where occasional showers are not a great problem if the food hasn't got wet and if drying can be resumed in open air the next day. Sun-drying can be done in parts of the Northwest east of the Cascade range, and in the Appalachians. Despite their heat, the humid areas of the South are not so good. It is worth noting that Cornell University's excellent 1977 bulletin does not include drying outdoors or in the sun. The authors' tacit instruction is *not* to try sun-drying in the Northeast.

What to Dry in Open-Air/Sun

Before we start, a warning: eggs, poultry, and meat—except for very lean beef, young lamb, or venison made into jerky—are not good for home drying; nor is fish, unless it is heavily salted cod, etc., that is more likely to be dried as a commercial venture. Reason: salmonella and staphylo-coccus bacteria thrive on these foods. There are also some vegetables whose storage life is comparatively short.

In the following list of sun-dryable produce, most of the fruits were exposed to extensive sulfuring before drying; with some of the "easier" fruits, consensus is that color and flavor are better when they finish drying in stacks in the shade, humidity permitting.

Fruits easier to sun-dry. Apples, apricots, cherries, coconut, dates, figs, guavas, nectarines, peaches, pears, plums, and prunes.

Fruits harder to sun-dry. Avocados, blackberries, bananas, breadfruit, dewberries, Loganberries, mameys (tropical apricots), and grapes.

Vegetables easier to sun-dry. Mature shell beans and peas, lentils and soybeans in the green state, chili (hot) peppers, sweet corn, sweet po-tatoes, cassava root, onion flakes, and soup mixture (shredded vegetables, and leaves and herbs for seasoning).

Vegetables harder to sun-dry. Asparagus, beets, broccoli, carrots, celery, greens (spinach, collards, beet and turnip tops, etc.), green/string/ snap beans ("leather britches" to old-timers), green (immature) peas, okra, green/sweet peppers, pimientos, pumpkin, squash. And tomatoes— but we'll tell how to do them indoors, too.

Drying Produce in Open-Air/Sun

Wash, peel, core, etc., and pre-treat according to individual instructions. Because vegetables must have more of their water removed than fruits do for safe drying, cut vegetables smaller than you cut fruits so they won't take too long to dry (being low-acid, vegetables are more likely to spoil during drying).

Spread prepared food on drying trays one layer deep ($\frac{1}{2}$ inch, or de-pending on size of the pieces); put over it a protective covering as de-scribed above; place trays in direct sun on a platform, trestles, or sloping roof—or on any sort of arrangement that allows air to circulate underneath them.

If you use clean sheets or the like to hold the food, a table, bench, or shed roof is a good place, but you lose the benefit of air circulating *under* the food.

Stir the food gently several times each day to let it dry evenly.

Before the dew rises after sundown, bring the trays indoors or stack them in a sheltered spot outdoors. If the night air is likely to remain very dry, the outdoor stack need not be covered; otherwise wait a little until the warmth of the sun has left the food, then drape a protecting sheet over the stack. Return the food to the direct sun the next morning.

At the end of the second day, start testing the food for dryness after it has cooled.

Stack-drying Produce in Shade

This variation of sun-drying relies on extremely dry air having considerable movement. This method gives a more even drying with less darkening than if the food was done entirely in direct sun; apricots, particularly, retain more of their natural color when shade-dried.

Prepare the food, cutting it in small pieces; put the trays in direct sun for one day or more—until the food is $\frac{2}{3}$ dry. Then stack the trays out of the sun but where they'll have the benefit of a full cross-draft, spacing them at least 6 inches apart with chocks of wood or bricks, etc. After several days the dried food is conditioned and packaged for storing.

Indoor Drying

Almost every food that sun-dries well can make a better product if it is dried more quickly, either in a dryer with separate heat source and blown air, or a well-managed oven. For some foods, especially the low-acid vegetables, processing in an indoor dryer is recommended even though outdoor drying conditions are reliable during the harvest months.

Herbs dry best hung in large kraft-paper bags (from the supermarket), tied by their stems, and the whole thing slung from a beam in a well-ventilated room.

Depending on the water content and size of the prepared food, and whether the dryer is loaded heavily or skimpily, good drying is possible within 12 hours in an indoor dryer.

On a smaller scale, the conventional oven of a cookstove can be made to perform as a dryer; the processing time is rather longer.

Microwaving note: do your herbs laid out between two sheets of paper toweling, at a high setting for 2 minutes, let sit one minute—then check. Repeat until leaves, when cool, may be brushed off any stems. Strip off leaves beforehand with herbs like basil, sage, etc.

Jerky meats, fresh or lightly salted, dry well in an oven (which is an indoor dryer of sorts)—often better than they do on trays in a regular drying box.

Salt fish is best done in open air, since a breeze outdoors on a sunny day is preferable to the limited ventilation afforded by a dryer or an oven.

Using a Dryer

Here you're increasing the speed of drying by use of temperatures higher than those reached outdoors in the sun, so be prepared to regulate heaters and shift trays around if you want the best results.

Rules-of-thumb for Indoor Dryer Times

The specifics for individual foods will say to start at a relatively lower temperature (this, to avoid case-hardening), then raising the heat after an hour or so, and lowering it again during the last one-third of drying time to prevent any "cooked" flavor, caramelizing, or scorching.

For drying, a conventional oven fluctuates in such swoops that it's simpler to set its thermostat at 140 F/60 C, and not change the heat unless the food may be tending to caramelize or cook toward the end of the drying time; then of course the temperature is lowered by 10 F/5.6 C. The convection oven may also be set initially at 140 F/60 C and let to carry on; with its fan, its times will be shorter than for the regular range oven.

This 140 F/60 C is the best across-the-board temperature if your dryer cannot be fine-tuned to changes, because it results in a safe product if not always a thrilling one.

The *How* of drying by artificial heat is simple if you keep in mind a few *Whys:*

In the usual home dryer always keep track of the temperature at the lowest tray, so you can use this heat as the base for judging the temperature higher up. And rotate the trays up or down every $\frac{1}{2}$ hour to correct this difference and ensure even drying.

Fresh food won't dry well if it is exposed to too much heat too soon. But for the majority of its total drying time the food must have enough heat to kill the growth cells of some spoilers, as well as to remove moisture that lets other ones thrive. This means that, no matter how low the temperature at which you start food in order to prevent case-hardening, etc., you have to raise the heat to a killing level and hold it there long enough to make it effective.

When the food has reached the $\frac{2}{3}$-dry stage, tend it with extra care to make sure it won't scorch. Keep rotating trays away from the heat source. If you need to, during the last 1 hour reduce the heat by 10 degrees or so.

Handling Food in a Dryer

Line or oil the trays—see the earlier comments on tray bottoms under Equipment for Drying; spread prepared food on them one layer deep if it's in large pieces, not more than $\frac{1}{2}$ inch deep if it's small. Place halved, pitted fruit with the cut side up (rich juice will have collected in the hollows if it was sulfured).

Stagger the trays on the slides: one pushed as far back as possible, the next one as far forward as possible, etc. (as in the sketch earlier).

Check the food every ½ hour, stirring it with your fingers, separating bits that are stuck together. Turn over large pieces halfway through the drying time—but wait until any juice in the hollows has disappeared before turning apricots, peaches, pears, etc. Pieces near the front and back ends of the trays usually start to dry first: move them to the center of the trays.

If you add fresh food to a load already in progress, put the new tray at the top of the stack.

Make needed room for fresh food by combining nearly dry material in deeper layers on trays in the center of the dryer; it can be finished here without worry, but keep stirring it.

Using an Oven

As far as you can, use an oven as you would a dryer, following general procedures and the specific instructions for each food.

Leave the upper (broiling) element of an electric oven turned off, and use only a low-temperature Bake setting for drying. With some electric ranges this broiling element stays partially on even with a Bake setting: if yours does this, simply put a cookie sheet on a rack in the uppermost shelf position to deflect the direct heat from the food being dried.

Most gas ovens have only one burner (at the bottom) for both baking and broiling. If your gas oven has an upper burner, don't turn it on.

Gas ovens are always vented, but electrics may not be. If your electric oven isn't vented, during drying time leave the door ajar at its first stop position.

If your oven isn't thermostatically controlled, hang an oven thermometer where you can see it on the shelf nearest the source of heat, leave the door ajar—and be prepared to hover more than usual over the food that's drying.

Preheat the oven to 140 F/60 C. *If the oven cannot be set this low, skip the lowest slide you would otherwise be using: keep the bottom tray at least 8 inches from the heat source.*

Don't overload the oven: with limited ventilation (even with a fan aimed toward the partly opened door) it can take as much fuel to dry a batch too big as might be used to dry two fairly modest batches.

Pre-drying Treatments for Produce

Before being dried at home by any method, fruits make a better product if they undergo one or more of the treatments given hereafter, while all vegetables are treated to stop the organic action that allows low-acid foods to spoil.

And, still speaking generally, a pre-drying treatment for fruits is optional, but the pre-drying treatment for vegetables is a must.

The optional treatments for fruit involve (1) temporary anti-oxidants, to hold their color while they're peeled/pitted/sliced; (2) blanching in steam or in syrup as a longer-range means of helping to save color and nutrients; (3) very quick blanching—either in boiling water or steam (lye is not recommended)—to remove or crack the skins; (4) exposing to the fumes of burning pure sulfur as longest-range protection.

The treatment for vegetables is steam-blanching. The quick dunk in boiling water that's used in freezing is not adequate to protect them against spoilage in drying; and the much longer boiling time needed here would waterlog the material, in addition to leaching away a number of its nutrients.

The following descriptions are given in the order that the treatments are likely to occur in handling produce for drying: they're not necessarily in order of importance.

Temporary Anti-Oxidant Treatment

Pure ascorbic acid is our best safe anti-oxidant, and is used a lot in preparing fruits for freezing. Use it here too. But with the difference that the solution will be somewhat stronger, and thus food coated with it can hold its color in transit in the open air for a longer time.

One cup of the solution will treat around 5 quarts of cut fruit, so prepare your amount accordingly. Sprinkle it over the fruit as you proceed with peeling, pitting, coring, slicing, etc., turning the pieces over and over gently to make sure each is coated thoroughly.

For apples: dissolve 3 teaspoons of pure crystalline ascorbic acid in each 1 cup of cold water.

For peaches, apricots, pears, nectarines: dissolve $1\frac{1}{2}$ teaspoons of pure crystalline ascorbic acid in each 1 cup of cold water.

If the variety of fruit you're working with is likely to become especially rusty-looking when the flesh is exposed to air, it's O.K. to increase the concentration of ascorbic acid as needed. The proportions above usually do the job.

The commercial anti-oxidant mixtures containing ascorbic acid don't work as effectively, volume for volume, as the pure Vitamin C does, but they're often easier to come by. Follow the directions for Cut Fruits on the package.

Blanching Fruits in Heavy Syrup

Blanching fruits in heavy syrup is little practiced by average householders, who generally want simplicity.

Treating Fruit Skins without Lye

A very quick dip in boiling water, *quite apart from the steam-blanching that helps keep the color and nutrients of certain cut fruits,* works well instead of any lye treatment. And it's safer for you and for your food, and the alkali in such soda compounds hurts many B vitamins and Vitamin C.

FOR CHECKING THE SKINS

Nature provides a wax-like coating on the skins of cherries, figs, grapes, prunes, and small dark plums, and certain firm berries like blueberries and huckleberries, and they all dry better if this waterproofing substance is removed beforehand. If the skins are cracked minutely (called "checking"), the fruit is unlikely to case-harden.

Gather not more than 1 pound of berries loosely in cheesecloth and hold them in 1 gallon of briskly boiling water for 30 seconds; lift out, dunk in plenty of fresh cold water to stop any further cooking action; shake off water, and carry on with the drying. If the amount of food—small in size or cut small—is kept to a maximum of 1 pound, and the water is at least 1 gallon and boiling its head off, there is virtually no lag between immersing the food and the return to a full rolling boil; so 30 seconds is feasible. At altitudes higher than 3500 ft/305 m, add boiling time to total 30 seconds for each additional 1000 ft/305 m.

Some people use food-grade pickling lime—calcium hydroxide—in boiling water to check the skins. Use 1 gallon of water, have only 1 pound of food; make up the solution according to the instructions on the package for firming pickles.

Steam-blanching before Drying

On the whole, vegetables to be dried are blanched in full steam at 212 F/ 100 C for longer time than they are blanched, either in steam or boiling water, before being frozen. The length of blanching time is given for each vegetable in the individual instructions, *as is a recommendation for high-altitude blanching.*

Put several inches of water in a large kettle that has a close-fitting lid; heat the water to boiling, and set over it—high enough to keep clear of the water—a rack or wire basket holding a layer of cut food not more than 2 inches deep. Cover, and let the food steam for half the time required; then test it to make sure that each piece is reached by steam. A sample from the center of the layer should be wilted and feel soft and heated through when it has been blanched enough.

In a pinch you can use a cheesecloth bag, skimpily loaded with food, and placed on the rack to steam. Be careful not to bunch the food so much that steam can't get at all of it easily.

Remove the food and spread it on paper toweling or clean cloths to remove the excess moisture while you steam the next load; lay toweling over it while it waits for further treatment or to go on the drying trays.

Microwave note for blanching: follow your oven's manual on blanching vegetables before freezing them.

Sulfuring

For many years sulfur has been used to preserve the color of drying fruits whose flesh darkens when exposed to air. The fruits generally treated with sulfur have been apples, apricots, nectarines, peaches, and pears; light-fleshed varieties of cherries, figs, plums, and prunes have also been treated with sulfur to prevent oxidation, though not so routinely.

In one method, the sulfur was applied by soaking produce in a sulfur solution. Such treatment is now banned as a preservative for fresh produce sold in supermarkets or offered at salad bars in restaurants.

Sulfur first melts—at around 240 F/116 C—becoming a brown goo before it ignites and burns with a clear blue flame that produces the acrid sulfur dioxide that penetrates evenly and is easy to judge the effect of. The usual amount to use is 1 level teaspoon burned for each 1 pound of prepared fruit.

Local drugstores had several kinds of dry sulfur, but we chose the "sublimed" variety—99½ percent pure, to be taken internally mixed with molasses (the classic folk tonic); it's a soft yellow powder with no taste and the faintest of scents that's nothing like the rotten-egg odor of hydrogen sulfide. The 2-ounce box was enough to do 16 to 18 pounds of prepared fruit.

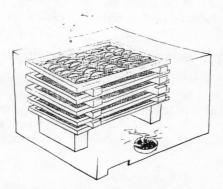

Sulfuring box in action: air-intake notch at bottom; small chimney hole at top to make a draft. The dish contains 1 teaspoon of powdered pure sulfur for each pound of prepared fruit. (*Drawing by Norman Rogers*)

How long to sulfur? Specific times are given in the instructions for individual fruits. Start to count sulfuring time *after* the sulfur has finished burning, which will take about 15 minutes. With the sulfuring box made airtight, you simply leave it inverted over the stacked trays for the required period to allow the sulfur dioxide fumes to reach all surfaces of the food.

For a sulfuring box you can use a stout, large carton of the sort that household appliances are shipped in. The box should be tall enough to cover an adequately spaced stack of up to 6 trays, and be about 12 inches longer than the trays from front to back so there'll be room for the sulfuring dish *beside* the stack. (See drawing.)

To unload the sulfur box, remove the trays from the top, being careful not to spill any juice that has collected in the hollows of the fruit.

Tests for Dryness in Produce

We rely on appearance and feel to judge dryness. Fruits generally can be considered adequately dry when no wetness can be squeezed from a piece of it when cut; and when it has become rather tough and pliable; and when a few pieces squeezed together fall apart when the pressure is released. "Leathery"—"suede-like"—"springy"—these are descriptions you'll see in the individual instructions. Several, such as figs and cherries, also are slightly "sticky."

Vegetables are generally "brittle" or "tough to brittle" when they're dry enough; an occasional one is "crisp." Again, instructions for specific vegetables will tell you what to look for.

Finally, foods still warm from the sun or hot from the dryer will seem softer, more pliable, more moist than they actually are. *So cool a test handful a few minutes before deciding it's done.*

Post-drying Treatments for Produce

Even after a sample from each tray of food has shown no moisture when cut and pressed, and feels the way its test says it should, you can't take for granted that the whole batch is uniformly dry. And especially if it's been dried outdoors do you need to get rid of any spoilers—air-borne micro-organisms or bugs you can see—that may have got to it somewhere along the line.

Conditioning

This makes sense particularly for food done in a dryer because there's often more chance of spotty results than in sun-drying, and you'll want moisture content equalized between under- and overdried pieces.

Cool the food on the trays, then pour it all into a large, open, nonporous container *that's not aluminum*—a big crock, enamel- or graniteware canner, even a washtub lined first with food-grade plastic and then with clean sheeting (wash-tubs are generally galvanized). Have the containers raised on trestles or tables, and in a warm, dry, airy, well-screened, animal-proof room.

Stir the food once a day—twice if you can manage to—for 10 days or 2 weeks, depending on the size of the pieces. It's O.K. to add freshly dried food to the conditioning batch, but naturally not if the food in a container is almost ready to store.

Fruits, usually being in larger pieces (and therefore more likely to need finishing off) than vegetables, need more conditioning time.

Pasteurizing

Pasteurizing is recommended strongly for killing insect eggs deposited on foods that have been dried in open-air/sun. It is effective as well for re-treating vegetables held in storage, although *if the vegetables show ANY SIGNS OF MOLD they should be destroyed*. Some molds produce af-latoxin, a dreaded food poisoning.

Don't bother cranking up the dryer for this, and don't do large amounts at a time: use an oven with a thermometer in it, and time the process.

Preheat the oven to 175 F/80 C. Spread the food loosely not more than 1 inch deep on the trays; don't do more than two trays' worth at the same time. Heat brittle-dried vegetables, cut small, for 10 minutes at 175 F/80 C; treat fruits—cut larger and therefore needing more time—for 15 minutes at 175 F/80 C.

Remove each pasteurized batch and spread it out to cool on clean toweling, etc. Cover lightly with cheesecloth to keep dried food clean. Package one batch while other batches are pasteurizing.

Storage

Hold your food in small quantities: 1-gallon glass jars with screwtop lids; or in 5-gallon *food-grade* freezer bags that are then stored in a metal container with a close-fitting lid. *Do not use heavy plastic trash or garbage bags,* or *plastic barrels,* or *galvanized metal barrels/trash cans* unless they are well lined with food-grade material that will not let any acid in the food come into contact with the metal.

Check your supplies on a frequent schedule, to make sure that no part of your food has become damp or contaminated. (The conditioning treat-ment is a great safeguard here; as is pasteurizing to destroy any insect eggs.)

When the dry food—conditioned well, and pasteurized if necessary—is thoroughly cool, it will go into its safe and critter-proof containers; if it is still warm, it is likely to sweat and cause trouble. Then the containers will be labeled and go to storage in a cool, dry place.

Three temperatures are pivotal in the storing of food: freezing; 48 F/9 C, when insects start to become active; and when fats melt, about 95

F/35 C. The lowest temperature short of freezing is hard to maintain, so it makes sense to consider 40 F/4 C a low easily held, and 60 F/16 C a reasonable top. Temperatures more than 70 F/21 C should be avoided.

Note: if you're dealing with only small amounts of dried food, especially cereal foods, an infestation of small bugs can be destroyed by putting the material loosely in a food-grade freezer bag much too large for it, and holding it in the freezer (at Zero F/ − 18 C) for several days.

Desiccants

For the average householder, the best way is to examine the food, repasteurizing it if necessary. Or you could use carbon dioxide—which is the "dry ice" that reduces oxygen in the stored contents and hence retards rancidity or spoilage in a container of food. The carbon dioxide is placed near the top of the contents, because, its gas being heavier than air, the fumes sink down through the food.

For controlling moisture in grain or grain products stored in large containers, a natural dessicant of proven worth is "diatomaceous earth," also known as *kieselguhr,* which absorbs up to 4 times its weight in water (moisture) without becoming fluid itself.

Silica gel is often recommended as a desiccant; it's gettable at craft-supply stores where it is bought by people who dry flowers. It seems quite expensive to us, but perhaps the cost is offset by the fact that it can be dried out in a slow oven and re-used often.

Simple, and cheaper, is to put in the bottom of a small glass storage jar a shallow layer of clean crushed concrete that has no added coloring or deodorizers: that's right, we're talking about catbox litter. It's not food-grade, so protect the food with coffee-filter paper trimmed enough larger than the circumference of the (small) jar to allow it to bend up the sides of the container for an inch or so. This paper will allow transfer of any moisture from the food down to the crushed concrete.

Fumigants

Some experts in home-storage of quite large amounts of grains, especially, use carbon tetrachloride—which we saw around in the old days to be used as a fabric cleaner and spot-remover. A few drops of this substance on a small wad of cotton is placed in the barrel of grain, etc. BE WARY OF THIS SUBSTANCE: in some states it is banned from sale to the general public. We mention it only because it was brought to *PFB*'s attention, and we cannot recommend it.

Cooking Dried Fruits

Pour boiling water over them in a saucepan *just to cover*—no more now: they shouldn't be drowned, and you can always add more if you need to—and simmer the fruit, covered, for 10 to 15 minutes, depending on the size of the pieces. Remove from heat and let cool, still covered. Sweeten to taste at the very end of cooking, or when removed from heat (sugar tends to toughen fruit fibers in cooking). For best flavor, chill the fruit overnight before serving.

If the fruit is to be "reconstituted" to use in a cooked dish (a pie or a cream dessert, say), put it in a bowl, add boiling water just to the top of the fruit; cover; and let it soak up the water for several hours, or until tender. Add water sparingly and only if the pieces seem still to be tough, because the liquid is full of good things and should be included in the recipe as if it were natural juice.

Cooking Dried Vegetables

Before being cooked, all vegetables except greens are soaked in cold water just to cover until they are nearly restored to their original texture. Never give them any more water than they can take up, and always cook them in the water they've soaked in.

Cover greens with enough boiling water to cover and simmer until tender.

DRYING FRUITS

See Pre-drying Treatments for Produce, earlier in this chapter, for proportions to use, in general, and criteria for dryness. Specifics are given with each fruit.

Apples

Best for drying are late-autumn or early-winter varieties. Peel, core, slice in ⅛-inch rings, coating slices with strong ascorbic-acid solution to hold color temporarily.

Dryer. Steam-blanch 5 minutes; press out excess moisture. Begin them at 130 F/54 C; raise gradually to 150 F/66 C after the first hour; when nearly dry, reduce to 140 F/60 C. Test dry. Condition. Package; store. Average total drying time: up to 6 hours, depending on size of slices.

Open-air/sun. If not steam-blanched for 5 minutes, sulfur for 60 minutes; if blanched, sulfur for 30 minutes. Proceed with drying. Test dry. Pasteurize. Package; store. *Solar dryer:* about 70 percent of open-air/sun time.

Room-drying. Prepare as above. Steam-blanch 5 minutes *and* sulfur 30 minutes; or sulfur only for 60 minutes. Thread on clean string, and festoon near the ceiling of a warm, dry, well-ventilated room (attic), or above a wood-burning stove in frequent use; or stack on trays with an electric fan blowing across them. *Protect against flies.* Test dry; pasteurize.

Dry test: leathery, suede-like; no moisture when cut and squeezed.

Apricots

Halve and stone. Hold against oxidizing with ascorbic-acid coating.

Dryer. Steam-blanch halves 15 minutes, slices 5 minutes. Press, start at 130 F/54 C, raise gradually after the first hour to 150 F/66 C. Reduce to 140 F/60 C for last hour or when nearly dry. Test dry. Condition; store. Average total drying time: up to 14 hours for halves, up to 6 hours for slices.

Open-air/sun. If steam-blanching as for a dryer, sulfur slices 30 minutes, halves 90 minutes. Not blanched, sulfur slices 1 hour, halves 2 hours. Remove halves carefully to drying trays so as not to spill juice in the hollows, and place cut-side up in the drying trays. Turn when all visible juice has disappeared. Test dry. Pasteurize; store. *Solar dryer:* about 70 percent of open-air/sun time.

Dry test. Leathery, pieces separate naturally after squeezing; no moisture when cut.

Fermenía's Apricots

The following method is general in high-country farms in the Rockies. *PFB* saw it practiced at Las Trampas pueblo in New Mexico (8000 ft/ 2438 m) by Sra. Fermenía López.

Wash whole, dunk in a solution of 2 tablespoons pickling salt to each 1 gallon water to hold color; tear open, pit, and press inside-out to expose greatest surface to drying.

Open-air/sun: on racks, full sun for several hours, then in roofed south-facing patio to finish. When leathery, condition (but don't pasteurize); store in airtight containers in cool place.

Fermenía presses each apricot half inside-out to expose the most surface. (*Photograph from Self Reliance Foundation*)

Berries, Firm

Check (crack) the skins of blueberries, huckleberries, currants, and cranberries, etc. by dipping for 15 to 30 seconds (depending on toughness of skin) in rapidly boiling water. Plunge into cold water. Remove excess moisture.

Dryer. Start at 120 F/49 C, increase to 130 F/54 C after one hour, then to 140 F/60 C; they will rattle on the trays when nearly dry. Keep at 140 F/60 C until dry. Test dry. Condition; store. Average total drying time: up to 4 hours.

Open-air/sun. Check the skins as above for the dryer. Remove excess moisture and put on trays one layer deep in the sun. Test dry. Pasteurize; store. *Solar dryer:* about 70 percent of open-air/sun time.

Dry test. Hard. No moisture when crushed.

Berries, Soft

There are so many better ways to use these—canned, frozen, in preserves—that there's not much use in drying them. Strawberries are especially blah and unrecognizable when dried.

Cherries

If not pitting, check their skins with a 15- to 30-second dunk in boiling water; cool.

Dryer. Press out moisture, start at 120 F/49 C for one hour, increase gradually to 145 F/63 C and hold there until nearly dry. Reduce to 135 F/57 C the last hour if danger of scorching. Test dry. Pasteurize. Cool and store. Total drying time: up to 6 hours.

Open-air/sun. Pit. Sulfur for 20 minutes. Dry. Test dry. Pasteurize. Cool and store. *Solar dryer:* about 70 percent of open-air/sun time.

Dry test. Leathery and sticky.

Figs

Small figs or ones that are partly dry on the tree may be dried whole. Large juicy figs are halved.

Dryer. Check in boiling water for 30 to 45 seconds. Cool quickly. If cut in half, steam-blanch for 20 minutes. Some people syrup-blanch whole figs before drying. To dry, start at 120 F., increase after the first hour to 145 F/63 C. When nearly dry, reduce to 130 F/54 C. Test dry. Condition. Cool and store. Total average drying time: up to 5 hours for halves.

Open-air/sun. Check the skins as above if drying whole. Sulfur light-colored varieties (like Kadota) for 1 hour before drying. If figs are to be halved, do not check the skins—instead, steam-blanch the halves for 20 minutes and then sulfur for 30 minutes. Test dry. Pasteurize. Cool and store. *Solar dryer:* about 70 percent of open-air/sun time.

Dry test. Leathery, with flesh pliable; slightly sticky to the touch, but they don't cling together after squeezing.

Grapes

Use only Thompson or other seedless varieties for drying. Check by dipping 15 to 30 seconds in boiling water and cooling immediately. Proceed as for whole Cherries.

Dryer. Proceed as for Cherries. Total average drying time: up to 8 hours.

Open-air/sun. Handle like Cherries, but don't sulfur. *Solar dryer:* about 70 percent of open-air/sun time.

Leathers (Peach, Etc.)

These sheets of pliable dried pulp may be made from virtually all fruits and berries, with peaches, apples, and wild blackberries leading the field (see also Tomatoes). The following is a general rule, so experiment with only small batches until you get the fresh, tart flavor you like. Three to 3½ cups of prepared fresh fruit will make approximately two good-sized leathers on cookie sheets—depending on the type of fruit and the size of the pieces. Added sweetening is not necessary, but helps bind the texture.

Use fully ripe fruit. Peel or not, core/stone, cut small; coat with an anti-oxidant, but the brief precooking should prevent some darkening. Measure prepared fruit, and add 1½ tablespoons sugar or honey for each 1 cup of cut fruit; an alternative is 1 tablespoon granulated fructose for each 1 cup of puréed fruit. Bring just to boiling, cook gently until tender. Remove from heat and, when the fruit is cool enough to handle, put it through a fine sieve or food mill.

Lay long sheets of foil or plastic *freezer*-wrap on wet cookie sheets—wet, so the foil/plastic will cling—allowing extra at ends and sides, and oil it well. Pour enough fruit pulp in the center of each sheet, tilting it to spread ¼ inch deep (it will dry much thinner), and to within 2 inches of the rims.

Dryer. Start at 130 F/54 C; raise to 145 F/63 C after the first hour and hold there until the surface is no longer tacky to the touch, or for 45 minutes. When nearly dry, reduce heat to 135 F/57 C. Test dry. Cool.

Drying time depends on juiciness of the fruit—usually about 2 to 3 hours. To make heavier leather, spread fresh pulp thinly on a layer that has lost all tackiness: building up on a nearly dry layer is better than working with a too-thick original layer.

To store, leave each sheet of leather on the plastic wrap on which it was dried and roll it up, tucking in the sides of the wrap as you go along. Overwrap each roll for further protection against moisture. Refrigerate until used, up to a couple of months; freeze for long-term storage.

Open-air/sun. Cover from dust and insects with cheesecloth held several inches away from the fresh fruit pulp, and place in direct sun. Bring inside at night. Protective cover can be left off when the leather is no longer tacky to the touch. Finish with a pasteurizing treatment at 145 F/ 63 C for 30 minutes. Total sun-drying time about 24 hours.

Dry test. Pliable and leathery, stretches slightly when torn; surface slick, with no drag when rubbed lightly with the fingertips.

Nectarines

Treat as for Apricots.

Peaches

Yellow-fleshed freestone varieties are the best for home-drying. Commercially dried peaches are halved, and seldom peeled. (For home-drying slices, however, peel.) Halve and stone the fruit; leave in halves or cut in slices. Scoop out any red pigment in the cavity (it darkens greatly during drying). Treat slices or halves with ascorbic-acid coat as you go along to hold color temporarily.

Dryer. Steam-blanch slices 8 minutes, unpeeled halves 15 to 20 minutes. Start drying at 130 F/54 C, increase gradually to 155 F/68 C after the first hour. Turn over halves when all visible juice has disappeared. Reduce to 140 F/60 C when nearly dry to prevent scorching. Average total drying time: up to 15 hours for halves and up to 6 hours for slices.

Open-air/sun. Prepare as for the dryer. If steam-blanching slices and halves as above, sulfur slices 30 minutes, halves for 90 minutes. If not blanched, sulfur 60 minutes and 2 hours, respectively. Be careful not to spill the juice in the hollows when transferring the halves to drying trays, where they're placed cut-side up. Proceed as for Apricots. Test dry. Pasteurize. Cool and store. *Solar dryer:* about 70 percent of open-air/sun time.

Dry test. Leathery, rather tough.

Pears

Best for drying are Bartletts, picked quite firm before they are ripe, then held at 70 F/21 C in boxes in a dry, airy place for about 1 week—when usually they're ready. Split lengthwise, remove core and woody vein, leave in halves (or slice and pare off skin). Coat cut fruit with ascorbic acid.

Dryer. Steam-blanch slices 5 minutes, halves 20 minutes. Start at 130 F/54 C, gradually increasing after the first hour to 150 F/66 C. Reduce to 140 F/60 C for last hour or when nearly dry. Test dry. Condition; cool and store. Average total drying time: up to 6 hours for slices, 15 hours for halves.

Open-air/sun. Sulfur as for Peaches; dry like Peaches. Test dry. Pasteurize; cool and store. *Solar dryer:* about 70 percent of open-air/sun time.

Dry test. Suede-like and springy. No moisture when cut and squeezed.

Plums and Prunes

Joe Carcione, network television's "Green Grocer," has pointed out that the Italian prune-plums have so much more natural sugar than other varieties that they dry well *whole* without fermenting; nor need they be pitted beforehand. Other kinds of plums should be pitted, then sliced or quartered in order to dry without spoiling. Check the skins with a 30–45-second dunk in boiling water. Cool.

Dryer. Steam-blanch 15 minutes if halved and stoned, 5 minutes if sliced. Start *slices* and *halves* at 130 F/54 C, gradually increase to 150 F/66 C after the first hour; reduce to 140 F/60 C when nearly dry. Start *whole,* checked fruit at 120 F/49 C, increase to 150 F/66 C gradually after the first hour; reduce to 140 F/60 C when nearly dry. Test dry. Condition; cool and store. Average total drying time for slices: up to 6 hours; halves, up to 8 hours; whole, up to 14.

Open-air/sun. Check the skins of whole fruit. Sulfur whole fruit for 2 hours. Sulfur slices and halves for 1 hour. Test dry. Pasteurize. Cool and store. *Solar dryer:* about 70 percent of open-air/sun time.

Dry test. Pliable, leathery. A handful will spring apart after squeezing.

DRYING VEGETABLES AND HERBS

Vegetables are partly precooked by blanching before being dried, and this step *is NOT optional:* it helps to stop enzymatic action that leads to spoilage.

Except for corn dried on the cob, all vegetables are pasteurized if their processing heat has not been high enough, or prolonged enough, to destroy spoilage organisms. Pasteurizing is particularly important for sun-dried vegetables.

Vegetables are cut smaller than fruits are, in order to shorten the drying process—for the faster the drying, the better the product. The approximate total drying times in a dryer are not given below, but they range from around 4 to 12 hours, depending on the texture and size of the pieces.

Beans—Green/Snap/String/Wax (Leather Britches)

String if necessary. Split pods of larger varieties lengthwise, so they dry faster. Steam for 15 to 20 minutes.

Dryer. Start *whole* at 120 F/49 C and increase to 150 F/66 C after the first hour; reduce to 130 F/54 C when nearly dry. For *split beans,* start at 130 F/54 C, increase to 150 F/66 C after first hour, and decrease to 130 F/54 C when nearly dry. Test. Condition; pasteurize. Cool and store.

Open-air/sun. Handle exactly as for the dryer. Test dry. Pasteurize certainly. Cool and store. *Solar dryer:* about 70 percent of open-air sun time.

Room-drying. Do not split. String through the upper ⅓ with clean string, keeping the beans about ½ inch apart. Hang in warm, dry, well-aired room. Test. Pasteurize certainly. Cool and store. (Old-timers would drape strings near the ceiling over the wood cookstove; they gave the name "leather britches" to these dried beans—probably because they take so long to cook tender.)

Dry test. Brittle.

Beans, Lima (and Shell)

Allow to become full-grown—beyond the stage you would when picking them for the table, or for freezing or canning—but before the pods are dry. Shell. Put in very shallow layers in the steaming basket and steam for 10 minutes. Spread thinly on trays.

Dryer. Start at 140 F/60 C, gradually increase to 160 F/71 C after the first hour; reduce to 130 F/54 C when nearly dry. Test dry. Condition; pasteurize. Cool and store.

Open-air/sun. Not as satisfactory for such a dense, low-acid vegetable as processing in a dryer is. However, follow preparation as for a dryer. Test dry. Condition if necessary; pasteurize certainly. Cool and store. *Solar dryer:* about 70 percent of open-air/sun time.

Dry test. Hard, brittle; break clean when broken.

Beets

Choose small beets, leave ½ inch of the tops lest they bleed; steam until cooked through—30 to 45 minutes. Cool; trim roots and crowns, peel. Slice crossways only ⅛ thick, OR shred on coarse knife of a vegetable grater: shredded dries more quickly but the cooking use is more limited.

Dryer. Put slices in at 120 F/49 C and increase to 150 F/66 C after first hour; reduce to 130 F/54 C when nearly dry. Put finer shreds in at 130 F/54 C. Increase gradually to 150 F/66 C after first hour; turn down to 140 F/60 C when nearly dry. Test dry. Condition; pasteurize. Cool and store.

Open-air/sun. Prepare as for dryer, but shreds are recommended here—better because they're faster. Test dry. Condition if necessary; pasteurize certainly. Cool and store. *Solar dryer:* about 70 percent of open-air/sun time.

Dry test. Slices very tough, but can be bent; shreds are brittle.

Broccoli

Trim and cut as for serving. Cut thin stalks lengthwise in quarters; split thicker stalks in eighths. Steam 8 minutes for thin pieces, 12 minutes for thicker.

Dryer. Start at 120 F/49 C, gradually increasing to 150 F/66 C after the first hour; reduce to 140 F/60 C when nearly dry. Test dry. Condition; pasteurize. Cool and store.

Open-air/sun. Prepare as for dryer. Test dry. Condition if necessary; pasteurize certainly. Cool and store. *Solar dryer:* about 70 percent of open-air/sun time.

Dry test. Brittle.

Cabbage

Drying is least feasible, but a little cabbage can be handy for soup. Storage life is short. Remove outer leaves; Quarter, cut out core, shred coarsely. Steam 8 to 10 minutes. All leaf vegetables mat on the trays during drying, so spread evenly and no more than ½ inch deep; dry half the weight per batch as for other foods.

Dryer. Start at 120 F/49 C, increase gradually to 140 F/60 C after the first hour; reduce to 130 F/54 C when nearly dry; thin part of leaves may scorch. Stir food often to prevent matting. Test dry. Condition if necessary; pasteurize. Cool, store.

Carrots

Choose crisp, tender carrots with no woodiness. Leave on ½ inch of the tops. Steam until cooked through but not mushy—about 20 to 30 minutes, depending on the size. Trim off tails, crowns with tops, and any whiskers. Cut in ⅛-inch rings, or shred.

Dryer. Proceed as for Beets, either sliced or shredded. Test dry. Condition; pasteurize. Cool and store.

Open-air/sun. Proceed as for Beets, either sliced or shredded. Test dry. Condition if necessary; pasteurize certainly. Cool and store. *Solar dryer:* about 70 percent of open-air/sun time.

Dry test. Slices very tough and leathery, but will bend; shreds are brittle.

Celery

For drying *leaves,* see Herbs. Split outer stalks lengthwise, leave small center ones whole; trim off leaves to dry as herb seasoning. Cut stalks across no larger than ¼-inch pieces. Steam 4 minutes.

Dryer. Start at 130 F/54 C, increase to 150 F/66 C after the first hour; reduce to 130 F/54 C when nearly dry. Test dry. Condition; pasteurize, because the maximum heat may not be long enough to stop spoilers. Cool and store.

Open-air/sun. Prepare as for the dryer. Test dry. Pasteurize certainly. Cool and store. *Solar dryer:* about 70 percent of open-air/sun time.

Dry test. Brittle chips.

Corn-on-the-Cob

Use popcorn and flint varieties for this. (Flint corn was the food-grain of the Colonists, who were taught by the Indians to use it. It is different from "dent" corn, which shrinks as it dries.) The kernels of both flint and popcorn remain plump when hard and dry.

These varieties are allowed to mature in the field and become partly dry in the husk on the stalks. Both are usually air-dried in the husk. However, in some hot countries the husks are peeled back from the partly dried ears and braided together or tied together.

Dry test for popcorn: rub off a little and pop it. If the result's satisfactory, then immediately put it into moisture/vapor-proof containers with tight closures, to prevent it from getting too dry to pop (the remaining moisture in the kernel is what makes it explode in heat).

Dry test for flint corn: brittle—it cracks when you whack it. Store in sound air- and moisture-proof barrels; but if you must hold it in large cloth bags, invert the bags every few weeks: this prevents any moisture from collecting on the bag where it touches the floor.

Corn, Parched

Correctly dried sweet corn is more than a stop-gap for the many people who consider it superior in flavor to canned corn. Any variety of sweet corn will do. Gather in the milk stage as if it were going straight to the table. Husk. Steam it on the cob for 15 minutes for more mature ears, 20 minutes for quite immature ears (the younger it is, the longer it takes to set the milk). It's a good idea to separate the corn into lots with older/larger and younger/smaller kernels so you can handle them uniformly. When cool enough to handle, cut it from the cob as for canning or freezing whole-kernel corn. Don't worry about the glumes and bits of silk: these are easily sifted out after the kernels are dry.

Dryer. Spread shallow on the trays. Start at 140 F/60 C; raise to 165 F/74 C gradually after the first hour; reduce to 140 F/60 C when nearly dry, or for the last hour. Stir frequently to keep it from lumping together as it dries. Test dry. Condition. (Pasteurizing is not necessary following processing in a dryer *if the temperature has been held as high as 165 F/74 C for an hour.*) The silk and glumes will separate to the bottom of the conditioning container; but if you don't condition, shake several cupfuls at a time in a colander whose holes are large enough to let glumes and silk through. Best stored in moisture/vapor-proof containers in small amounts.

Open-air/sun. Prepare exactly as for the dryer. Stir frequently to avoid lumping. Pasteurize certainly. Shake free of glumes and silk. Package and store. *Solar dryer:* about 70 percent of open-air/sun time.

To cook. Rinse in cold water, drain; cover with fresh cold water and let stand overnight. Add water to cover, salt to taste, and boil gently until kernels are tender—about 30 minutes—stirring often and adding a bit more water as needed to keep from scorching. Drain off excess water, season with cream, butter, pepper.

Dry test. Brittle, glassy, and semi-transparent; a piece cracks clean when broken.

Garlic

Treat it like onions if you *must* dry it.

Herbs

This category includes celery leaves as well as the greenery from all aromatic herbs—basil, parsley, sage, tarragon: whatever you like.

All such seasonings are *air-dried* at temperatures never more than 100 F/38 C (higher, and they lose the oils we value for flavor); and as much light as possible should be excluded during the process. Also, see your microwave manual.

Gather on a sunny morning, take only plants that have started to bloom. Cut with plenty of stem, strip tough leaves from lower 6 inches. Wash stalks with leaves in clear water, let drain on paper toweling.

Bag-drying. Collect 6 to 12 stems loosely together, and over the bunched leaves put a commodius brown-paper bag—one large enough so the herbs will not touch the sides. Tie the mouth of the bag loosely around the stems 2 inches from their ends, and hang the whole business high up in a warm, dry, airy room. When the leaves have become brittle, knock them from the stems and package in airtight containers and store away from light.

Dry test. Readily crumbled. Rub to pulverize.

Mixed Vegetables

These are never dried in combination: drying times and temperatures vary too much between types of vegetables. Dry vegetables and seasoning separately, *then* combine them in small packets to suit your taste and future use.

Mushrooms

Only young, unbruised, absolutely fresh mushrooms should be dried. Preferably wipe clean with damp cloth. Remove stems, slice caps in ⅛-

inch strips—cut stems across in ⅛-inch rings—and treat them with ascorbic-acid coating as you work. Steam for 12 to 15 minutes.

Dryer. Start at 130 F/54 C, increase gradually to 150 F/66 C after the first hour; reduce to 140 F/60 C when nearly dry. Test dry. Condition; pasteurize. Cool and store.

Sliced caps and stems process at the same temperature sequences, but stem pieces usually take longer.

Open-air/sun. Prepare as for the dryer. Test dry. Condition if necessary; pasteurize certainly. Cool and store. *Solar dryer:* about 70 percent of open-air/sun time.

Dry test. Brittle.

Onions

Dried onions do not hold so long as other vegetables do in storage (they are like carrots and cabbage, above). Peel; slice carefully to uniform thickness. No steaming needed.

Dryer. Put them in at 140 F/60 C and keep them there until nearly dry, watching carefully that thinner pieces are not browning. Reduce to 130 F/54 C for the last hour if necessary. Test dry. Condition. Cool and store.

Open-air/sun. Prepare as for the dryer. Test dry. Pasteurize. Cool and store. *Solar dryer:* about 70 percent of open-air/sun time.

Dry test. Light-colored, but brittle.

Peas, Black-eyed

Treat like Beans (Shell), above.

Peas, Green

Choose young, tender peas as you'd serve them fresh from the garden. From there on, treat them like Shell Beans, above.

Dry test. Shriveled and hard; shatter when hit with a hammer.

Peppers, Hot (Chili)

Choose mature, dark-red pods. Thread them on a string through the stalks, and hang them in the sun on a south wall. When dry, the pods will be shrunken, dark, and may be bent without snapping.

Peppers, Sweet (Green or Bell)

Split, core, remove seeds; quarter. Steam 10 to 12 minutes.

Dryer. Start at 120 F/49 C, gradually increase to 150 F/66 C after the first hour; reduce to 140 F/60 C when nearly dry (if any are thin-walled, reduce to 130 F/54 C toward the end, and keep stirring them well). Test dry. Condition. Cool and store.

Open-air/sun. Prepare as for the dryer. Test dry. Condition if necessary; pasteurize certainly. Cool and store. *Solar dryer:* about 70 percent of open-air/sun time.

Dry test. Crisp and brittle.

Potatoes, Sweet (and Yams)

Only firm, smooth sweet potatoes or yams should be used. Steam whole and unpeeled until cooked through but not mushy, about 30 to 40 minutes. Trim, peel; cut in ⅛-inch slices, or shred.

Dryer. Proceed as for sliced or shredded Beets. Test dry. Condition; pasteurize. Cool and store.

Open-air/sun. Prepare as for dryer. Test dry. Condition if necessary; pasteurize certainly. Cool and store. *Solar dryer:* about 70 percent of open-air/sun time.

Dry test. Slices extremely leathery, not pliable; shreds, brittle.

Potatoes, White ("Irish")

These root-cellar too well to bother drying. But dry like Turnips, below.

Pumpkin

Deep-orange varieties with thick, solid flesh make the best product. There's not much use in drying in chunks, because they're to be mashed after cooking. Take them directly from the garden (they shouldn't be conditioned as for root-cellaring). Split in half, then cut in manageable pieces for peeling and removing seeds and all pith. Shred with the coarse blade of a vegetable grater (less than ⅛ inch thick). In shallow layers in the basket, steam for 6 minutes.

Dryer. Proceed as for shredded Beets, above. Test dry. Condition; pasteurize if length of maximum processing heat isn't enough to stop spoilers. Cool and store.

Open-air/sun drying. Prepare as for the dryer. Test dry. Pasteurize certainly. Cool and store. *Solar dryer:* about 70 percent of open-air/sun time.

Dry test. Brittle chips.

Rutabagas and Turnips

Seldom dried, but treat like shredded Carrots.

Spinach (and Other Greens)

Use only young, tender, crisp leaves. Place loosely in the steaming basket and steam for 4 to 6 minutes, or until well wilted. Remove coarse midribs; cut larger leaves in half. Spread sparsely on drying trays, keeping overlaps to a minimum.

Dryer. Start at 140 F/60 C, increase to 150 F/66 C after the first hour; if necessary, reduce to 140 F/60 C when nearly dry, to avoid browning. Test dry. Condition. Cool and store.

Open-air/sun drying. Prepare as for the dryer. Test dry. Pasteurize certainly. Cool and store. *Solar dryer:* about 70 percent of open-air/sun time.

Dry test. Easily crumbled.

Squash (all Varieties)

Treat like Pumpkin.

Drying Tomatoes

The newest commercially put-by food to reach celebrity status at this writing is the imported sun-dried tomato—a dark-red morsel usually salted, tough, and expensive. Almost always from Italy, where it is used much as North American cooks use their home-canned tomatoes, it is the plum/pasta/Roma type, the chunky little oblong without much juice but mighty in flavor. Since the mid-eighties it has superseded the classic "canner" in our catalogs.

To reconstitute unpeeled salted halves: cover with hot water, let stand until soft and plumped. If the water is not too salty, cook them in it for

Home-dried tomatoes (*right*) are plumper, with brighter color. (*Photograph by Jeffery V. Baird*)

sauces and soups, etc. To hold for snipping—used like pimientos or olives for a garnish—remove from soak-water, rinse if you like, pat dry, and put in a storage jar with olive oil to cover.

For Esther Swift's version: peeled and unsalted, they are more pliable and need no freshening. She stores them in small bags in the freezer; they are thawed and eaten out-of-hand as a treat; or used as above.

Leather (made as with Apricots) is cut small, soaked and simmered. For sauce, titivate with onion, garlic, peppers, as liked.

Italian Style, Unpeeled

Wash well, halve lengthwise; remove stem-base and heavy midrib. Salt to remove moisture from tissues: spread flattened halves on a platter, cut-side up, sprinkle 1 teaspoon canning salt for each 1 pound of tomatoes; stack several layers, weight with a plate for an hour. Steam-blanch 4 minutes.

Dryer. Treat like Apricots, condensing time to half, about 4–5 hours.

Open-air/sun. Treat like Apricots; time will be nearly halved.

Dry test. Pliable but not soft. (Store-bought ones are quite tough, but they have traveled far.)

Esther Swift's Style

She dries hers held above a wood-burning stove on racks covered with nylon net (easier to clean than cotton, etc.), halves set cut-side up until the centers are no longer juicy, then she turns them over. Drying time: up to 2 days, less if stove is banked for all night.

Dryer. If using stainless steel netting, treat it with a nonstick cooking spray. Set dryer for 140 F/60 C; when centers are no longer tacky, lower to 130 F/54 C at the ⅔-done test. Total time: about 4½ hours.

Dry test. Springy when squeezed, softer than unpeeled style.

DRYING MEAT AND FISH

We shan't give blow-by-blow instructions for making jerky as the Mountain Men did, or drying codfish with the expertise of a Newfoundland native. Here are the basic steps. Work in small batches, with complete sanitation; don't cut corners. Refrigerate or freeze the finished product: high-protein foods like these invite spoilage.

Using Dried Meat and Dried Fish

Jerky traditionally was shaved off (or gnawed off) and eaten as is, because it was a staple for overland wanderers who were traveling light and far from assured supplies of fresh meat. (Helpful ins-and-outs of concentrated journey food are to be found in Horace Kephart's *Camping & Woodcraft;* see "pemmican" especially.) Today many versions of it appear in stick form as snacks, for either the Long Trail or a cocktail party.

Dried salt fish—the type described below—is always freshened by soaking beforehand, either in cold water or fresh milk; the soaking liquid is discarded here, because of the extremely high salt content. Such fish were standard fare even in the hinterlands far from salt water.

Drying Meat (Jerky)

Jerked meat is roughly ¼ the weight of its fresh raw state.

Preferred meats for jerking are mature beef and venison (elk is too fatty), and only the lean muscle is used. Partially freeze meat if possible to make slicing easier. Cut lengthwise of the grain in strips as long as possible, 1 inch wide and ½ inch thick.

Dry test. Brittle, as a green stick: it won't snap clean, as a dry stick does. Be sure to test it *after* it cools, because it's pliable when still warm, even though enough moisture is out of it.

Unsalted Jerky

This does not mean unseasoned—there's a bit of salt for flavor—but the meat is not salted heavily to draw out moisture or to act mildly as a preservative.

Lay cut strips on a cutting-board, and with a blunt-rimmed saucer or a meat mallet, pound the following seasonings (or your own variations thereof) into both sides of the meat: salt, pepper, garlic powder, your favorite herb. Use not more than 1 teaspoon salt for each 1 pound of fresh meat, and the other seasonings according to your taste.

Arrange seasoned strips ½ inch apart on wire racks treated with nonstick cooking spray. Put them in a preheated 150 F/66 C oven, and immediately turn the heat back to 120 F/49 C. Spread aluminum foil on the bottom of the oven to catch drippings. If your oven is not vented, leave its door ajar at the first stop position. After 5 or 6 hours turn the strips over; continue drying at the same temperature for 4 hours more, when you check for dryness. When dry enough, jerky is shriveled and black, and is brittle when cooled.

Wrap the sticks of jerky in moisture/vapor-proof material, put the packages in a stout container with a close-fitting lid, and store below 40 F/4 C in the refrigerator (or freeze it). Reconstitute by simmering in water to cover.

Salted Jerky

Dry this in the sun; or, if you're emulating the frontiersmen, hang them 4 feet above a very slow, non-smoking fire that's not much more than a bed of coals.

Prepare a brine of 2½ cups of pickling salt for 3 quarts of water, and in it soak the cut strips of meat for 1 or 2 days. Remove and wipe dry.

OVER COALS

Before you're ready to begin drying the salted meat, start a fire of hardwood and let it burn down to coals. Feed the fire with small hardwood so carefully that juice does not ooze out from the excess heat, or the meat starts to cook.

Depending on conditions, drying could take 24 hours. Test for dryness; package and store in refrigerator or freezer.

IN OPEN AIR

A method used on old-time hunting trips deep into the High Plains. Choose a time when you'll have good—but not roasting—sun, dry air day and night, and a gentle breeze. Hang the salted strips from a drying frame such as described above (of course with no fire), and leave them there until they become brittle-dry.

A Basic Procedure for Drying Fish

Drying fish at home is not something to be undertaken lightly. The fish must undergo a long dry-salting period before it is put out to dry; and, since home-drying is best done outdoors in the shade, the procedure requires a trustworthy breeze, fairly low humidity, and critter-proof holding tubs and racks.

Dry any *lean* fish (cod of course is the classic). Coat all surfaces of each fish with pure pickling salt, using 1 pound of salt for 2 pounds of fish, and stack the opened fish flesh-side up on a slatted wooden rack outdoors. Don't make the stacks more than 12 layers deep, with the top layer skin-side up. Leave them stacked from 1 to 2 weeks, depending on the height of fish and the dryness of the air. Brine made by the salt and fish juices will drain away. Move the pile inside each night and weight it down to press out more brine.

Scrub the fish again to remove the salt, and put them on wooden frames outdoors to complete the necessary removal of moisture from their tissues. Hang or spread the fish on cross-pieces in an open shed with good ventilation; direct sun on the fish can start it to sunburn (cook) at only 75 F/24 C. Bring fish in at night, re-piling to ensure even drying; re-spread on the racks more often with skin-side up.

To store, cut in manageable chunks if the fish are large; wrap in moisture/vapor-proof plastic; pack in tight wooden boxes, and store in a dry, cool place 32–40 F (Zero–4 C).

Dry test. No imprint is left when the fleshy part of a fish is pinched between thumb and forefinger.

22
ROOT-CELLARING

Of all the time-tested ways of putting food by, only wintering-over in cold storage at home is less satisfactory today than it was a century or more ago. The reason is simple. All the technological advances we're so pleased with in construction and heating have given us cozy, dry basements instead of cool, damp cellars, and the chilly shed off the pantry has given way to a warm passageway between carport and kitchen.

Therefore, this section is telling how to re-create conditions that several generations of North Americans have devoted themselves to improving. It includes some indoor areas that are warmer and drier than the traditional outbuilding or cellar with stone walls and a packed earthen floor, and it also includes some arrangements outdoors that are a good deal more rough-and-ready.

To root-cellar is to store for the winter a variety of fresh, whole, raw vegetables and fruits that have not been processed in any way to increase their keeping qualities. This means that such foods must be held for use during the winter—and some even longer, into the next growing season—without being subjected to an unnatural amount of heat or of cold or of dryness.

HOW IT WORKS

Used commonly, root-cellaring means to hold these foods for several months after their normal harvest in a cold, rather moist atmosphere that will not allow them to freeze or to complete their natural cycle to decomposition.

The freezing points and warmth tolerances of produce vary. The range to shoot for generally, though, is 32–40 F/Zero–4 C—the effective span for refrigeration—with only a couple of vegetables needing warmer storage to keep their texture over the months. In this range the growth of spoilage micro-organisms and the rate of enzymatic action (which causes overripening and eventual rotting) are slowed down a great deal.

Good home root-cellaring involves some control of the amount of air the produce is exposed to, since winter air is often let in to keep the temperature down. But fresh whole fruits and vegetables respire after they're harvested (some more than others: apples seem almost to *pant* in storage), so the breathing of many types is reduced by layering them with clean dry leaves, sand, moss, earth, etc., or even by wrapping each individually in paper. These measures of course aren't as effective as those of commercial refrigerated storage, which rely in part on drastic reduction of the oxygen in the air supply, but they work well enough for the more limited results expected from home methods. We'll be describing a variety of storage arrangements in a minute.

The beauty of root-cellaring is that it deals only with whole vegetables and fruits and there are no hidden dangers: if it doesn't work, we know by looking and touching and smelling that the stuff has spoiled, and we don't eat it. On the other hand, it's something that sounds a lot more feasible than it may really turn out to be.

First, the householder must learn something about the idiosyncrasies of the fruits and vegetables he plans to store on a fairly large scale: for example, apples and potatoes—the most popular things to carry over through winter—can't be stored near each other, and the odor of turnips and cabbages in the basement can penetrate up into the living quarters, and squashes want to be warmer than carrots do.

Then he casts around for the right sort of storage. And the solution may cost more than its value to his over-all food program, especially if it's a structure more elaborate or permanent than the family's make-up warrants. But aren't there the less pretentious outdoor pits, or the more casual barrels sunk in the face of a bank? Yes; and they're fun to use— except in deep-snow country when they can be a worry to get at.

The late Samuel Ogden of Landgrove, Vermont, organic gardener and noted Green Mountain countryman, warned the newcomer to cold-climate root-cellaring to avoid three things: (1) counting too heavily on cold storage; (2) having too much diversity; (3) and having the food inaccessible in bad weather.

Equipment for Root-Cellaring

Storage place, indoor or outdoor.
Clean wooden boxes/lugs/crates or barrels; or stout large cardboard cartons (for produce that wants to be dry, not damp).

Plenty of clean paper for wrapping individually, or shredding.
Plenty of clean dry leaves, sphagnum, peat moss, or sand.
A tub of sand to keep moistened to provide extra humidity if needed.
Simple wall thermometer certainly; humidity gauge (optional).

INDOOR STORAGE

The Classic Root Cellar Downstairs

There are fewer of these to be found as the years go by, even in the old houses in our part of the country. Usually in the corner of the original cellar-hole, they have two outside walls of masonry (part of the foundation), the floor is packed earth, and any partitions are designed more to support shelving than to keep out warmth from a nonexistent furnace. They incorporate at least one of the small windows that provide cross-ventilation for the whole cellar to keep overhead floor joists from rotting; propped open occasionally during the winter, it's the answer for regulating temperature and humidity.

Darken the window(s)—potatoes turn green in light when they're stored, and this isn't good. If necessary, keep clearing snow from the areaway that's below ground leading to the window. And check the whole thing for places where field mice can get in and feast on your crops during the lean winter months, and stop them up.

Using a Bulkhead

Many middle-aged houses have an outside entrance to the cellar: a flight of concrete steps down to the cellar wall, in which a wide door is hung to give access to the cellar. The top entrance to the steps—the hatch— is a door laid at an angle 45 degrees to the ground.

On the stairway, which probably is closed from the outside during the coldest months anyway, you can store barrels/boxes of produce. You could put up rough temporary wooden side walls along the steps; but certainly lay planks on the steps to set your containers on. Insulate the door into the cellar proper with glass batts. If you need to, keep a pail of dampened sand on one of the steps to add humidity. You're likely to be propping the hatchway door open a few inches from time to time to help maintain proper temperature on the steps. This means shoveling snow from the bulkhead, and piling it back on when this outside door is closed again.

A Dry Shed

This takes the place of the garage advocated by some people—but not by us: too much oil and gasoline odor (some produce soaks stray odors up like a sponge), and far too great a quantity of lead-filled emissions from running motors. And anyway, temperature is often uncontrollable.

Instead of using the garage, partition off storage space in the wood-floored shed leading into the kitchen, if you have an old house in the country. Or segregate a storage area in a cold, seldom used passageway.

Storage areas like these are usually not fit for such long-term storage as the basement root cellar or store room is.

Up Attic

An old-fashioned attic generally is the last place in the house to cool off naturally as cold weather sets in; and unless the roof is well insulated, the attic temperature rises on sunny winter days. This fluctuation doesn't matter much for some foods, however (see the chart), though it does for onions, say. The answer is to wall-off, and ceil, a northeast corner for anything that needs maintained low temperature and dryness. Then you can put pumpkins and such near the stairway leading to the attic—and leave the hall door downstairs open whenever you need to.

In Picnic Chests/Hampers

Harvest carrots, beets, turnips, etc., late in the fall. The handling described for carrots works well for the others too.

Cut tops off carrots, leaving about ½ inch of stem. Wash away garden soil, then wipe fairly dry—"fairly dry" because you want a little moisture; but not wet, lest the vegetables mold. Sack them 4 to 5 pounds at a crack in the largest *food-grade* freezer bags (see Using Space-Age Plastics in Chapter 4, for the warning against using trash or garbage bags in direct contact with food). Press out excess air from each filled bag, twist the top and tie it tight with string, rubber bands, coated wire.

Pack the bags in a polystyrene chest of the inexpensive sort you use to carry picnic food on ice, and store the chest in any cold spot, like an unheated roughly walled-off corner of the cellar or an enclosed sunless porch. Keep the lid tightly on the chest. For phasing out of such plastics, see the note for Chapter 22 in the Appendix.

Late Apples in Small Metal Barrels

To use small metal trash barrels, you'll need to provide some insulation so they will equal the efficiency of milk cans (which are becoming "collectibles" for nostalgia buffs, and so are fairly scarce). Put 3 inches of dry sawdust in the bottom of the barrel and pack 1 inch of sawdust between the outside apples and the metal sides. Electrician's tape or other heavy-duty plastic self-adhesive strips may be wound tightly around the upper edge of the barrel, just below the rim, to create friction with the cover and ensure that it fits as smoothly tight as a milk can's lid does.

SMALL-SCALE OUTDOOR STORAGE

There's a good deal of information around that contains ideas for full-dress outdoor buildings for root-cellaring. Of these we suggest the USDA *Home and Garden Bulletin No. 119, Storing Vegetables and Fruits in Basements, Cellars, Outbuildings, and Pits*—available from your County Agent. He can steer you to the most recent USDA and Commerce publications; don't forget your public library's catalog, or *Readers' Guide to Periodical Literature*. As the practicality of homesteading has been developed, so has grown the amount of material published on virtually every phase of preserving crops.

Some Mild-Climate Pits, Etc.

The USDA bulletin and the other sources describe several easy-to-make and cheap outdoor storage facilities, all either on well-drained ground or sunk only several inches below the surface. See the chart for which produce likes the conditions they provide.

However, such arrangements can be counted on *only in places with fairly mild winters* that have no great extremes in temperature. At any rate, make a number of small storage places, fill them with only one type of produce to each space, and be prepared to bring the entire contents of a store-place indoors for short storage once the space is opened.

Mild-Climate Cone "Pits"

Most instructions call a storage place like this a pit, but it's really a conical mound above ground. To make it, lay down a bed of straw or leaves, etc.; pile the vegetables *or* fruits (don't mix them together) on the bedding; cover the pile well with a layer of the bedding material. With a shovel, pat earth on the straw/leaf layer to hold it down, extending a "chimney" of the straw to what will be the top of the cone to help ventilate and control the humidity of the innards of the mound. Use a piece of board

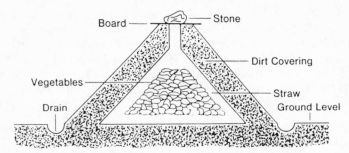

A mild-climate cone "pit." (*Drawing by Irving Perkins Associates*)

weighted by a stone to act as a cap for the ventilator. Surround the "pit" with a small ditch that drains away surface water.

As colder weather comes, add to the protective layer of earth, even finishing with a layer of coarse manure in January.

Mild-Climate Covered Barrel

Still called a "pit" is a barrel laid on its side on an insulating bed of straw, chopped cornstalks, leaves, etc. Put only one type of produce in the barrel on bedding of straw/leaves. Prop a cover over the mouth of the barrel; cover all with a layer of straw, etc., and earth on top to hold it down.

Outdoor Frame for Celery, Etc.

Dig a trench 1 foot deep and 2 feet wide, and long enough to hold all the celery you plan to store. Pull the celery, leaving soil on the roots, and promptly pack the clumps upright in ranks 3 to 5 plants wide. Water the roots as you range the plants in the trench. Leave the trench open until the tops dry out, then cover it with a slanted roof. This you make by setting on edge a 12-inch board along one side of the trench, to act as an upper support for the cross-piece of board, etc., that you lay athwart the trench. Cover this pitched roof with straw and earth.

Walter Needham's Cold-Climate Pit

As he told in *A Book of Country Things,* he was raised in rural Vermont by a grandfather for whom a candle-mold was a labor-saving device. So whenever we want to learn about totally practical methods of pioneer living in the cold country, we turn to Walter Needham. He was the first to point out that the conical "pit" wouldn't do an adequate job in 20-below Zero Fahrenheit weather. This is his alternative:

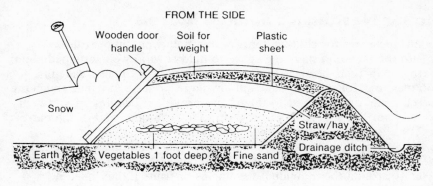

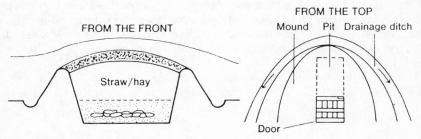

Cold-climate storage in a pit. (*Drawing by Irving Perkins Associates*)

Choose the place for your pit on a rise of ground to avoid seepage. There, shovel out a pit about 1½ to 2 feet deep and 4 feet wide at the bottom, throwing dirt up all around to build a rim that will turn water away; dig a V-shaped drainage ditch around it for extra protection (see sketches). Take out any stones near the sides of the pit because frost will carry from one stone to another in rocky ground. The pit needn't go below the deep frost-line if such frost conductors are removed. Pack the bottom of the pit with dry mortar sand 2 to 3 inches deep: the loam, having retained moisture, will freeze; the sand holds the food away from the loam.

On the layer of sand make a layer of vegetables not more than 1 foot deep; cover the vegetables with more fine sand, dribbling it in the crevices, to fill the pit nearly to ground level. Cover the sand with straw or Nature-dried hardwood leaves, or mulch hay (hay that got rained on before it cured), mounded to shed the weather. Hold down this cover with a thin layer of sod—or, nowadays, plastic sheeting weighted down with 1 to 2 inches of earth. Cover one end of the mound with a door laid on its side and slanted back almost like a bulkhead entrance. In winter you'll move the door away to dig in for the vegetables, and, as they're taken out, move the door back along the mound.

This root-pit is best for beets, carrots, turnips, and potatoes.

Walter Needham's Sunken Barrels

Again, these are for cold-winter areas with uneven temperature.

Into the face of a bank dig space to hold several well-scrubbed metal barrels with their heads removed—one barrel for apples, say; one for potatoes, one for turnips. Take out any large stones that would touch the barrels and conduct frost to them, and provide a bedding of straw/dry leaves, etc., for the barrels to rest on. Slant the open end of the barrels slightly downward, so water will tend to run out.

Put straw or whatever in the barrels for the produce to lie on, and fill the barrels from back to front, using dry leaves or similar material to pack casually around the individual vegetables or fruits if they need it.

Over the opening put a snug cover propped against it—a stout wooden "door" with a *wooden* handle (did you ever have the skin of your palm freeze on to metal in bitterly cold weather?). Dig a shallow V-shaped drainage ditch to carry surface water away from the barrels (as in the sketch).

The snow will be added protection in the deepest cold of the winter. Shovel it back against the door after removing food from the barrels.

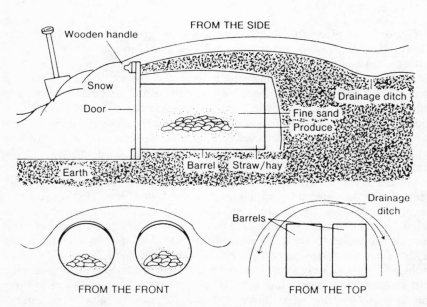

Cold-climate storage in sunken barrels. (*Drawing by Irving Perkins Associates*)

How to Ripen Fruit

Apples keep longer if they don't touch each other, and apples and potatoes should never be stored close together. *Reason:* apples respire more than most fruits do; they seem to give off extra amounts of the gas ethylene—along with other gases—and this peculiarity allows them to help pears, peaches, and tomatoes to ripen. Use a heavy brown kraft-paper bag from the supermarket, punch about half a dozen small holes in it (to let some of the gases escape); put pears, peaches, or tomatoes in the bag without crowding, *and include one sound ripe apple.* Bend over the top of the bag several times and hold it with a paperclip; set the bag on a shelf in the pantry where it will not be too warm. Check every day to make sure that no soft/rotten/mold spots are appearing, and that things are ripening well. It's amazing what two days will do for ripening these fruits.

Avocados ripen well in a brown paper bag; so do bananas; so do small melons. The brown paper is just porous enough to allow an exchange of fresh air and the gases from the fruit.

The popular plastic "ripening bowl" developed, if we remember right, by a scientist at the University of California, Davis, department of agriculture, performs like a brown paper bag—but in a more deliberate, sophisticated, and predictable fashion.

Root-Cellaring Fruits

Only several of the most popular fruits root-cellar well; and of these, apples retain their texture and flavor longest, with several varieties of pears next in storage life.

Like vegetables, fruits to be stored over the winter should be harvested as late as possible in the season, and be as chilled as you can get them before they're put in their storage containers (it will take even a properly cold root cellar a good deal of time to remove the field heat from a box of warm apples).

Because they absorb odors from potatoes, turnips, and other "strong" vegetables, fruits should have their own special section partitioned off in the root cellar if they are stored in quantity; otherwise put them in another area where the conditions simulate those of a root cellar (see the chart), or keep them as far from the offending vegetables as possible.

We recommend clean, stout cartons, wooden boxes or splitwood fruit baskets over the classic apple barrels for storing fruits for a small family. Metal barrels are best used for fruits in underground storage, with the barrel well insulated from frost in the earth.

Some fruits are individually wrapped for best keeping, but all should be bedded on a layer of insulating—and protecting—straw, hay, clean dry leaves, with the straw, etc., between each layer of fruit, and several inches of bedding on top of the container.

RECOMMENDED CONDITIONS FOR OVER-THE-WINTER COLD STORAGE

PRODUCE	FOOD FREEZES AT (F)	TYPE OF STORAGE	IDEAL TEMPERATURE (F)	RELATIVE HUMIDITY (%)	AIR CIRCULATION	AVERAGE STORAGE LIFE
Fruits:						
Apples	29.0	RC-F	at 32	MM: 80-90	moderate	4-6 months
Grapefruit	29.8	RC-F	at 32	MM: 80-90	slight	1-1½ months
Grapes	28.1	RC-F	at 32	MM: 80-90	slight	1-2 months
Pears	29.2	RC-F	at 32	MM/M: 85-90	slight	2-7 months
Vegetables:						
Beans, dried	won't	DS; A	32-50	D: 70	moderate	12+ months
Beets	(c. 30)	BSR; P/B; RC	32-40	M: 90-95	slight	4-5 months
Cabbage	30.4	P/B; RC; DS	at 32	MM/M: 85-90	slight	late F-W
Carrots	(c. 30)	BSR; P/B; RC	32-40	M: 90-95	slight	6 months
Cauliflower	30.3	RC	at 32	MM: 80-90	slight	1½-2 months
Celery	31.6	BSR; frame; RC	at 32	MM/M: 85-90	slight	late F-W
Chinese cabbage	(c. 31.9)	BSR; frame; RC	32-34	VM: 95-98	slight	3-4 months
Dried seed, live	won't	A	32-40	D: 70	slight	12+ months
Endive	31.9	frame	at 32	MM/M: 85-90	moderate	2-3 months
Horseradish	(c. 30.4)	BSR; P/B; RC	at 32	M: 90-95	slight	4-6 months
Kale	(c. 31.9)	frame	at 32	VM: 95-98	moderate	1 month

Kohlrabi	(c. 30)	BSR; P/B; RC	32–40	M: 90–95	slight	2–3 months
Leeks	(c. 31.9)	BSR; P/B; RC	at 32	MM: 80–90	moderate	1–3 months
Onions	30.6	DS; A	at 32	D: 70	moderate	F–W
Parsnips	30.4	BSR; P/B; RC	at 32	M: 90–95	slight	F–W
Peas, dried	won't	DS; A	32–50	D: 70	moderate	12+ months
Peppers	30.7	BSR; RC; DS	45–50	MM: 80–90	slight	$\frac{1}{2}$–1 month
Popcorn	won't	A	to 75	D: 70	slight	12+ months
Potatoes	30.9	BSR; P/B; RC	35–40	MM: 80–90	slight	F–W
Pumpkins	30.5	BSR; A	at 55	MD/D: 70–75	moderate	F–W
Salsify	(c. 30.4)	BSR; P/B; RC	at 32	M: 90–95	slight	4–5 months
Squash	30.5	BSR; A; RC	at 55	VD: 50–70	moderate	F–W
Sweet potatoes	29.7	BSR; DS	55–60	MD/D: 70–75	moderate	F–W
Tomatoes, green	31.0	BSR; DS	55–70	MM: 80–90	moderate	1–1½ months
Turnips	(c. 30)	P/B; RC	at 32	M: 90–95	slight	2–4 months
Winter radishes	(c. 30)	BSR; P/B; RC	at 32	M: 90–95	slight	2–4 months

A = attic: BSR = basement store room: DS = dry shed: frame = coldframe: P = outdoor pit: B = buried barrel: P/B = pit or barrel: RC = root cellar: RC-F = root cellar for fruit: D = dry: VD = very dry: MD/D = moderately dry to dry: M = moist: MM = moderately moist: MM/M = moderately moist to moist: VM = very moist: F = fall: W = winter. This tabular form could not accommodate metric conversions alongside Fahrenheit figures. Therefore, we refer you to the relevant metric conversions in Chapter 3, as simpler for you. Figures in parenthesis are approximate.

371

All fruits need checking periodically for spoilage. If you're afraid your fruit will deteriorate faster than you can eat it fresh, have a midwinter preserve-making session. They're fun on cold lowery days.

Apples

Best keepers: *late* varieties, notably Winesap, Yellow Newton, Northern Spy; then Jonathan, McIntosh in New England, Cortland, Delicious. Pick when mature but still hard, and store only perfect fruit. Apples kept in quantity in home cold storage usually will be "aged" from Christmas on.

Apples breathe during storage, so put them in the fruit room of a root cellar so they don't give off their odor (or moisture) to vegetables. Wrap individually in paper (to cut down their oxygen intake); put them in stout cartons, boxes, barrels that can be covered, and have been insulated with straw, hay or clean dry leaves. If you use large food-grade plastic bags or liners for the boxes, etc., cut ¼-inch breathing holes in about 12 places in each bag. They also may be stored in hay- or straw-lined pits or in buried barrels covered with straw and soil, etc.

Check periodically and remove any apples that show signs of spoiling. See the chart for ideal conditions.

Grapefruit and Oranges

Store unwrapped in stout open cartons or boxes in the fruit room of a root cellar (see chart for conditions). Inspect often for spoilage, removing spoiled ones and wiping their mold off sound fruit they've touched.

Grapes

Catawbas keep best, then Tokays and Concords. Pick mature but before fully ripe.

Grapes absorb odors from other produce, so give them their own corner of the root-cellar fruit room (see chart for conditions). Hold in stout cartons or boxes lined with a cushion of straw, etc., with straw between each layer; don't burden the bottom bunches with more than three layers above them, fitting the bunches in gently. Cover with a layer of straw. Check often for spoilage.

Pears

Best keeper of the dessert varieties is Anjou, with Bosc and Comice popular among the shorter keepers. (Bartlett and Kieffer ripen more quickly and earlier: the former is especially good for canning, the latter for spicing whole or used in preserves; see Chapters 7 and 19.)

Pick mature but still green and hard. Hold loosely in boxes in a dry, well-aired place at 50–70 F/10–21 C for a week before storing. Then store them like apples. See chart for conditions.

Warning: pears that have started ripening above 75 F/24 C during the interim between picking and storage, or are root-cellared at too high a temperature, will spoil, often breaking down or rotting inside near the core while the outside looks sound.

Root-Cellaring Vegetables

Root-cellared vegetables freeze sooner than fruits do, as a rule; and, if you store a variety beyond the commonest root crops—beets, carrots, potatoes, turnips, and rutabagas—you need several different kinds of storage conditions. See the chart again, and the individual instructions below.

Wooden crates and movable bins, splitwood baskets, stout cartons—all make good containers for indoor storage. Insulating and layering materials are straw, hay, clean leaves, sphagnum and peat moss, and dry sand. The moist sand suggested for certain vegetables shouldn't be at all puddly-wet: if it's cold to the touch and falls apart when squeezed, leaving a few particles stuck to your hand, it should be the right degree of dampness.

Don't fill containers so deeply that the produce at the bottom is ignored in the periodic examinations for spoilage. And forgo building permanent bins that can't be moved outside for between-season scrubbing and sunning—stout shelving for the containers at convenient heights off the floor is a much better use of storage space.

Beans (Shell), Dried

Cool the finished beans and package in plastic bags which you then put in large, covered, insect- and mouse-proof containers. See chart for conditions.

Beets

Harvest in late fall after nights are 30 F/−1 C (they withstand frosts in the field) but when the soil is dry. Do not wash. Leave tails and ½ inch of crown when removing the tops. Pack in bins, boxes, or crates between layers of moist sand, peat, or moss; or bag like Carrots in the largest size of food-grade plastic freezer bags; cut ½-inch breathing holes in about 12 places. See chart.

Incidental intelligence: A noted Shaker cook always bakes her large beets (then takes off their tips and tails, peels them, cuts them for serving): she says they retain their juiciness and flavor best this way. Try it!

Cabbage, Late

Cabbage is not harmed by freezing in the field if it's thawed slowly in moist sand in the root cellar and not allowed to refreeze. Late cabbage can be stored effectively in several ways. (1) Roots and any damaged outer leaves are removed and the heads are wrapped closely in newspaper before being put in bins or boxes in an outdoor root cellar (the odor is more noticeable when they are wrapped than when covered with sand or soil). (2) With roots removed, the heads are covered with moist soil or sand in a bin in the root cellar. (3) In pit storage, stem and root are left on and they are placed head-side down. Straw, hay, or clean, dry leaves may be packed between the heads for added protection and the whole business covered with soil. (4) The outer leaves are removed and cabbages are hung upside down in a dry place at normal room temperature for several days or until they "paper over." Then they are hung upside down in the root cellar.

Warning: cabbages have one of the strongest odors of all vegetables, so don't store them where the smell can waft through the house. See chart.

Carrots

Carrots may stay in the garden after the first frosts. After digging, handle like Beets. See chart for conditions.

Cauliflower

Another hardy vegetable that can withstand early frosts. Cut off the root and leave plenty of protecting outer leaves; store in boxes or baskets with loose moist sand around and covering the heads. See chart for conditions.

Celery

Celery should not be stored near turnips and cabbages, which taint its flavor.

Pull the plant, root and all; leave the tops on. Do not wash. Place the roots firmly in moist sand or soil, pressing it well around the roots. Water the covered roots to keep them moist *but do not water the leaves*.

The procedure for celery may be followed in a trench, a coldframe-bed, or in a corner of the root-cellar floor that has been partitioned off to a height of 6 inches. The closer the celery is stood upright, wherever it's stored, the better. See chart for conditions.

Chinese Cabbage

Pull and treat like Celery. See chart for conditions.

Dried Seed (Live)

So long as it is kept quite dry, live seed won't germinate. Store in food-grade plastic bags that are then put in a large, mouse-proof covered container; or in canning jars that are wrapped in newspaper to keep out the light. It can't be hurt by natural low temperatures: see chart for conditions.

Endive

Pull as for Celery. Do not trim, but tie all the leaves close together to keep out light and air so the inner leaves will bleach. Set upright and close together with moist soil around the roots, again as for Celery. See chart for conditions.

Horseradish

One of the three vegetables that winters-over beautifully in the garden *when kept frozen*. Mulch carefully until the weather is cold enough to freeze it, then uncover to permit freezing and, when it has frozen in the ground, mulch heavily to prevent thawing. For root-cellaring, prepare and handle like Beets. See chart for conditions.

Kale

Treat like Celery right on down the line. See chart for conditions.

Kohlrabi

Handle like Beets; see chart.

Leeks

See Celery and chart.

Onions

Pull onions after the tops have fallen over, turned yellow, and have started to dry—but examine for thrips (which can cause premature wilting, etc.).

Bruised or thick-necked onions don't store well.

Onions grown from sets are stored in a cool, very dry place on trays made of chicken wire and the tops pointing down through the mesh.

Onions must be conditioned—allowed to "paper over"—in rows in the field; turn them several times so their outsides dry evenly. Smaller amounts may be surface-dried on racks in a dry, airy place under cover; or the tops may be braided and the bunches hung in a dry room. After they are conditioned, trim the tops and hang the onions in net bags or baskets in a dry, airy storage place. See chart for conditions.

Parsnips

Actually *improved* by wintering frozen in the garden (and not allowed to thaw), but may be root-cellared if necessary. Treat like Beets or Horse-radish. See chart for conditions.

Peppers

Careful control of temperature and moisture is imperative in storing peppers (see chart): they decay if they get too damp or the temperature goes below 40 to 45 F/4 to 7 C.

Pick before the first frost; sort for firmness; wash and dry thoroughly—handling carefully because they bruise easily.

Put them one layer deep in shallow wooden boxes or cartons lined with food-grade plastic in which you cut about twelve ¼-inch holes; close the top of the plastic. Even under ideal conditions the storage life is limited.

Popcorn

See Beans (Shell), dried; see chart.

Potatoes, Early

Don't harvest after/during heavy rains, or on a hot day. Dig them carefully early in the morning when the temperature is no more than 70 F/21 C. Condition them for 2 weeks at 60–70 F/16–21 C in moist air to allow any injuries to heal: early potatoes will not heal if they are conditioned in windy or sunny places. After conditioning, store at 60 F/16 C for 4 to 6 weeks. These early varieties do not keep long, and spoil readily held at over 80 F/27 C.

Potatoes

Late potatoes are much better keepers than early varieties. Dig carefully. Hold them in moist air about 2 weeks between 60 to 75 F/16 to 24 C to condition: do *not* leave them out in the sun and wind. Put them, not too deep, in crates, boxes or bins stored in a dark indoor or outdoor root cellar; cover to keep away all light (to prevent their turning green, which could mean the presence of solanine, not good).

After several months' storage, potatoes held at 35 F/2 C may become sweet. If they do, remove them to storage at 70 F/21 C for a week or so before using them. Potato sprouts must be removed whenever they appear, especially toward the end of winter. Early sprouting indicates poor storage conditions. See chart for conditions.

Warning: when potatoes have green areas under their skins it means they have been exposed to the sun when they should have been covered well. Result: solanine, a poisonous alkaloid that must be cut away (a reason why country cooks always gouged out around the sprouts on potatoes). Potatoes are a member of the deadly nightshade family.

And remember: keep potatoes and apples well separated from each other—potatoes make apples musty.

Pumpkins

Harvest before frost, leaving on a few inches of stem. Condition at 80 F/ 27 C for about 2 weeks to harden the rind and heal surface injuries. Store them in fairly dry air at about 55 F/13 C (see chart). Watch the temperature carefully: too warm, and they get stringy; and pumpkins (and squashes) suffer chill damage in storage below 50 F/10 C—they're not for outdoor cellars or pits. Just because they are big and tough doesn't mean they can be handled roughly, so place them in rows on shelves, not dumped in a pile in a corner.

Radishes, Winter

Handle like beets; see chart.

Salsify

The third vegetable (with parsnips and horseradish) that winters-over to advantage in the garden—so long as it remains frozen. If salsify must be stored, dig it when the soil is dry late in the season but before it freezes. Handle and root-cellar like Beets (and see chart).

Squashes

Condition and store like Pumpkins—but drier (see chart).

Sweet Potatoes

If a killing frost comes before you can dig them, cut the plants off at soil level, so decay in the vines can't penetrate down into the tubers.

Sweet potatoes are really quite tender, so handle them gently: sort and crate them in the field. Condition at 80 to 85 F/27 to 29 C for 10 days to 2 weeks near a furnace or a warm chimney, maintaining high humidity by covering the stacked crates (which have wooden strips between for spacing) with plastic sheeting or a clean tarpaulin, etc. Then store in fairly dry and warm conditions (see chart). Like Pumpkins and Squashes, they damage from chill below 50 F/10 C.

Tomatoes, Mature Green

For storing, harvest late but before the first hard frost, and only from vigorous plants. Wash gently, remove stems, dry; sort out all that show any reddening and store these separately.

Pack no more than two layers deep with dry leaves, hay, straw, or shredded paper (plastic bags with air-holes are more likely to cause decay). Sort every week to separate faster-ripening tomatoes. See the chart.

Turnips

These and rutabagas withstand fall frosts better than most other root crops, but don't let them freeze/thaw/freeze. Storage odor can penetrate up from the basement, so store them by themselves outdoors (see chart for conditions).

Handle like Beets; pack in moist sand, peat, etc.

Waxing Turnips and Rutabagas

A number of readers have asked about the feasibility, and then the procedure, for waxing rutabagas and turnips the way commercial growers do, to prevent the vegetables' odor from spreading and to protect the skins. As to feasibility: not really practical for the usual householder. The vegetables must be dunked in the melted wax (it's plain old paraffin wax, the sort on hand to seal jellies and jams), and there must be enough of it in liquified form. And it must be just the right temperature—too cool and it globs, too hot and it scalds. And the air in the room must be just right as well, lest the wax break off before it penetrates the pores of the skins. (Remember, too, that rodents like paraffin wax.)

Better to use the most clinging of the food-grade plastic wraps, bought in extra-large rolls from your farm-supply/garden-and-seed store. Press the wrap tightly around each separate vegetable; you can bag several together, or put them in a large carton or bin.

If you do want to try waxing, though, be sure the rutabagas or turnips are clean and absolutely dry before you dunk them.

23
PUTTING BY PRESENTS FOR CHRISTMAS

By the middle of May, *PFB* has started making presents for the following Christmas, and we'll keep adding to the trove on up to the day itself. Nor are the presents designed only for the big holidays: they're great fun to have on hand for favorite hostesses in the drearier winter months, or for shut-ins. Or for your own pleasure—always doubled by sharing.

Aside from maple syrup made in the sugarhouse at the foot of the low mowing, the first crop is rhubarb. For some souls it is truly a passion. Older friends, especially, enjoy it on hot buttermilk biscuits as old-time Scottish Rhubarb Ginger Jam—a comfort on a snowy Sunday morning more than half a year away. Give a foil pan of biscuits along with the jam. (What did we ever do before disposable foil dishes came along?)

CONVERSIONS FOR PUTTING BY CHRISTMAS PRESENTS

Do look at the conversions for metrics, with workable roundings-off, and for altitude—both in Chapter 3— and apply them. **Note: this chapter has Water-Bath processing—adjust for your altitude.**

Rhubarb Ginger Marmalade

Nine ½-pint jars

You can put this by from May through June—or when the rhubarb crop is available in your area. From an Agriculture Canada bulletin, it's the first written rule *PFB* has seen for a variation of a favorite old British recipe (translated, reluctantly, from metrics given in the original).

    4½ pounds rhubarb, cut in 1-inch pieces
    6⅓ cups sugar
    4 medium oranges
    5 teaspoons ground ginger

Mix rhubarb and sugar and let stand covered overnight to start the juices. Peel oranges by scoring the rind with a knife and removing it in tidy narrow strips; cut rind in inch-long pieces, cover with water, and simmer until tender (about 15 minutes); drain and save. Chop orange flesh, removing any seeds. Combine rhubarb, sugar, ginger, and orange pulp in a stainless steel or enameled kettle. Bring to full rolling boil, stirring constantly, until just before jelly stage, until jam holds its shape on the spoon or passes the refrigerator test (see Chapter 18). Remove kettle from heat and add orange rind; stir, skim. Ladle into hot ½-pint jars, leaving ¼ inch of headroom; cap with two-piece screwband closure, give 10-minute bath at 195 F/91 C. Remove jars from kettle and cool upright.

Buttermilk Biscuits

1 dozen good-sized biscuits

These can be made anywhere from a week to 10 days ahead of time and frozen in their aluminum baking tins, wrapped in Saran-type wrap and packaged in a plastic freezer bag.

    2 cups sifted flour (made from soft winter wheat is most delicate)
    ½ teaspoon salt
    1 teaspoons baking powder
    1 teaspoon baking soda
    1½ teaspoons sugar (for light browning)
    4 tablespoons (2 ounces in weight) lard or vegetable shortening
    ¾ to 1 cup fresh cultured buttermilk

Stir dry ingredients together. Rub in shortening until the mixture is mealy. Stir in buttermilk until all dry ingredients are gathered, and turn out onto a floured board. Knead gently a couple of passes, roll lightly to ¾ inch thick. Cut with floured cutter, place barely touching in a greased pan. Bake at 425 F for 15 minutes, or until golden. Let biscuits cool upside-down on a cookie rack if they aren't going to be eaten right away; package individually.

For a gala breakfast (so rich it could double as a shortcake): roll and cut biscuits, let them rest. Meanwhile, put ½ inch of maple syrup in the bottom of the baking pan, add 2 tablespoons butter and ¾ cup broken nutmeats. Set on a burner over medium heat, and boil gently, stirring, until the syrup is slightly reduced and sticky-looking. Remove from heat and immediately lay cut biscuits on the maple-nut mixture so that biscuits barely touch, and put into the oven. Bake 15 minutes, remove from oven; let sit 2 minutes, then turn over upside-down onto a serving rack. Eat hot, with cold, cold rhubarb sauce on the side.

For the Children

Then mini-harvests and preserving bouts follow on each other's heels as the preserving season takes over midsummer. There are jams and jellies, and relishes, and late autumn specialties as they come along—ending with treats expressly for the children's holiday.

A specialty of any house at Christmas is Gingerbread Men—to tuck into a hamper, string on the tree, serve with a light custard dessert to take the curse off too many heavy sweets.

Gingerbread Men
About 4 dozen small figures

These cookies can be made months ahead of time and frozen. Or the dough can be made months ahead of time—freeze it in a flattened ball wrapped closely in saran-type wrap. Tuck a label inside a freezer bag with the wrapped dough, remove air from the bag, and tie as shown in Chapter 13; store in the door of your upright freezer or in the basket of your chest freezer. Remove the dough from the freezer 2 days before you plan to make the cookies, and thaw 1 day in the refrigerator and then on your counter—the dough should be stiff.

½ cup butter (and it must be butter or the texture suffers)
1 cup light brown sugar, packed firmly
¼ teaspoon baking soda, dissolved in molasses
⅓ cup molasses (half dark and half light works well)
1 large unbeaten egg
½ teaspoon each: freshly grated nutmeg, ground cloves, ground cinnamon
2 cups all-purpose flour

Blend all together—electric mixer or a good processor is handy here; knead for a few minutes, wrap as an airtight packet and let it rest in the refrigerator for several hours. Knead well again, until it's as elastic as very dense bread dough. Chill or freeze. To cook: thaw, roll very thin, cut, and bake at 325 F for 8 to 10 minutes; watch lest the cookies spoil from overbrowning.

If the cookies are small and can be used as tree ornaments, put a hole in their heads to hold a bit of bright ribbon to hang them by. Before removing the hot cookies from their baking sheets, twist the blunt end of a kitchen matchstick in the center of each head until you feel the cookie pan; much better than trying to puncture the cooled cookie (which will break), or bake each with a match-end sticking up in the air.

Decorate with eyes, buttons, etc., using a toothpick dipped into a paste of ¼ cup confectioner's sugar moistened with enough lemon juice or water to make it blob (it will dry hard soon). Children enjoy making the faces on these traditional cookies.

For Our Elders

All small portions—especially the ½-pint jars with calico hoods—are welcome in a carton or basket. Here's a time to remember sugar-free jams and jellies (from Chapter 18), because so many of our older people may be flirting with diabetes. Do up some ½-pint jars of Dilly Beans (Chapter 19)—and include some Boiled Cider (perhaps even a small individual Steamed Pudding, which comes later); and make some tiny tarts of Green Tomato Mincemeat (Chapter 19)—it's lighter and therefore better for some appetites than the suet one with venison or beef—with a little cognac stirred into the filling before it's put in the pie pastry.

A holiday display of put-by presents. (*Photograph by Jeffery V. Baird*)

If you have occasion to mail your tiny tarts, use a plastic egg carton. Nestle each tart in its petits-fours paper (got from Maid of Scandinavia) in the holders on the bottom, lay a strip of saran film over them, and then lay another tart on it face down (but with its own little paper cup). Finally, close the carton, tape it together, and you have a splendid mailer.

Boiled Apple Cider
Two ½-pint jars

This can be made when fresh cider is available in your area (be sure to use *freshly pressed* cider). This precious liquid is a tart, clear, sweet taste of autumn. You might suggest that the recipient of this gift tries a teaspoon or two drizzled over vanilla ice cream or custard. Elegant as a topping, charming as an ingredient, a little goes a long way. Figure 1 gallon of freshly pressed cider to produce about 14 fluid ounces of Boiled Cider: *not* a procedure for newly wall-papered kitchens. Northern Spys are our favorites for this.

In a tall stainless steel or enameled kettle, boil cider over high heat, skimming all the while and cooking as fast as you can, as with maple sap being evaporated into syrup. Late in the season (or if the cider has been frozen) there will be too many soft curds of coagulated material to skim away easily, so pour the whole batch quickly through a good-sized jelly bag. Straining it now will go fast because it is still so thin. Rinse out the kettle, return clarified cider to it, and keep boiling hard until it lightly coats a metal spoon. Watch for a change in the bubbles, and try the Sheet Test for jelly. If drops do *not* run along the spoon's edge and combine, the cider is ready. A minute longer it will tear off in a lump, and it has gone too far. At this point add ½ cup of boiling water ¼ cup at a time, test, and pour into hot ½-pint jars, leaving up to ½ inch of headroom if need be, to fill two jars. Wipe sealing rims, cap with prepared disk lid, tighten screwband, firmly tight. Finish in a Boiling–Water Bath in a small stockpot that also holds "dummy" jars of water to complete the canner load: process for 15 minutes at a gentle boil, to drive air from the headroom and effect a good seal. Remove jars; cool upright.

All-purpose Pie Pastry
2 crusts

This pastry can be made and frozen in 1-crust balls up to 3 months ahead of time. If you will be using the pastry for tartlets, freeze pastry in balls the size of large shooting marbles, wrapped individually then all together in a freezer bag. Come time to make your tarts, simply press and flatten each "marble" in the cup of your tart pan. Fully assembled tartlets can be frozen, then removed from pans and repackaged for freezing up to one month ahead of the holiday.

The basic ingredients may be titivated with more sugar, almond or vanilla extract, an extra egg yolk if the batch is doubled, etc. A combination of unsalted sweet butter and good shortening—never oil—does best, but cholesterol-worrisome diets use less egg yolk, and corn-oil margarine as the shortening.

> 2 cups all-purpose flour
> $\frac{2}{3}$ cups lard, butter, or combination with vegetable shortening
> $\frac{1}{2}$ teaspoon salt *optional*
> 1 tablespoon sugar—more for a sweeter pastry—for a golden top crust
> 1 large egg
> $1\frac{1}{2}$ teaspoons vinegar
> Ice water to bring egg-vinegar-water to $\frac{1}{2}$ cup total

Combine dry ingredients, cut in shortening quickly until all is mealy, pour in the $\frac{1}{2}$ cup egg-vinegar-ice water mixture. When dough gathers, knead it a couple of quick passes, separate into 1-crust balls, let it rest, chilled, until rolled out.

Green Tomato Mincemeat tartlets packed in egg cartons—ideal for mailing or shipping. (*Photograph by Jeffery V. Baird*)

Include in the gift basket some ½-pint and pint jars of Pie Filling (from Chapter 12). Then, come the cold winter days, the keeper of a small household can make some pies. And tuck in some individual aluminum pie pans, 4½ inches across: these will take most of a pint of filling—dessert for two.

To ensure a better bake with the little individual pans, puncture them *from the inside out,* using an icepick or a similar small, sharp instrument.

Thank heaven for foil pans! Poke holes in the bottom to get a better bake in 12 to 15 minutes for biscuits. (*Photograph by Jeffery V. Baird*)

Raspberry "Vinegar," a Cockaigne
About 3½ pints

This beautiful quencher can be made in late summer or early fall when raspberries are abundant. But if your fresh raspberries are too precious, it can be made anytime of the year simply by asking your garden center (e.g., Agway) or county Farm Bureau when a frozen fruit sale will be held in your area. These sales are wonderful opportunities to order quantities of fresh frozen fruits at a very inexpensive price, for just such things as this "vinegar"—or for your own mid-winter pies.

5 pounds frozen raspberries, of course *not* sweetened
5 cups freshly pressed apple cider vinegar
About 8 cups sugar

Frozen, the berries are well "broken down," and the vinegar therefore can go into them in one addition, rather than in two separate additions as with fresh berries. In a large enameled or stainless steel bowl, pour vinegar over thawed berries. Stir, cover, and refrigerate overnight. Stir again the next day, refrigerate again. The third day, bring berries and vinegar to the boiling point in an enameled kettle, cook gently no more than 3 minutes, pour immediately through a stout jelly bag or maple-syrup-straining felt. Measure extracted juice, add ⅔ to ¾ cup sugar (or to taste) for each 1 cup of juice, stir, bring to a simmer, stirring and skimming. Pour hot into hot pint and ½-pint canning jars, leaving ½ inch of headroom. Cap with scalded disk lid held in place with a screwband. Process in a Hot–Water Bath to cover—190 F/88 C—for 15 minutes. Remove jars, cool upright away from drafts.

Use ¼ cup over cracked ice in a tall glass, fill with still or sparkling water, and stir—a holiday toast if an older person doesn't take hard liquor.

For Your Part of the Party

What to take to a holiday get-together that's going to have its share of Christmas cookies and eggnog and fruitcake and specialty dips? Hot Pepper Jam, that's what. Turn a squat ½-pint jar out in the center of a dispensable waxed holiday paper platter. Around it spread some good, softened cream cheese. As a vehicle for this crowd-pleaser, make Pita Triangles instead of buying wheat crackers: cut apart the largest loaves of Syrian white flat bread, spread open the halves, and tear them into pieces. The result should be triangles about 2½ to 3 inches to a side. Jumble them in the middle of a 275 F oven for 20 or 30 minutes or so, turning if they seem to need it. The result is crisp, tasty bits to use for any special spread.

Whether a holiday party is buffet or a full sit-down dinner, a colorful contribution of Dot Robbins's Christmas Pickle or, perhaps, Dilly Brussels Sprouts, are sure to be welcome—as would be the tasty put-by stand-by, Corn Relish (see Chapter 19).

Hot Pepper Jam
About three½-pint jars

Make this jam in August, when jalapeños have come in and green peppers are getting quite inexpensive.

There are tiny bits of green and hot peppers floating in this—not much like conventional jam, but they make a more interesting texture than if they were strained out. Keep batches small, because you're adding all the components for gelling that the peppers lack—pectin, acid, and sugar. Do up several batches' worth of the peppers and measure out what you need as you go along. Remember that jalapeños are savagely hot: avoid fumes, wear rubber gloves when you do them all at once, open and cut them under water, and don't touch face or eyes.

1 cup ground sweet bell pepper, including juice (1 quite large pepper)
2 tablespoons finely ground jalapeño peppers (convincing but tolerable; your family may opt for more jalapeños in later batches)
¾ cup cider vinegar
½ cup water
⅛ teaspoon salt
2 tablespoons strained lemon juice
1 box standard powdered pectin (1¾ ounces)
2½ cups granulated sugar
2 drops of green food color if liked

Wash and seed the sweet pepper, mince fine in a blender or food processor (easier if whirred with part of the vinegar); measure a good cup of the pepper and its juice into a heavy stainless steel or enameled pot. Holding two or three jalapeños under clean water, remove stems and seeds, split and mince fine: try not to inhale fumes, and take care to wash hands carefully after handling. Measure 2 tablespoons (plus 1 more if you're hardy) into the pot with the sweet pepper; add and stir together the remaining vinegar, water, salt, lemon juice, and pectin—*powdered pectin is added* to fruit juice *before sugar.* Set the pot on medium-high heat and quickly bring to boiling; add sugar and bring again to a full rolling boil that cannot be stirred down; boil hard for 1 minute, remove from heat, skim froth, stir. Ladle into clean hot ½-pint jars, leaving ¼ inch of headroom. Thoroughly wipe sealing rim of jars with fresh paper towel, put on prepared disk lid, and screw the band down firmly tight. Process in a covered pasteurizing bath at 185 F/85 C for 10 minutes; remove jars from canner and set on a dry folded towel where they can cool upright and naturally.

Dilly Brussels Sprouts
7 pint jars

This can be put by in late August or September—or whenever you see a good buy on Brussels sprouts.

Prepare 7 wide-mouth pint jars as for Dilly Green Beans in Chapter 19, using one fresh or frozen whole dill head, one clove fresh garlic, and ¼ teaspoon crushed dried hot red pepper in each jar. Soak Brussels sprouts for 10 minutes in a cold brine of 1 tablespoon salt to each 4 cups of water to drive out any bugs; rinse well: peel and trim to a uniform 1-inch diameter. In the bottom of each sprout cut an X deeply up into its core to allow the pickling solution to reach well inside the vegetable. Pack each jar solidly with raw sprouts to within ¾ inch of the top, tunking the bottom of the jar to settle the contents. Prepare boiling salted pickling liquid, pour over packed sprouts, leaving ½ inch of headroom. With a slender plastic blade remove any trapped air from jar; wipe sealing rim, cap with prepared disk lid and screwband firmly tight. Process in a Boiling–Water Bath for 10 minutes. Remove jars, cool upright.

Judge the amount of pickling solution as you go along. If, toward the end, you think you're running a little short, fill the remaining couple of jars equally and top up each with straight vinegar—*never* with water, lest you destroy the safe balance of acetic acid and low-acid vegetable.

Dot Robbins's Christmas Pickle
7 pints

8 or 9 large *ripe* cucumbers
7 cups white sugar
2 cups white vinegar
½ teaspoon oil of cloves
½ teaspoon oil of cinnamon
3 ten-ounce jars of maraschino cherries

First day: peel cucumbers, remove seeds, cut in 1-inch pieces, and put them in a large enameled or stainless steel kettle. Add water to cover and boil gently until barely tender (about 10 minutes). Remove, drain well, and put in a large glass or crockery bowl. Combine sugar, vinegar, and oils of clove and cinnamon in the empty kettle, bring to a boil, and then pour over the cucumber pieces in the bowl. Cover and let stand overnight at room temperature.

Second day: drain off syrup and bring it to a boil. Pour over the cucumbers and again let stand overnight.

Third day: put cucumber pieces, with syrup, in the kettle, bring all to a boil, and add three jars maraschino cherries and their juice. Return to heat, and when it is boiling again ladle into hot pint jars, leaving ½ inch of headroom. Adjust lids and process in a Boiling–Water Bath (212 F/100 C) for 10 minutes after the canner has returned to a full boil. Remove and complete seals if necessary.

The spice essences are usually carried in natural-food stores.

For the Larder

Always appreciated at holiday time are presents that can be stored away and brought out long after the Christmas season has passed, when special foods and treats are certain spirit-lifters and reminders of friendship. Mint Sauce, spicy Peppercorn Jelly, or pretty Wine Jelly make fine gifts—as do unusual easy-to-create Vinegars, bottled and beribboned. You might also present a package of Dried Tomatoes (see Chapter 21). These are currently the rage in specialty shops, and are a real treat for anyone who has purchased them there for a dear price.

Wine Jelly
4 small jars

Put this by *whenever*. It's easy and simple and can even be made last-minute.

- 3 cups sugar
- 2 cups any drinkable table wine (part may be Sherry)
- 2 three-ounce foil pouch liquid fruit pectin

Measure sugar and wine into the top of double boiler; mix well. Place over, but do not let touch, rapidly boiling water; stir 3 to 4 minutes, or till sugar is dissolved. Remove from heat. At once stir in fruit pectin and mix well. Skim off any foam. Quickly pour hot jelly into hot $\frac{1}{2}$-pint canning jars—the squat Kerr jars are ideal here—leaving $\frac{1}{8}$ inch of headroom. Cap with disk and band turned firmly tight.

Finish 10 minutes in a Hot–Water Bath at 190 F/88 C. Remove jars; let cool upright.

Peppercorn Jelly
Five $\frac{1}{2}$-pint jars—old-style squat ones are best

Paula Reds are *PFB*'s favorite apple for this recipe. If you're using Paula Reds, plan to do this jelly early on—they're among the season's earliest apples, and they aren't around for long. If Paula Reds aren't available, any good jelly apple—especially with pink juice—will do.

- $3\frac{1}{2}$ pounds Paula Red apples (early variety, with lovely pink juice)
- $\frac{1}{2}$ cup white wine (Chablis is good here)
- 2 tablespoons lemon juice
- 2 to $2\frac{1}{2}$ cups water
- 4 cups white sugar
- 2 tablespoons (with their juice) of bottled pickled green peppercorns (half a $3\frac{1}{2}$-ounce bottle)

Wash and cut up apples, but do not peel or core. Put them in a stainless, steel or enameled pot, add the lemon juice and water to cover. Cook, stirring, until apples are mush; strain through a stout jelly bag. Measure 4 cups of the pretty juice into the washed pot, add the wine and the peppercorns. Bring to a simmer, skim out the peppercorns, and hold them in a cup (you'll put them back in a minute). Add sugar to the hot juice; bring to boiling and cook until it tests Done for jelly: it tears in a sheet off the spoon, or a blob of it wrinkles on an icy plate when pushed with your finger. Off heat, promptly return pep-percorns to the jelly, remove any froth, pour into the wide-mouth ½-pint jars that turn their contents out on a plate so nicely. Leave a scant ¼ inch of headroom, wipe sealing rim of jars, cap. Finish in a 180 F/82 C simmering Hot–Water Bath for 15 minutes. Remove jars, cool upright and naturally. During the first 10 minutes, twirl each jar twice to distribute peppercorns.

Mint Sauce
In ½-pint jars

Plan to make this in July or August, when mint is tender and before it's gone to blossom. It's the real thing, and worth doing plenty of. Proportions are the key: how much you make at a crack depends upon how much fresh, good mint you have. *PFB*'s favorite is apple mint, tender leaves well washed and dried off in a salad spinner:

> 2 parts fresh, washed, well-packed-down mint leaves
> 3 parts white vinegar
> 1½ parts white sugar
>   Pinch of salt
> ¼ teaspoon crystalline citric acid

Chop mint fine in a food processor with steel blade in place, or in a blender: use enough white vinegar to help the process, but *do not* purée (you want tiny bits of leaf); set aside. Bring to boil in a stainless steel or enamelware pot the sugar and vinegar—you should have about 2+ parts of vinegar re-maining, to go with the 1½ parts of sugar. When the mixture has boiled hard for 3 minutes, add finely chopped mint, salt, and citric acid. Bring again to boil, pour immediately into clean, hot ½-pint jars, leaving ¼ inch of headroom; cap. Process gently at 190 F/88 C for 10 minutes. Remove jars; let cool upright and naturally.

Specialty Vinegars

When nasturtium and chive blossoms come in—somewhere between July and mid-August, depending on your area—take time to make these wonderful, bright-tasting vinegars. You may change the proportions and the ingredients as you go along. Use wide-mouth jars, or cadge 1-gallon wide-mouth jars from your friends at the neighborhood diner.

Nasturtium Vinegar: Use distilled white vinegar, and for every 2 quarts of vinegar, use 1 cup firmly packed nasturtium blossoms that have been swished in salted water to get rid of bugs. Rinse in fresh water and dry in a salad spinner, then pack blossoms down hard in a measuring cup. Add ½ teaspoon salt, perhaps 2 shallots (garlic and onion are too strong for the delicacy of the blossoms), and a 2-inch strip of orange zest. Shake all together, put saran film over the mouth and screw on the top, and forget it for at least one month.

Then strain out the blossoms, and to clarify the vinegar (mostly of pollen) pour vinegar piecemeal through an automatic coffee filter. This makes a charmer of a present packaged in small bottles with bayonet caps that snack juices have come in. The color of this finished vinegar is a lovely jewel-like red-orange.

Chive Vinegar: Follow the same direction for chive blossoms as for nasturtium blossoms. Omit shallots and any herbs; add more orange zest, or try lemon zest. This vinegar will be a lovely pinkish lavender with the faintest aroma of chive.

Finale

Finally *PFB* offers a Steamed Holiday Pudding—a put-by Christmas treat that is truly top-of-the-line, the pièce de résistance for children and elders, or as a hostess gift, or, most especially, as the grand finale for any holiday gathering.

Because both this Holiday Pudding and the Spicy Fruit Pudding freeze wonderfully, they can be made long ahead of time—even well before Thanksgiving—when you're likely to have a few more free minutes.

Directions for Steaming the Pudding

In a Pressure saucepan or Pressure Canner put $1\frac{1}{2}$ or $2\frac{1}{2}$ quarts of water, respectively. Put in the covered mold, which need not be weighted or propped since it has so little water to float in; put on the lid of the saucepan/canner and tighten—but do not stop the vent with a deadweight gauge or other closure. Allow the canner to vent steam strongly for 20 minutes, then close the vent with the relevant weight, and process at 10 pounds for 50 minutes.

At a small dinner, use two 1-pound shortening cans, well greased, filled no more than $\frac{2}{3}$ full (or slightly less, to make equal puddings). Following general procedure above, vent for 15 minutes and process at 10 pounds for 25 minutes.

For individual puddings in a Pressure saucepan or canner this recipe yields only 6 to 8 small molds, but the larger of these may be halved and served cut side down, with spirits flaming and toppings as described below. In tapered $\frac{1}{2}$-pint canning jars, with lids on firmly (or in No. 1 cans, covered with heavy-duty foil held in place with wire twisted tight) and in a Pressure saucepan: vent 10 minutes, process at 10 pounds for 20 minutes. Slightly larger molds like a tapered peanut-butter jar or wide-mouth pint canning jar, in an 8-quart canner: vent 20 minutes, process at 10 pounds for 25 minutes.

The present: Steamed Fruit Pudding for 10 (or 1, right). Delicious with Boiled Cider Sauce. (*Photograph by Jeffery V. Baird*)

Steamed Holiday Pudding
10-plus servings

This freezes like a dream: when cool, wrap carefully with a "drugstore fold" in heavy-duty aluminum foil, freeze. Heat wrapped pudding(s) in a 350 F oven at start of the dinner.

   2 tablespoons melted butter or margarine for greasing the mold
   1 cup grated raw carrots (about 3 medium)
   1 cup grated raw potatoes (about 2 large)
   1 teaspoon baking soda
   $\frac{1}{2}$ cup butter or margarine, melted ($\frac{1}{4}$-pound stick)
   1 cup brown sugar
   $1\frac{1}{4}$ cups all-purpose flour
   $\frac{1}{2}$ teaspoon salt (optional: which is half what the old rule asks for)
   1 teaspoon ground cinnamon
   $\frac{1}{2}$ teaspoon freshly grated nutmeg
   $\frac{1}{2}$ teaspoon ground cloves
   1 cup seeded raisins (OR 1 cup firmly packed chopped seedless raisins)
   1 cup heavy cream for whipped topping (optional)

Best is the classic charlotte mold with a cover, or a straight-sided casserole or souffle dish that can be covered with thick foil that's bound tight with wire. First, grease your mold. Melt extra butter, and with a pastry brush paint the inside of the mold generously, paying special attention to any swirled designs in the bottom and to all surfaces of the center tube if it has one. Put the molds in a cold place to let the butter congeal while you make the batter.

The steel blade of a food processor does a good job with the vegetables here, cutting them fine and freeing the juice well. In a bowl combine grated carrots and potatoes, dissolve the baking soda in the vegetables' juices. Stir in the sugar well, and add the ½ cup melted butter. Sift flour with spices and optional salt, and add; dust raisins with a bit of flour and stir in. Pour the batter into greased mold to no more than ⅔ full; cover tightly and steam as directed in first example of Steaming, above. When pudding is done, take cover off mold and let it sit for 5 minutes, then up-end on a rack if the pudding is to be cooled temporarily, or on a butter salver if to be served warm with a flavored whipped cream or hard sauce if liked. Or pour brandy over it, set aflame for a flourish then add topping as it is cut and served.

Spicy Fruit Pudding
10 to 12 individual puddings

Many thanks to Betty Wenstadt of Presto Industries—good friends to pre-servers everywhere. (Original recipe called for candied fruit: the change is *PFB*'s.) Use a large Pressure Canner here.

    1 cup mixed dried fruit
    ¼ cup walnut meats, broken
    ¼ cup pecans, broken
    ½ cup dates, chopped
    ¼ cup raisins
    1 tablespoon flour
    ¼ cup shortening
    2 tablespoons sugar
    3 tablespoons honey
    1 egg (large)
    ⅜ cup flour
    ¼ teaspoon salt
    ¼ teaspoon baking powder
    ¼ teaspoon ground allspice
    ⅛ teaspoon ground cloves
    ⅛ teaspoon freshly grated nutmeg
    1½ tablespoons orange juice
    2 cups water

Dredge fruit and nuts with the 1 tablespoon flour. Cream shortening and sugar, add honey; add egg and beat until smooth. Sift all dry ingredients and work them into the sweet mixture, added alternately with the orange juice.

Combine the batter with floured fruits and nuts, mixing well. By heaping spoonfuls, put about $1\frac{1}{2}$ inches of batter in each tapered wide-mouth pint canning jar, well buttered; cap with disk lid and tightened screwband. In your Pressure Canner with 2 inches of warmed water, set jars on the rack adding a cake-cooling rack on top of them to hold the second layer, which you stagger so no jar is obstructed from below. Close canner's lid, vent strongly 10 minutes, process 20 minutes at 10 psig (15 psig above 5000 ft/1524 m). Follow general instructions above for drying, unmolding, etc., and packaging.

Boiled Cider Sauce
Two ½-pint jars

This Sauce uses the Boiled Cider you put by earlier in the autumn. It can be made at the last minute—or give this recipe along with a present of Boiled Cider. It's a good nonalcoholic topping for Holiday Steamed Pudding.

$1\frac{1}{2}$ cups sugar
$\frac{1}{2}$ cup butter (or good margarine)
2 tablespoons modified cornstarch
$\frac{1}{3}$ cup + 3 tablespoons water
$\frac{3}{4}$ cup Boiled Cider

In a heavy pot, combine butter, sugar, $\frac{1}{3}$ cup of water, and the Boiled Cider. Bring to a simmer over Low heat, stirring. Make a "slurry" of the modified cornstarch and 3 tablespoons of water, in a smooth paste. Dilute with several spoonfuls of the hot cider mixture, return all to the pot, and cook until glossy, clear, and thickened. Pour immediately into hot ½-pint canning jars, leaving a scant $\frac{1}{4}$ inch of headroom. Wipe sealing rim, put on disk lid, tighten screwband firmly. Process 10 minutes in a Hot–Water Bath at 190 F/88 C; remove jars; let cool upright.

APPENDIX:
WHERE TO FIND THINGS

In their sequence, the letters C-E-S can give help to every householder in the United States on any matter of preserving food. They stand for Cooperative Extension Service, and mean that the U.S. Department of Agriculture is ready to instruct and advise anybody, anywhere, about any branch of food technology, including how to preserve food safely at home. Each state land-grant college or university has a "USDA-CES," one of whose virtues is that it has no axe to grind. None. Except, of course, to sharpen our awareness of safety.

From People

The CES person most homemakers deal with is their county's Extension Home Economist, who used to be called a County Agent in Home Economics (hence our continual "Ask your county agent"), and who will handle your questions. Chances are that Extension has filled gaps in the agent's education as well as in *PFB*'s: at a giant state university an undergraduate course in food preservation—offered toward a degree in home economics—may be only one credit, and an elective unit at that.

A sprawling metropolis may not have agents listed for every county it embraces geographically, but it is likely to have an Expanded Food and Nutrition Program (EFNEP) center, with branches in boroughs or settlement houses. Try them.

In the telephone directory your most likely listing will be plain "Extension Service." Or look under your county agencies, and your state university. The Yellow Pages may list county agents under "Vocational and Educational Guidance."

There is an annual *County Agents Directory* published by a commercial press. It's expensive to buy, but a good research librarian can track it down; and of course the state CES headquarters will have it.

More General

Best bet for USDA H&G (Home & Garden) bulletins still circulating is the Consumer Information Center, Department EE, Pueblo, Colorado 81009, and ask to be put on the mailing list for their *Consumer Information Catalog*—free, and it comes out quarterly. There are extensive catalog-lists of publications sold by the Superintendent of Documents, U.S. Government Printing Office, Washington, D.C. 20402. Send the correct amount of payment (money-order, check, stamps if they'll be accepted, but never coins or currency). Be prepared for backing-and-forthing for weeks if the material is temporarily unavailable or the price has gone up. Deal directly with the documents distribution people in Colorado, above, whenever you can.

In 1978 four USDA agencies—the Agricultural Research Service, the Cooperative Research Service, the Extension Service, and the National Agricultural Library—merged to become the Science and Education Administration (SEA), U.S. Department of Agriculture. A number of the resultant publications following this consolidation of agencies are still listed as prepared by the Agricultural Research Service. This is a nuisance for the consumer but perhaps a help to government computers. Anyway, look for the publications *by number* as well as by title; where the numbers have been changed, the familiar old one may appear either on the back cover with printing/edition information, or on the inside front cover (federal publications are not so picky as the book trade is about citing earlier publication data).

In Canada, visit or write your provincial agriculture departments, which issue attractive materials that are notably to the point and well written; their measurements are metric. Several of Ontario's free materials are favorites.

At this writing, Agriculture Canada is phasing out its food preservation publications. A pity: the Dominion does fine research and possesses a crisp, non-gobbledygook style. Currently, though, you can get information from Mrs. Pauline Kloseveych, Food Development Division, Agriculture Canada, Ottawa, Ontario, K1A 0C7.

We refer often to the *Morbidity and Mortality Weekly Report* from the Centers for Disease Control in Atlanta, Georgia, now printed and distributed by the Massachusetts Medical Society, C.S.P.O. Box 9120, Waltham, Massachusetts, 02254-9120.

Of the materials on nutrition that *PFB* receives, the best is *Diet and Nutrition News* from Tufts University. It has depth and scholarship, and what seems to be unfailing good sense.

Chapter by Chapter

Chapter 1: What Is It?

The six methods of food preservation described in the Master Food Preserver workbook of the University of California, Davis (led by Kathryn Boor, 1987), are credited to the University of Illinois MFP program manual.

The canning-jar, etc., figures for the United States in 1985 were worked out by Sue Hovey of Kerr Glass Manufacturing Company, Chicago, and Joan Randle of Ball Corporation, Muncie, Indiana, and *PFB*.

Chapter 2: Why Foods Spoil

USDA *H&G Bulletin No. 162, Keeping Food Safe to Eat.*

The *FDA Consumer,* official magazine of the Food and Drug Administration, Rockville, Maryland 20857.

The introductions to preserving food at home in all accredited manuals.

Food Safety for the Family, 533 K, from the Consumer Information Center, Pueblo, Colorado 81009.

Chapter 3: Altitude and Metrics

Altitude. The Colorado State University pamphlets are especially good for their simplicity and quickness, among them: *Pamphlet 41, High Altitude Food Preparation* (Revised 1977), by Pat Kendall, Extension Specialist, Food Science and Nutrition, Colorado State University, Fort Collins 80523. (Inquire about price from Bulletin Room.)

National Presto Industries, Inc., pamphlet AD79–3449B, *Canning at 15 Pounds Pressure* (1978), has, on its page 7, a detailed chart for high-temperature processing of vegetables, meats, and seafood at altitudes ranging from under 3000 feet/914 meters to over 7000 ft/2134 m (but less than 10,000 ft/3048 m). For copies, write to Consumer Services division of this maker of Pressure Canners at simply Eau Claire, Wisconsin 54701. Mrs. Betty Wenstadt is Home Economist for Presto in charge of their test kitchens and is generous with information.

Also extremely valuable for their sensible recognition of the problems "out in the field" are the New Mexico State University, Las Cruces 88001, bulletins, especially *Guide E-313, Home Canning Pressures and Processing Times,* revised 1978 by Mae Martha Johnson, Extension Nutrition Specialist.

Metrics. Where to begin? *Readers' Guide to Periodical Literature* in the reference department of your local public library, for openers; then *Subject Guide to Books in Print,* same department. The Pueblo Information Center (see above) lists government publications.

One source of metric kitchen scales is the Hanson Company of Shubuta, Mississippi 39360. They also make small by-the-ounce ones that they call "recipe" scales.

Chapter 4: Fair Warning

On irradiation, the *Readers' Guide,* cited above, is probably the quickest way to locate pros and cons. And the Pueblo center, again mentioned above.

Much of our material came from long-established consumer groups. In April 1987, the Atomic Industry Forum, Inc. (7101 Wisconsin Avenue, Bethesda, Maryland 20814-4891) published *Background Information* updating the use of low-dose radiation on certain foods as an alternative to the use of chemicals.

Chapter 5: Common Ingredients and How to Use Them

Artificial sweeteners. One of the best and briefest discussions of the Delaney Clause, responsible for the FDA's stand on saccharin (and more substances) is the *Food Facts from Rutgers, Cancer: Laws and Chemistry,* Cook College, April–October 1981; P.O. Box 231, New Brunswick, New Jersey 08903. From time to time the Centers for Disease Control include items on aspartame (industrially, NutraSweet; for household use, Equal); see the *Morbidity and Mortality Weekly Report* (for an index, write to MMWR's address, listed earlier).

Other addresses for common ingredients: the Morton Salt Company, 110 North Wacker Drive, Chicago, Illinois 60606, David Strietelmeier, Technical Director; and for information on food-grade chemicals and other substances, Fisher Scientific Company, 461 Riverside Avenue, P.O. Box 379, Medford, Massachusetts 02155.

For thickening, the possible newcomer to everyday kitchens is a modified, waxy maize (corn) starch called Clearjel A by its maker, National Starch and Chemical Company, 10 Finderne Avenue, P.O. Box 6500, Bridgewater, New Jersey 08807. Supplied to the frozen food industry among other commercial users, it is a better performer than "mochiko" because it will not break down under prolonged heat processing, and it will not get opaque or chunky but become pleasantly firm. It's a prime ingredient in the Pie Fillings in Chapter 12. The catch is that the company sells it in 500-pound bags. The thing to do is to mount a campaign to persuade a favorite specialty store to stock it at reasonable cost.

However, Dacus, Inc., P.O. Drawer 2067, Tupelo, Mississippi 38803, is willing to try distributing Clearjel A along with its established line of Mrs. Wage's pickling and gelling products *if* they get enough individual orders to warrant adding it to their line. Jim McGee is the person to write to at Dacus. Or homemakers' groups could pool money and order from a jobber; Paul Smith of National Starch, above, can tell you the jobbers in your area; write to him at Food Starch Division, Sales. (It is Mr. Smith who described an ideal gravy to *PFB* as adding Clearjel A in a "cold-water slurry stirred into hot pan drippings.")

Meanwhile it's a good idea to have mochiko on hand, and if you can't find it, write to Ms. Jeane Pate of Koda Farms, South Dos Palos, California 93665 for jobbers/retailers in your area. Koda Farms does *not* sell mochiko retail.

Chapter 6: The Canning Methods

The general reference books include: of course, *The Complete Guide to Home Canning* (1988), USDA Center for Excellence in Home Food Preservation, which supersedes four major USDA H&G bulletins on home canning (*No. 8, Home Canning of Fruits and Vegetables; No. 56, How to Make Jellies, Jams, and Preserves at Home; No. 92, Making Pickles and Relishes at Home;* and *No. 106, Home Canning of Meat and Poultry*). The Pueblo, Colorado, *Consumer Information Catalog* (above) can tell you how much it is; so can the Superintendent of Documents, Washington, D.C.; and don't forget to ask your county agent.

Ball Blue Book, the Guide to Home Canning and Freezing, gettable wherever canning supplies from Ball are sold. The *Kerr Home Canning and Freezing Book:* again, wherever canning supplies are sold. The *Bernardin Home Canning Guide,* where Bernardin lids are featured in the United States, and wherever the range of jars and closures are sold in the Dominion (inquire of the home economist at Bernardin of Canada, 120 The East Mall, Toronto, Ontario M82 5V5, Canada).

The Almanac of the Canning, Freezing, Preserving Industries (1986), Edward E. Judge & Sons, Inc., P.O. Box 866, Westminster, Maryland 21157.

The *NEFCO Canning Book* (Nutrition Education and Food Conservation), Dixie Canner Equipment Company, Inc., Athens, Georgia 30601. Edited by William C. Hurst, Ph.D., Food Scientist, University of Georgia. It deals with canning at home or in a community canning kitchen, using either pint or quart glass jars, or steel (so-called "tin") cans, sealed with a special mechanism. *PFB* relies on Dr. Hurst and the National Food Processors Association for processing procedures for *cans*.

So Easy to Preserve by Susan Reynolds, et al., Cooperative Extension Service, the University of Georgia (1984)—a favorite with *PFB* because of its style and organization.

Worthwhile pamphlets/bulletins start with Dr. Pat Kendall's high-altitude publications on canning fruits and vegetables, from Colorado State University, Fort Collins 80523; ask at the Bulletin Room. Notably sensible.

Also for altitude canning, National Presto Industries, Inc., pamphlet AD78-3449C for ½-pint, pint, 1½-pint, and quart jars, at 15 psig, showing the processing times needed as the altitude increases. *Valuable for small households that have 4- or 6-quart cookers, using ½-pint and pint jars.* Presto warns that the times at 15 psig given apply *only to altitudes up to 3000 ft/914 m.*

Canning Canadian Fruits and Vegetables /La mise en conserve de fruits et legumes du Canada (1982, 1985), publication 1560/E, Communications Branch, Agriculture Canada, Ottawa K1A 0C7. Once again, inquire from Mrs. Pauline Kloseveych for information.

Mirro & Foley Companies, P.O. Box 409, Manitowoc, Wisconsin 54220, make Pressure Canners, as well as pressure saucepans; their entire line is deadweight gauges. Mary Karbon, of the customer relations department, is helpful.

National Presto Industries (mentioned above) has Mrs. Betty Wenstadt and Jo Ann O'Gara as experts in their test kitchen. Presto makes both canners and saucepans, and has dial gauge as well as deadweight gauge canners. The Presto people have helped fund much research, including the 15-psig research at the University of Minnesota, and work in home-canning of seafood. Eau Claire, Wisconsin 54701 will reach them.

Joan Randle of Ball Corporation (345 South High Street, Muncie, Indiana 47305-2326) and Sue Hovey of Kerr Glass Manufacturing Company (2444 West 16th Street, Chicago, Illinois 60608) are always ready to help.

Ball and Kerr are the long-established makers of canning jars in the United States. Anchor Glass Company, Dominion Glass, and Bernardin all make jars in Canada. Anchor no longer makes jars in the United States: just closures. Bernardin's Canadian range is widest of all, and includes an innovative plastic *storage* lid, for use in freezing and other storage (the company warns against *canning* with it).

Dixie Canner Equipment Company is a leading supplier of machinery for community canning kitchens that use metal cans with crimped-lid closures (although their equipment also includes handling glass). They will send their *NEFCO Cannery* booklet describing the criteria for all sizes of canning kitchens. Write to P.O. Box 1348, Athens, Georgia 30603-1348 (phone: 404/549-1914). They do not supply cans.

Freund Can Company, 197 West 84th Street, Chicago, Illinois 60620, has a wide selection of cans (and sells in less than giant carloads). They also have a manually operated can sealer.

Ives-Way Products, Inc., Saratoga Lane, Buffalo Grove, Illinois 60090, is another possible source.

Specializing in small kitchens is Joel M. Jackson's Food Preservation Systems, 1604 Old New Windsor Road, New Windsor, Maryland 21776 (phone: 301/635-2765). Mr. Jackson developed the Ball Corporation's community canning kitchens some years ago; they were given to the World Ministry of Brethren Churches, and now are his project. He works mainly with neighborhood groups—churches, community-action organizations, soup kitchens, etc. Called community canning centers, there are 200 of them in the United States alone, with others in Canada.

And finally, a promised word on the atmospheric steam canner: in mid-October 1987, the USDA's Center for Excellence research team analyzed data from a battery of tests on the atmospheric canner. The researchers reported that this canner is strikingly poor in "kill power"/lethality when it is treated as a substitute for the traditional B–W Bath.

Thermal tests showed that inner temperatures in jars of strong-acid food processed in this canner could reach the same temperature in a B–W Bath. Examining the contents of the processed jars showed, however, that spoilage micro-organisms had not been destroyed by atmospheric steam. Therefore, *PFB* assumes the following: this canner cannot be used with B–W Bath processing times; new times must be worked out for the atmospheric canner if it is to be safe; and the folkway procedures for operating this canner that have been published casually and by cataloguers are not only inadequate but potentially dangerous as well.

At the end of October, after talking with the Center's experts, *PFB* also believes that, when/if the atmospheric canner is cleared for use, it will be reserved for *finishing* baths to ensure seals, or for strong-acid food that has been pre-cooked.

Chapters 7–12: Canning Specific Foods

Refer to *The Complete Guide to Home Canning* for the latest USDA–CES research on processing traditional foods; where there is a discrepancy in handling or processing compared with outdated material, use the *Guide*'s processing instructions. Apply latest research in using old-time jelly/jam, etc., recipes.

Canadian bulletins from Agriculture Canada are *Jams, Jellies, and Other Preserves,* publication 1753/E, and *Canning Meat at Home,* publication 5187/E, 1984.

See also the following: USDA *Farmers' Bulletin No. 2265, Pork Slaughtering, Cutting, Preserving, and Cooking on the Farm* (1978).

USDA *Farmers' Bulletin No. 2209 . . . Beef on the Farm* (1969).

USDA *Farmers' Bulletin No. 2152 . . . Lamb and Mutton,* etc. (1967).

USDA *H&G Bulletin No. 106, Home Canning of Meat and Poultry* (1970).

The Missouri, North Dakota, and Georgia bulletins on dressing, cutting, preserving, and cooking wild game are the most lively ones we've seen in a long time. Meanwhile Audrey Alley Gorton's *The Venison Book—How to Dress, Cut up and Cook Your Deer* (first published in 1957 and still going strong) is published by The Stephen Greene Press Inc., 15 Muzzey Street, Lexington, Massachusetts 02173.

Publication PNW 194, Canning Seafood (1979), Pacific Northwest Extension Publication (Cooperative Extension Services of the Universities of Idaho, Oregon, and Washington).

The chapter "Canning Convenience Foods" is well served by Presto's Form AD79–3552B—a leaflet titled *Creative Canning* that deals with 4- or 6-quart canners up to 3000 ft/914 m. *PFB* is devoted to this 8-pager because we're making an effort to help small families of two members.

Chapter 13: Getting and Using a Freezer

All bulletins describe sensibly how to use and manage a freezer. *PFB*'s favorite freezer wraps are the saran-types, like Dow's, because of their greater moisture/vapor-proof capabilities.

Food-grade polyvinyl chloride is used in supermarkets and packaging centers and is on the market from several makers to be used at home. We don't use the self-locking freezer bags a great deal, but again prefer Dow's for what so far seems to be a superior design to withstand knockabout handling in the freezer. All such bags for us, however, pop open a corner if roughly handled.

Chapters 14–17: Freezing Specific Foods

There have been so many updates in freezing publications that we say again, "Ask your county agent,"—especially for regional specialities like freshwater fish or seafood.

Again, Dr. Kendall of Colorado State University Extension is worth reading for freezing vegetables and combination main dishes: both subjects because of her recommendations for dealing with high altitude.

Of special interest for smaller families and for those who have microwave ovens are several new bulletins dealing with freezing convenience foods. (Microwaving, by the way, is used here only for thawing or preparing to serve the food; *PFB* does not undertake to go into microwave *cooking*.)

Frozen Foods, publication 504, from the Ministry of Food and Agriculture, Ontario, Canada. Agriculture Canada also has *Freezing Foods,* publication 892/E.

See also *Freezing Cooked and Prepared Foods at Home,* Louisiana State University CES, Baton Rouge; and *Freezing Prepared Foods at Home,* leaflet 2751, Division of Agricultural Sciences, University of California.

Chapter 18: Jellies, Jams, and Other Sweet Things

In addition to the publications superseded by the Home Canning Guide, we like *Jellies, Jams, and Preserves,* Louisiana State University CES, recently updated. See also *Jellies, Jams, and Other Preserves* by Agriculture Canada, publication 1753/E.

The *Ball Blue Book* and the Kerr book, as well as Bernardin's, have new and fun fruit combinations.

Chapter 19: Pickles, Relishes, and Other Spicy Things

Our favorite continues to be *Safe Methods of Preparing Pickles, Relishes, Chutneys,* from the Division of Agricultural Sciences, University of California. This publication is the clearest exposition of the how/why of using a pasteurizing bath to finish the seal, and for warnings about the use of oil in making home-pickled products.

Also, we use *The Principles of Making Pickles and Relishes* by Dr. Ruth Patrick of the Louisiana State CES; and PFB thanks to Dr. Henry Fleming and Dr. Roger F. McFeeters, both of North Carolina State University Department of Food Science.

Chapter 20: Curing with Salt and Smoke

A Complete Guide to Home Meat Curing, Morton Salt Company, 110 North Wacker Drive, Chicago, Illinois 60606.

USDA *Farmers' Bulletin No. 2265, Pork Slaughtering . . . Preserving . . .* (1978).

B-2259, *Home Smoking and Pickling of Fish,* University of Wisconsin, Agricultural Bulletin Building, 1535 Observatory Drive, Madison, Wisconsin 53706.

Publication P-325, Smoking Fish (1979), Co-operative Extension Service, University of Alaska and the USDA.

Chapter 21: Drying

USDA *H&G Bulletin No. 217, Drying Foods at Home.*

Drying Foods at Home, publication of the Agriculture Extension Service, University of Minnesota.

Home-Drying of Foods, Information Bulletin 120 by Klippstein and Humphrey of Cornell, is gettable from the Mailing Room, Building 7, Research Park, Cornell University, Ithaca, New York 14853, for a price of 40 cents per copy.

A consistent favorite (and do what you can to get it back in print) is *Sun-Dry Your Fruits and Vegetables,* from the USDA. It was written way back in 1958 and intended as "a guide for home economists around the world."

Chapter 22: Root-Cellaring

USDA *H&G Bulletin No. 119, Storing Vegetables and Fruits in Basements, Cellars, Outbuildings and Pits* (1973).

PFB praised using Styrofoam picnic chests for small-scale storage. A clean, sound chest of this expanded rigid polystyrene plastic is a boon, and save them because they are likely to become scarce in some regions. Reason: they are not biodegradable, and, because their manufacture often uses chlorofluorocarbons that damage severely the ozone layer protecting the earth's atmosphere, a number of states are considering legislation against making them.

Chapter 23: Putting By Presents for Christmas

One of the happiest booklets on this subject that we know of is *Gifts from Your Kitchen* by Elenora Hines, RD, Food, Nutrition, and Health Specialist, Tuskegee University Cooperative Extension Program, 108 Extension Building, Tuskegee University, Tuskegee, Alabama 36088. Never mind that it comes from so far away from *PFB*'s home territory: its warmth, gentleness, inventiveness, and ability to "make do" with charm are lessons especially worthy at Christmas-time.

For professional-style wrappings, boxes, etc., write for a Maid of Scandinavia catalog, 3244 Raleigh Avenue, Minneapolis, Minnesota 55416. It has the widest selection we know of for petits fours cups, boxes, doilies, food decorations, etc., priced reasonably and offered in small quantities to the homemaker.

Williams-Sonoma, Mail Order Department, P.O. Box 7456, San Francisco, California 94120-7456: their catalogs come several times a year, and occasionally have wrappings, etc., for food gifts. Again, you don't have to buy large amounts.

Finally: if you are considering how to sell your jellies and jams and pickles, etc., in roadside shops or at farmers' markets, get in touch with your state's Department of Agriculture, and often the Product Development Department. The people there can tell you what's needed to meet health or distribution laws.

INDEX